Lecture Notes in
Computer Science

T0238348

Lecture Notes in Computer Science 625

Edited by G. Goos and J. Hartmanis

Advisory Board: W. Brauer   D. Gries   J. Stoer

W. Vogler

Modular Construction and Partial Order Semantics of Petri Nets

Springer-Verlag

Berlin Heidelberg New York
London Paris Tokyo
Hong Kong Barcelona
Budapest

Series Editors

Gerhard Goos
Universität Karlsruhe
Postfach 69 80
Vincenz-Priessnitz-Straße 1
W-7500 Karlsruhe, FRG

Juris Hartmanis
Department of Computer Science
Cornell University
5149 Upson Hall
Ithaca, NY 14853, USA

Author

Walter Vogler
Institut für Informatik, Technische Universität München
Postfach 20 24 20, W-8000 München 2, FRG

CR Subject Classification (1991): D.2.2, F.1.2, F.4.3

ISBN 3-540-55767-9 Springer-Verlag Berlin Heidelberg New York
ISBN 0-387-55767-9 Springer-Verlag New York Berlin Heidelberg

Typesetting: Camera ready by author/editor
Printing and binding: Druckhaus Beltz, Hemsbach/Bergstr.
45/3140-543210 - Printed on acid-free paper

Foreword

The concept of the computer and the ways to use it as well as the understanding of what (theoretical) computer science is or should be have undergone rather deep changes since the middle of the 1930s when the first real computers were designed and built and the first theoretical concepts of computers and computability, of programs and their description were developed and studied. Traditional computer science is based on the paradigm of sequential computations to evaluate functions, briefly sketchable by the metaphor of a single individual considered as a calculator (see Turing's argumentation for the definition of his machine model). Today, the computer serves much more diverse and complex purposes; it has to be considered as a component in a distributed, interactive system composed of humans, computers, machines, and other artificial or natural dynamic systems, such that notions like communication, coordination, collaboration, cooperation, etc., play a more and more important role - and the new metaphor now is a group of individuals engaged in some or all of the four "co"-activities just mentioned (for more details on this, see Lect. Notes in Computer Science 555).

This new view of informatics is not only due to technological developments and application requirements, but also to research in theoretical informatics, in particular in the field of Petri nets. Petri nets allow us to model (distributed, concurrent) systems by a formalism which separately represents (local) actions, (local) states and the (local) interrelations between the holdings of states and the executions of actions (which means that the structure as well as the dynamics are described in the same formalism).

The formalism of Petri nets allows for very detailed descriptions on an operational level. This makes it necessary to develop methodologies and techniques for modular construction of Petri nets and for appropriate behaviour descriptions. Since there are also several other formalisms for concurrent systems, there also exist many modular construction techniques and behaviour notions for such systems - and in particular there has been much discussion on the adequacy of semantic notions, in particular (since Petri introduced the idea in 1976) on the necessity of using partial-order semantics. In this book, Vogler studies this issue in depth - he starts with the most basic techniques of Petri net construction and with requirements on the nets to be constructed, and then shows which notions are minimally required. This way, he can for example prove that failures semantics (which originally was developed for TCSP) is just the right notion to characterize the constructability of deadlock-free nets by TCSP-like parallel composition. The most important contribution perhaps is Vogler's study of action refinement - he was the first to provide formal results on the adequacy of partial order semantics and branching equivalences for the study of action refinement, and moreover he even shows that only a restricted type of partial orders (the interval orders) are necessary.

Vogler's approach allows him to systematically study the broad spectrum of construction, semantics and equivalence notions in such a way that many of their inherent properties and their interrelationships become much clearer, and many more or less philosophical discussions of the past are now obsolete. On the basis of this (and selected work by other authors), it should now be possible to study problems of practical applications of these notions within concrete methodologies of system construction.

München, June 1992   Wilfried Brauer

Preface

Petri nets are a well-known model for parallel systems, used for both applications and theoretical studies. Like any formal model, they can be used for specification, modelling and analysis; Petri nets in particular offer a graphical representation and a clear view of concurrency. For the design of large systems, modular construction is indispensible; hence, considerable effort has been spent on studying the modular construction of Petri nets. This book presents some contributions to this research area.

In bottom-up design, nets are put together and the intention is to determine the behaviour of the composed system from the behaviour of its components; as operators for the combination of nets we consider parallel composition with synchronous and with asynchronous communication. For the top-down design, we study the refinement of the elementary parts of nets, i.e. of places and transitions. A refinement step is performed with one of two possible intentions in mind. Either the refined net is expected to have the same behaviour as the unrefined net, in which case we speak of a behaviour-preserving refinement; or we expect that refining two nets with the same behaviour leads to nets that have the same behaviour again, in which case we speak of an equivalence-preserving refinement; the equivalence-preserving refinement of transitions is also called action refinement.

This book presents behaviour descriptions that support these modular construction methods of nets. Many such descriptions are possible. Therefore, special care is taken to justify the descriptions presented here by showing what is called full abstractness; i.e., when considering some construction method, we not only present a suitable behaviour description, but also show that it makes exactly those distinctions of nets that are necessary to support the given method and to take into account some simple feature of behaviour like deadlock-freeness. For example, failure semantics is the right behaviour description for constructing deadlock-free nets using parallel composition with synchronous communication. As one of the highlights, we show that in order to support action refinement and to take into account failure semantics some form of partial order semantics is necessary.

This work is a revised version of my *Habilitationsschrift* written at the Technische Universität München. It would not have been possible without the support I have received from many people. First of all, my thanks go to Professor W. Brauer for the good working atmosphere he has created in his group and for his helpful advice and valuable comments over the last few years. I also would like to thank Professors M. Broy and M. Nielsen, who acted as further referees of my *Habilitationsschrift*. I am particularly grateful to Professor R. Halin, who guided my way through graph theory before I changed over to computer science.

I have profited from numerous discussions with many people, and I am especially grateful to all my former and present colleagues from Hamburg and München. For many years, Dirk Taubner has shared an office and his knowledge especially on failure semantics with me, and our discussions have helped me a lot. I am also greatly indebted to Eike Best, Jörg Desel, Volker Diekert, Rob van Glabbeek, Robert Gold, Ulla Goltz, Astrid

Kiehn, Wolfgang Reisig, Thomas Tensi, and Rolf Walter.

Work on this book was partially supported by the Deutsche Forschungsgemeinschaft, Sonderforschungsbereich 342: Methoden und Werkzeuge zur Nutzung paralleler Rechnerarchitekturen, TU München, and the ESPRIT Basic Research Action No. 3148 DEMON (Design Methods Based on Nets).

Last not least, I thank Harald Hadwiger and Dieter Stein, who have helped me enormously to transform my notes into a LaTeX document.

München, June 1992   Walter Vogler

Contents

Chapter 1

Introduction

A concurrent system, such as a network of processors, an operating system, or a manufacturing system, consists of several partly autonomous components, which run in parallel and influence each other by interactions. Thus, to determine the behaviour of the system it is not enough to know how each of its components in isolation transforms initially given input objects like data or raw materials into output objects produced at the end. In the case of a sequential system this would be sufficient; for example, the behaviour of a procedure in a sequential program can be defined as a function from memory states to memory states. But for a concurrent system, we also have to know how each component reacts to outside influences and how it influences its environment while it is running. In fact, the same applies to the entire system in so far as it interacts with its environment and is thus itself a component of a larger system. Furthermore, the activities of the components may be unsynchronized; as a consequence the interaction of the system components may have the effect that the entire system behaves nondeterministically, even if its components are deterministic. For example, if two senders share a channel, the behaviour of the system may depend on the timing of their messages.

In view of these complications, the design of concurrent systems particularly requires a formal method. In the design process an informal conception is transformed into a formal system model, and the first benefit of the formal model is that its development helps to uncover deficiences and ambiguities in the informal conception. Once the model is completed, it can be analyzed formally and relevant properties can be verified, or at least the concurrent system can be tested by means of a simulation.

In the approach we adopt in this book, the behaviour of a concurrent system is described in terms of the actions it can perform. Here an action is any activity that we view as a conceptual entity; in particular, it may be an act of communication. A simple behaviour description of this kind is the set of all possible sequences of actions. But this semantics has been criticized in two points.

First, the concurrent execution of actions is seen as equivalent to arbitrary interleaving, i.e. to executing these activites in an arbitrary order. Thus concurrency is simply reduced to some form of nondeterminism. Such a semantics is called an interleaving semantics. Alternatively, one could try to represent concurrency explicitly, e.g. by describing a system run by a partial order of actions. Such a semantics would be 'truly concurrent'.

Secondly, the above semantics gives no information about the nondeterministic choices that have been made during a system run, about the branching structure of the system behaviour. Such a semantics is called a linear-time semantics. In order to give a branching-time semantics one must not only compare system runs, but also consider how internal conflicts are resolved. Prominent branching-time semantics are failure semantics [BHR84] and bisimulation [Par81,Mil83].

These considerations show that there is more than one way of defining the behaviour of a system. Which definition is chosen depends on the system properties that are regarded as relevant. It is even feasible to view a simple property as the behaviour of a system, e.g. the semantics of a system possibly is just to be or not to be free of deadlocks. Thus we cannot expect to find *the* behaviour of a concurrent system. Instead we can compare various semantics and study their properties. A most important requirement for a semantics is that it support the modular construction of systems.

To reduce and manage system complexity we have to design large systems in a modular fashion either bottom-up by composing subsystems with known behaviour in such a way that we can determine the behaviour of the whole system from its parts, or top-down by refining parts of a rough model by more detailed system descriptions. In the latter case we either ensure that the behaviour is essentially preserved or proceed again in such a way that we can determine the behaviour of the refined system from that of the rough model and the refinement, in which case we speak of an equivalence-preserving refinement. In all these cases, systems are constructed from building blocks, and a semantics supports the modular construction of systems if it describes the behaviour of the building blocks, i.e. their interfaces, in such a way that we can control the behaviour of the entire system as just described.

This book contributes to the theory of designing concurrent systems with Petri nets. A Petri net is a formal system model based on concepts from automata theory, linear algebra and graph theory. Besides the general advantages of a formal model and the verification methods based on linear algebra, Petri nets are additionally attractive since – as graph-theoretic objects – they have a graphical representation. Already in the design process this graphical representation offers a visual impression of the concurrent system and how it is built from subsystems and distributed in space; it gives a clear image of concurrency, sequentiality and conflict, both on the concrete visual level and on the abstract graph-theoretic level.

In particular, the visualization of concurrency makes it very natural to consider concurrency as a feature that deserves a proper presentation on the semantic level. Petri net theory has a long tradition in studying 'true concurrency' in the semantics of concurrent systems. Most often 'true concurrency' is captured by giving a semantics based on partial orders, and partial orders also invite a graph-theoretic representation as Hasse diagrams. On the other hand, the branching structure of systems has not been given much attention.

It must also be mentioned that modularity as described above has been a somewhat weak point of Petri net theory. A Petri net is defined as a whole and not in the first instance obtained by composing subnets; correspondingly its semantics, i.e. the firing rule or a derivative of it, does not rely on the semantics of some subnets, although the firing rule is local in character.

This is totally different in process algebras like CCS [Mil80,Mil89], TCSP [BHR84,

Hoa85] or ACP [BK84]. Here systems are described by process terms, which are by nature built from subterms. Naturally, the semantics of a process term is obtained from the semantics of its subterms, no matter whether the semantics is an operational semantics defined according to the structured operational approach of [Plo81] or a denotational semantics. Thus process algebras are a priori compositional.

Traditionally, in process algebras concurrency has been reduced to interleaving; this may be due to their roots in algebra. Often it has been argued that interleaving is simpler than 'true concurrency' and just as expressive, i.e. sufficient for any practical purpose. Instead, the emphasis has been on studying the branching structure of processes. Thus, while neglecting the dimension of interleaving versus 'true concurrency', process algebra has concentrated on the orthogonal dimension of linear-time versus branching-time semantics, and vice versa for Petri net theory.

In recent years, both these approaches have increasingly influenced each other, and a lot of effort has been made to combine their respective merits. Partial order semantics for process terms have been developed, see e.g. [BC87,DDNM88,Old89,NEL89,Ace89]. Furthermore, semantics in terms of Petri nets have been given to process algebras, such that a process term, which is an operator applied to some subterms, is translated to a net that is an appropriate composition of nets related to those subterms; see e.g. [GV87,Gol87, Gol88b,Tau89]. This allows one to give some partial order semantics to process algebras by translating a term to a net and taking (one of) its partial order semantics. Viewed the other way, those nets that are translations of process terms form a restricted class of nets for which several compositional semantics can be given if we apply results from process algebra. Similarly, several authors have suggested solving the compositionality problem of Petri nets by working with restricted, more structured classes of nets that are built from very simple nets, see e.g. the state-net-decomposable nets studied in [Bes88a], and see [BDC92] for a survey.

Other authors have concentrated on solving the compositionality problem for the unrestricted class of all nets. They have suggested various transformations that are behaviour-preserving in some sense, and various composition operators such that the behaviour of a composed net (in some sense) can be determined from the behaviour of its components, see e.g. [And83,Bau88,Ber87,DCDMS87,Gra81,Mül85,Sou91,SM83,Val79,Vos87]. This is the area in which this book is located.

The system models in this book are labelled place/transition-nets without capacities, i.e. place/transition-nets where the transitions are labelled with actions. As indicated above, these are uninterpreted names of activities. Their use allows one to abstract from details of a system description that are of no importance for the user of the system. Transitions with the same label represent the same activity in different internal situations. Very important is the use of λ-labelled transitions, which represent internal activities that are invisible for the user; thus we can abstract from activities that are important on a low-level system description when we consider the system behaviour on a higher level.

For the modular construction of nets we concentrate on two sorts of operators: parallel composition with synchronous or asynchronous communication for the bottom-up design of nets, and refinement of transitions and places for the top-down design. These operators are especially interesting since they are also graphically meaningful. Given two nets, parallel composition corresponds to composition by merging transitions in the synchronous case and to merging places in the asynchronous case. Given one net a refinement replaces a basic net element, i.e. a place or a transition, by some net, thus giving a more detailed description of a local state or an activity. Other operators are only touched upon, e.g. hiding, which allows one to abstract from details by turning some visible actions into internal actions.

Typical for the approach of this book is the situation in Chapter 3. We define an operator ∥ (more precisely a family of operators indexed by a set of actions that are to be synchronized), and then we want to find out which nets, if combined with any environment via ∥, can be exchanged without changing the behaviour of the composed net; we call such nets *externally equivalent* following [Bau88]. At the same time we want to explore which behaviour notions are suitable, since there is no general agreement about this point. This makes the situation slightly obscure: either we should fix our behaviour notion, and then we can try to characterize externally equivalent nets; or we should fix the exchanges of nets we want to carry out, and then we can try to find out which behaviour is preserved by such exchanges. Fortunately, our approach leads to quite a satisfactory solution for this matrix of problems. First, we fix a very simple sort of behaviour, namely we just distinguish deadlock-free nets from those that can deadlock. Then we give an *internal* characterization of the corresponding external equivalence, i.e. we determine when nets can be exchanged in any environment without referring in our characterization to all possible environment nets; namely, nets are externally equivalent if and only if they have the same failure semantics. When proving this we find that exchanging failure-equivalent nets preserves the failure semantics of the composed net. Thus our composition operator together with failure semantics is compositional in the sense that we can determine the behaviour of a composed net from the behaviour of its components. At the same time we discover that exchanging failure-equivalent nets preserves behaviour in a much stronger sense than originally required; thus our equivalence works for a whole range of behaviour notions.

External equivalence is closely related to testing equivalence in the sense of [DNH84], which refers to a notion of observability. In principle the reasoning for the above external equivalence can also be expressed in terms of observability. But in this book the argument is not that deadlock or divergence (infinite internal looping) are observable in some sense, but rather that these are important features of behaviour that we must control in the modular construction of a system. External equivalence is also closely related to full abstractness [Mil77], which is a stronger requirement: it considers the exchange of two nets in any context built by applying the operators under consideration possibly many times, while we consider contexts where we have only one application of an operator. In other words, an internal semantics characterizing an external equivalence describes the interface of a building block such that this description is sufficient to deduce the relevant behaviour of a *system* constructed from two building blocks; on the other hand, a fully abstract semantics describes the interface of a building block such that we can determine the interface of a *building block* constructed from two building blocks, and thus we can determine the relevant behaviour of a system constructed from any number of building blocks. Naturally, the latter is to be preferred in general. But in all natural cases we consider, external and fully abstract equivalences coincide, and thus our results are stronger if we start with the weaker requirement, i.e. if we start with the study of external equivalences.

Our results can also be seen as a justification of failure semantics; they show that failure equivalence is just the right equivalence, if we want an equivalence that is compositional with respect to ‖ and are mainly interested in the deadlocking behaviour of systems. This view is especially interesting in Chapter 5, where we consider action refinement. This operator refines some action a by a more detailed description of this activity; it replaces *every* a-labelled transition by a copy of some net. Regarding the dispute of interleaving versus 'true concurrency', our results in Chapter 5 show that for a congruence for action refinement that respects e.g. failure equivalence the power of partial order semantics is needed. Thus we justify the use of 'true concurrency'.

As explained above, by modular construction we understand either composition or refinement, and the latter is subdivided into behaviour- and equivalence-preserving refinement. Composition is studied in Chapters 3, 4, and 7. Chapters 3 and 7 are concerned with synchronous communication, where Chapter 7 studies nets with capacities contrary to the model we use in general. Asynchronous communication is treated in Chapter 4. These chapters develop suitable interface descriptions for the composition of systems from building blocks. Behaviour-preserving refinement is studied in Chapter 4; modules are characterized that are suitable for replacing a transition or a place in any context, and these results are obtained by putting behaviour-preserving refinement in the framework of composition. Chapters 5 and 6 are devoted to equivalence-preserving refinement; Chapter 5 is concerned with linear-time and failure semantics, Chapter 6 with various types of bisimulation. Here the emphasis is on the behaviour of the rough model. We develop interface descriptions for the rough model that together with the inserted refinement nets allow us to deduce the interface description of a partly refined model and finally the relevant behaviour of the detailed system.

In more detail, we proceed as follows. Chapter 2 introduces Petri nets, where Section 2.1 briefly reviews the basic notions. Section 2.2 is devoted to the linear-time partial order semantics of Petri nets; we define the well-known processes (of nets) and partial words and adapt them to labelled nets. Complementarily, Section 2.3 describes some points in the linear-time/branching-time spectrum for the interleaving case; we define two failure-type semantics, one taking account of divergence and the other not, and some versions of bisimulation.

In Chapter 3 we study parallel composition with synchronization of actions from some given set; as described above, we show that the two types of failure semantics we have introduced are just right for a compositional semantics if we are mainly interested in deadlock-free or deadlock- and divergence-free systems. In Section 3.3 we study some modifications. One concerns the treatment of infinite system runs, and we touch upon the problem of fairness. We consider an adaption to safe nets and to the case where we are interested in liveness (in the Petri net sense) instead of deadlock-freeness. In Section 3.4 we mention the further operators hiding, relabelling, and the choice operator.

If we restrict ourselves to the exchange of nets that are in some sense deterministic, we can improve our results. Not only do we get more favourable decidability results, we also can show that our simple requirement, that the exchange of equivalent nets preserves deadlock-freeness, guarantees behaviour preservation in a much stronger sense; such an exchange results in a net that is bisimilar to the original net; see also [Eng85]. This is presented in Chapter 4, where in particular we show how these results can be applied to the refinement of places. Dually, we initiate the study of a parallel composition operator with asynchronous communication in Section 4.3; in Section 4.4 we explore which features make a net deterministic in this context, and show how the behaviour-preserving refinement of transitions fits into this framework.

Often we cannot expect that the refinement of transitions preserves behaviour, since the refined net may be able to perform some new actions that were not present in the unrefined net. In this case we would like to be able to determine the behaviour of the refined from that of the unrefined net. Thus equivalent nets, i.e. nets with the same behaviour, should be refined to nets that are equivalent again; in other words, the equivalence should be a congruence for action refinement. Such equivalence-preserving action refinements are studied in Chapter 5 and Chapter 6. In Chapter 5 we introduce a technique for action refinement and discuss, which refinement nets are suitable in this context. We show that partial order semantics is useful for defining congruences with respect to action refinement in Section 5.3, and we introduce a partial order semantics based on interval orders. This turns out to be just the right semantics in the sense that interval semiwords can be used to define three semantics that are fully abstract for action refinement and language-, failure- and failure/divergence-semantics respectively. We show this in Section 5.4, where we use interval words, a more or less sequential presentation of interval semiwords. The translation between these two descriptions of system runs is presented in Section 5.5 together with some decidability results.

In Chapter 6 we discuss congruences for action refinement of bisimulation-type. While partial orders are immediately useful for linear-time congruences, pomset bisimulation [BC87], a straightforward combination of partial order semantics and bisimulation, has turned out to fail for this purpose [BDKP91,GG89b]. History-preserving bisimulation, a more intricate combination of partial order semantics and bisimulation, is a congruence [BDKP91,GG89b]; but even this fails unless we restrict the use of internal actions. On the other hand, ST-bisimulation [GV87], which makes no explicit use of partial orders but is in fact closely related to interval semiwords, gives a congruence. We show that the ST-idea can be used to lift in a uniform way bisimulation, pomset bisimulation, history-preserving bisimulation and the newly introduced partial-word bisimulation to congruences with respect to action refinement without any restriction on the use of internal actions. At least in the first three cases we can also show full abstractness results.

For these considerations we restrict ourselves to event structures [NPW81], which can be seen as a special class of Petri nets, which are in particular acyclic. In many ways, this makes event structures theoretically easier to work with; but they have the considerable disadvantage that they have to be infinite in order to describe an infinite behaviour. The corresponding advantage of general Petri nets is somewhat lost when we work with history-preserving bisimulation. This type of bisimulation gives a detailed account of the interplay of causality and branching, and it has turned up in various papers – not only in the context of action refinement; but unlike the usual bisimulation it relates system runs instead of system states, and thus it necessarily refers to infinitely many objects if we are concerned with infinite behaviour. In Section 6.4 we give an alternative definition of a bisimulation for safe nets without internal transitions; in this definition each system state is described by a marking together with a pre-order on its tokens, where the pre-order contains information on causality in the token generation. We can show that this *OM*-bisimulation (*OM* = ordered marking) gives the same equivalence as history-preserving bisimulation. As a corollary we obtain that history-preserving bisimulation is decidable.

Chapter 7 looks at partial order semantics from another angle. Continuing a research initiated in [HRT89], we consider compositionality for nets *with* capacities. Here we are not concerned with full abstractness; instead we consider quality criteria of partial order semantics based on net transformations that are very natural in the presence of capacities. We give several characterizations and define a new partial order semantics that seems to be the natural choice in this framework, since it is the minimal semantics satisfying all our criteria.

In total, we will present more than forty different net semantics. As explained above, it must be left to the reader to choose the right one for a specific application. For example, if interleaving semantics is all the reader is interested in, attention can be restricted to Chapters 3 and 4 after reading appropriate portions of Chapter 2, but then a hierarchical design by action refinement cannot be accomplished. Conversely, if the reader wants to know a good reason for partial order semantics or if action refinement is the main operator of interest, then the reader should turn to Chapters 5 and 6.

In many applications it will turn out that only safe nets are needed. In this case Chapter 7 can be left out, which is only interesting if we have varying capacities. Also, in some cases our decidability results are based on the decidability of the reachability problem, but they become quite simple in the case of safe nets. In this book, emphasis is put not only on showing that some semantics is sufficient to guarantee certain properties for the modular net construction, but also on the necessity of the distinctions made by the semantics. These necessity results can fail if the nets under consideration are restricted to some subclass; therefore it is important that we keep an eye on this practically important class of safe nets, sometimes developing appropriate variations as in Section 3.3.

For applications it may look like a severe restriction that the actions, which label transitions, are just uninterpreted names; but in principle we can also deal with arbitrary data. For example, if we have an action '*input(n)*', where n is meant to be a natural number – although formally '*input(n)*' is just a meaningless name –, then we can use actions '*input(n)$_1$*','*input(n)$_2$*', ..., one for each value of n. Of course, in this way the input of n requires infinitely many transitions, and correspondingly the variable n is modelled by infinitely many places, one for each value of n. This approach is perfectly sufficient for theoretical investigations as we present them here. For practical applications, most often some form of high-level net will be more adequate; see e.g. [Gen87,Jen87,Rei90]. High-level nets can be seen as an abbreviation for place/transition-nets, and thus one can expect that our results carry over; often these place/transition nets are – possibly infinite – safe nets, which underlines the importance of safe nets. On the other hand, using high-level nets it is often desirable to work on a symbolic level, and here a lot of work remains to be done in order to transfer our results.

Chapter 2

Petri Nets and Their Semantics

This chapter gives a brief introduction to Petri nets and some aspects of their semantics. This introduction concentrates on those notions that are needed in this book and is rather mathematically minded. For further introductory information on Petri nets and especially on their applications, the reader is referred to e.g. [Pet81,Rei85,Bau90]. The first section defines Petri nets as we use them in this book, namely place/transition-nets whose transitions are labelled with actions. It also introduces the firing rule for transitions, steps, and sequences of transitions and steps.

In the second section, it is explained how concurrency of actions in a system run can be described explicitly. We introduce partial orders and three Petri net semantics based on partial orders: the well-known processes of nets, partial words and semiwords. We compare these among each other and also with firing sequences and firing step sequences; in particular, we present a short proof for the result of A. Kiehn that each least sequential partial word is the event structure of a process.

In the third section, we return to interleaving semantics, which identifies the concurrent execution of actions with their execution in arbitrary order, and consider the branching behaviour of nets. Branching means that a net, when performing some action sequence, may have the choice between different firing sequences that all correspond to the given action sequence, but nevertheless may lead to different states and different future behaviour. To take into account this branching behaviour, one often considers failure semantics or bisimulation, and we present two versions of failure semantics and four versions of bisimulation.

2.1   Basic notions of Petri nets

This section briefly introduces Petri nets and the basics of their dynamics. In this book, we will deal with Petri nets (place/transition-nets) whose transitions are labelled with actions from some infinite alphabet Σ or with the empty sequence λ. These actions are left uninterpreted; the labelling only indicates that a λ-labelled transition represents an *internal* action, which is not visible for an external observer, and that two transitions with the same label from Σ represent the same *visible* action occurring in different internal situations. (In the literature, an internal action is often denoted by τ instead of λ.)

Thus a *Petri net* $N = (S, T, W, M_N, lab)$ (or just a *net* for short) consists of

- disjoint (not necessarily finite) sets S of *places* and T of *transitions*

- the *weight function* $W : S \times T \cup T \times S \to \mathbb{N}_0$, which assigns to each pair (x, y) the corresponding *arc weight*

- the *initial marking* $M_N : S \to \mathbb{N}_0$

- the *labelling* $lab : T \to \Sigma \cup \{\lambda\}$.

The initial marking will always be indexed by the net, the other components only if it is necessary to distinguish the components of one net from those of another net. Usually, $S_{N'}, T_{N'}, \ldots, S_{N_1}, \ldots$ etc. will be abbreviated to $S', T', \ldots, S_1, \ldots$ etc.

Nets can be considered as labelled, directed, bipartite graphs: the vertices are the places and transitions, and we have an arc from x to y if $W(x, y) > 0$. Graphically, places are circles, transitions are boxes and arcs are arrows between them. If the identity of a place or a transition is given, then it is written next to the circle or box. The value of M_N is written inside the circle or indicated by an appropriate number of black dots, so-called *tokens*, inside the circle; the value of *lab* is written inside the box; an arrow from x to y is drawn if $W(x, y) > 0$, and it is labelled with $W(x, y)$ if $W(x, y) > 1$. We say that a net has *arc weight* 1, if W takes values in $\{0, 1\}$ only. The visible actions of a net form its *alphabet* $\alpha(N) = lab(T) - \{\lambda\}$. Figure 2.1 shows a net with arc weight 1.

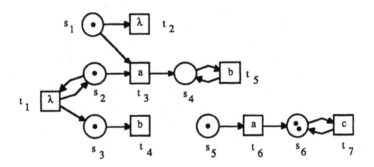

Figure 2.1

If we model a system by a net, then the transitions model the active parts of the system, the places the passive parts, and the markings describe the system states. The next definitions describe how the states are changed, they describe the so-called token game.

- For $x \in S \cup T$ the *preset* of x is ${}^\bullet x = \{y \in S \cup T \mid W(y, x) > 0\}$, the *postset* of x is $x^\bullet = \{y \in S \cup T \mid W(x, y) > 0\}$. For sets $X \subseteq S \cup T$ we put ${}^\bullet X = \bigcup_{x \in X} {}^\bullet x$ and $X^\bullet = \bigcup_{x \in X} x^\bullet$.

- A *multiset* over a set X is a function $\mu : X \to \mathbb{N}_0$; it is finite, if it has finite support, i.e. if $\mu(x) \neq 0$ for finitely many $x \in X$; it is empty, if $\mu(x) = 0$ for all x. The set of finite non-empty multisets over X is denoted by $\mathcal{M}(X)$. We write $x \in \mu$, if $\mu(x) \neq 0$.

On multisets we have the usual pointwise addition and scalar multiplication with scalars from $\mathbb{N}_0$. Furthermore, we identify $x \in X$ with the multiset that is 1 for x and 0 everywhere else. Thus especially $X \subseteq \mathcal{M}(X)$, and e.g. $\mu = 2x + 3y$ satisfies $\mu(x) = 2$, $\mu(y) = 3$, and $\mu(z) = 0$ for all $z \in X$ with $x \neq z \neq y$.

- A *marking* is a multiset over S, a *step* is a finite, non-empty multiset over T.

- A step μ is *enabled* under a marking M, denoted by $M[\mu\rangle$, if $\sum_{t \in \mu} \mu(t) \cdot W(s,t) \leq M(s)$ for all $s \in S$. Obviously, it suffices to check this inequality for all $s \in \bigcup_{t \in \mu} {}^{\bullet}t$.

  If $M[\mu\rangle$ and $M'(s) = M(s) + \sum_{t \in \mu} \mu(t)(W(t,s) - W(s,t))$, then we denote this by $M[\mu\rangle M'$ and say that μ can *occur* or *fire* under M yielding the *follower marking* M'. We also say that the transitions of μ can fire *concurrently*.

  Since transitions are special steps, this also defines $M[t\rangle$ and $M[t\rangle M'$ for $t \in T$ and markings M and M'.

- For a set X we denote by X^* the finite *sequences* over X as usual, X^ω denotes the infinite sequences, and $X^\infty = X^* \cup X^\omega$. The set of subsets is denoted by $\mathcal{P}(X)$.

- The above definition of enabling and occurrence can be extended to sequences as follows:

  A sequence $w \in \mathcal{M}(T)^*$ is *enabled* under a marking M, denoted by $M[w\rangle$, and yields the follower marking M' if it *occurs*, denoted by $M[w\rangle M'$, if

  - $w = \lambda$ and $M = M'$

  or

  - $w = w'\mu$ with $\mu \in \mathcal{M}(T)$, $M[w'\rangle M''$ and $M''[\mu\rangle M'$ for some marking M''.

  A sequence $w \in \mathcal{M}(T)^\omega$ is enabled under a marking M, also denoted by $M[w\rangle$, if $M[v\rangle$ for every finite prefix v of w.

- A *firing sequence* of N is some $w \in T^\infty$ with $M_N[w\rangle$, a *firing step sequence* of N is some $w \in \mathcal{M}(T)^\infty$ with $M_N[w\rangle$. We denote the set of finite (arbitrary, infinite) firing sequences of N by $FS(N)$ ($FS^\infty(N)$, $FS^\omega(N)$), and the sets of firing step sequences analogously by $FSS(N)$, $FSS^\infty(N)$ and $FSS^\omega(N)$.

- Any function $f : X \to Z \dot\cup \{\lambda\}$ can be extended as usual to a function f^* on X^* by $f^*(\lambda) = \lambda$ and $f^*(wx) = f^*(w)f(x)$ for $w \in X^*$, $x \in X$. Especially, if $f(x) = \lambda$ then $f^*(wx) = f^*(w)$. Thus f^* takes a sequence over X to the corresponding sequence of images under f, thereby deleting λ-labelled elements. This way we can regard f^* as being defined on infinite sequences as well. Similarly, we can define f^* on $\mathcal{M}(X)$ as follows:

  – $f^*(\mu) = \lambda$ if $f(x) = \lambda$ for all $x \in \mu$
  – $f^*(\mu) = \sum_{f(x) \neq \lambda} \mu(x) \cdot f(x)$.

Furthermore, we extend f^* to $\mathcal{M}(X)^\infty$ in the obvious way.

Especially, *lab** takes steps and step sequences to multisets of actions and sequences of such multisets, thereby deleting internal transitions. With this function, we can lift the enabledness and firing definitions to the level of actions, as explained below. Consequently, we can describe the behaviour of Petri nets on two levels: on the lower level we describe the behaviour in terms of the concrete transitions; on the higher level, which is based on the lower level, we describe the behaviour as it appears to an outside observer, i.e. in terms of the visible actions that are performed. These two levels of description we will meet several times in this book.

  • A sequence $v \in \mathcal{M}(\Sigma)^\infty$ is *image-enabled* under a marking M, denoted by $M[v\rangle\rangle$, if there is some $w \in \mathcal{M}(T)^\infty$ with $M[w\rangle$ and $lab^*(w) = v$. If additionally $M[w\rangle M'$, we write $M[v\rangle\rangle M'$. If $M = M_N$, then the sequence v is called an *image firing step sequence* and w is its *underlying* firing step sequence. (More precisely, w is one of the firing step sequences underlying v.) If w is a firing sequence, then v is an *image firing sequence* with *underlying* firing sequence w.

    The set $L(N) = lab^*(FS(N))$ is also called the *language* of N, furthermore we have $L^\omega(N) = lab^*(FS^\omega(N))$, $L^\infty(N) = lab^*(FS^\infty(N))$, and analogously for step sequences the sets $SL(N)$, the *step language* of N, $SL^\omega(N)$, and $SL^\infty(N)$ are defined. We call two nets N_1 and N_2 *language-equivalent* (*step-language-equivalent*) if $L(N_1) = L(N_2)$ ($SL(N_1) = SL(N_2)$).

  • For a marking M the set $[M\rangle$ of markings *reachable* from M is defined as $\{M' \mid \exists w \in T^* : M[w\rangle M'\}$. The set $[M_N\rangle$ is the set of reachable markings of N. Two nets N_1 and N_2 are *marking-equivalent* on a set P of common places, if $\{M|_P \mid M \in [M_{N_1}\rangle\} = \{M|_P \mid M \in [M_{N_2}\rangle\}$, where $\mid$ denotes the restriction of functions or relations.

Consider the net N of Figure 2.1. Under the initial marking M_N the step $2t_7 + t_1$ is enabled, the follower marking M has just one more token on s_3. Consequently, $2c$ is image-enabled, and we have $M_N[2c\rangle\rangle M$. On the other hand, $t_1 + t_3$ is not enabled under M_N, since this step would need two tokens on s_2. Also $2t_6$ is not enabled under M_N, but $2a$ is nevertheless image-enabled since $M_N[t_3 + t_6\rangle$. Furthermore we have $M_N[2b\rangle\rangle$, since $M_N[t_1(2t_4)\rangle$. Similarly, we have $M_N[b^\omega\rangle\rangle$, where b^ω is an infinite sequence of b's, since $M_N[(t_1 t_4)^\omega\rangle$. We can also see, that for all reachable markings M we have e.g. $M(s_5) + M(s_6) = 3$.

We have some more basic notions:

  • A net is *T-restricted*, if for all transitions t we have ${}^\bullet t \neq \emptyset \neq t^\bullet$.

  • If $S' \subseteq S$ and $T' \subseteq T$, then the *subnet* of N *induced* by $S' \cup T'$ has place set S' and transition set T', and the arc weight, initial marking and labelling are those of N

restricted to $S' \times T' \cup T' \times S'$, S' and T'. If $X \subseteq S \cup T$, then $N - X$ is the subnet of N induced by $(S \cup T) - X$.

A net or a subnet is a *loop* if it consists of a place s and a transition t such that $s \in {}^\bullet t \cap t^\bullet$. The default value for the arc weights of a loop is 1, i.e. if a net or a subnet is simply described as a loop, then we mean that both arcs have weight 1.

- A place s is *m-bounded* for $m \in \mathbb{N}_0$ if $M(s) \leq m$ for all reachable markings M. A net is *bounded*, if for every place s there exists some $m \in \mathbb{N}_0$ such that s is m-bounded. It is *safe*, if every place is 1-bounded. We also say that a marking M is *safe* if $M(s) \leq 1$ for all $s \in S$.

- A transition t is *live*, if $\forall M \in [M_N\rangle \; \exists M' \in [M\rangle : M'[t\rangle$. An action $a \in \Sigma$ is *image-live*, if $\forall M \in [M_N\rangle \; \exists M' \in [M\rangle : M'[a\rangle\rangle$. A net is *live*, if all its transitions are live. A transition is *dead*, if it is not enabled under any reachable marking.

- As we have already defined above, two not necessarily distinct transitions t_1 and t_2 are concurrently enabled under some marking M if $M[t_1 + t_2\rangle$. A net is *sequential*, if there are no transitions t_1, t_2 that are concurrently enabled under some reachable marking. A transition is *self-concurrent*, if $M[2t\rangle$ for some reachable marking M. An action $a \in \Sigma$ is *autoconcurrent*, if $M[2a\rangle\rangle$ for some reachable marking M. A net is free of self- or autoconcurrency, if no transition or no action is self- or autoconcurrent.

- Two distinct transitions t_1 and t_2 are in *conflict* under some marking M if $M[t_1\rangle$ and $M[t_2\rangle$, but not $M[t_1 + t_2\rangle$. A net is *conflict-free*, if there are no distinct transitions t_1, t_2 that are in conflict under some reachable marking.

Consider again the net N of Figure 2.1. N is not T-restricted. The place s_4 is safe, i.e. 1-bounded, s_6 is 3-bounded, while s_3 is unbounded. The transition t_7 is live, and b is image-live, although no b-labelled transition is live. In fact, no transition except t_7 is live, and no transition is dead. We have already seen that t_3 and t_6 are concurrently enabled, thus N is not sequential. Furthermore t_1 and t_3 are in conflict, thus N is not conflict-free. The transition t_7 is self-concurrent, and a is autoconcurrent, although no a-labelled transition is self-concurrent.

- The *reachability graph* $\mathcal{R}(N)$ of a net N is an arc-labelled, rooted, directed graph; its vertex set is $[M_N\rangle$ and M_N is a distinguished vertex, the root; whenever for $M, M' \in [M_N\rangle$ and $t \in T$ we have $M[t\rangle M'$, then there is an arc from M to M' labelled $lab(t)$.

If $\mathcal{R}(N)$ is finite, we can view it as a finite automaton with M_N as initial state and all reachable markings as final states. The language of N is just the language accepted by the finite automaton $\mathcal{R}(N)$.

Arc-labelled, rooted, directed graphs, so-called transition systems, are often used to describe directly the behaviour of concurrent systems or to give an operational semantics to process algebras like CCS [Mil80], TCSP [BHR84] or ACP [BK84]. For this purpose it is important that transition systems may be infinite. Also for Petri nets some operational semantics can be given by means of a transition system, as we have seen. At the same time transition systems can be seen as special Petri nets of arc weight 1: the vertices are the places, the root is the only marked place carrying one token, and the arcs are the (unbranched) transitions. In this way many of our results apply to transition systems as well, and one reason why we study infinite nets in this book is that we want to cover infinite transition systems, too. (Not all our results apply directly to transition systems; the reason is that the net operators we use do not necessarily give a transition system when applied to nets that correspond to transition systems.)

Another model for concurrent systems are prime event structures with binary conflicts [NPW81], which we just call event structures for short. Also event structures can be seen as special Petri nets of arc weight 1. These nets are acyclic and safe, i.e. every transition can fire at most once; they are also T-restricted, and if they describe a possibly infinite behaviour, they must be infinite. See Chapter 6 for details.

- A net is *image-finite* if for every reachable marking M and every $a \in \Sigma \cup \{\lambda\}$ there are only finitely many M' with $M[a\rangle\rangle M'$.

- A net is *enabling-finite* if for every reachable marking M there are only finitely many transitions enabled under M, i.e. if $\mathcal{R}(N)$ is locally finite in the sense that for each vertex there are only finitely many outgoing arcs.

In general, the nets considered in this book do not have a capacity function (capacities will be studied in Chapter 7), thus there is no a priori bound enforced for the number of tokens on a place.

- We say that the labelling of a net N is *injective for* some $A \subseteq \Sigma$, if $lab(t) = lab(t') \in A$ implies $t = t'$. We say that a net is *injectively labelled*, if its labelling is injective for Σ. It is λ-*free*, if $\lambda \notin lab(T)$.

Very often unlabelled nets are studied in the literature; these can be seen as special labelled nets, where the labelling of the transitions is the identity and therefore omitted. With the above notions, unlabelled nets coincide (up to isomorphism) with λ-free, injectively labelled nets.

In general, we will not distinguish between isomorphic nets (nor between isomorphic partial orders etc.).

2.2   Partial order semantics of nets

Very often the behaviour of a net is described by its language. For some purposes this may be adequate. But consider the nets of Figure 2.2.

These nets are language-equivalent, but obviously the first net can perform actions a and b independently of each other, while the second cannot. In many situations this will make quite a difference. Presumably, the first net will be faster, and this ability of speeding up execution of independent actions is one important reason to study parallel systems. On the other hand, a and b may be an abstract view of quite complex activities,

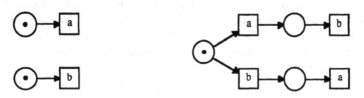

Figure 2.2

each consisting of several subactivities which might even interfere with each other. If this is the case, the first net is an abstract view of a system which could run into complications the second system is safe of. (We will come to this in Chapter 5.)

Thus, an *interleaving semantics* (like the language of a net), which equates the concurrent execution of actions with their execution in an arbitrary order, is often inadequate for concurrent systems.

The nets of Figure 2.2 are already distinguished by step-language equivalence. The implicit assumption behind the step language semantics is, that there exists a global clock and each action just takes one unit of time as measured by this clock. Both parts of this assumption may fail in a distributed system. The nets of Figure 2.3 can both start a and c concurrently, in the first net b can occur when a has finished, while in the second net b has to wait for c, too. This makes quite a difference if c takes more time than a or if, in the absence of a global clock, we do not know how the finish of a is related to the finish of c. Thus, a semantics that models 'true concurrency' should distinguish the nets of Figure 2.3, which are step-language-equivalent.

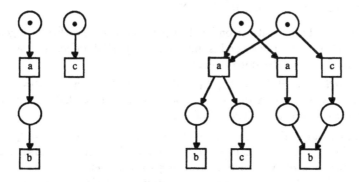

Figure 2.3

Petri net theory has a long tradition of studying 'true concurrency' by way of partial order semantics. In this book, we will restrict our attention to finite partial orders. (For a more general approach see e.g. [BF88].)

- A (finite) *partial order* $(X, <)$ consists of a finite set X and a transitive, irreflexive relation $<$. As usual, we write $x \leq y$ for $x < y$ or $x = y$. Often $<$ is called a strict partial order, while $\leq$ is a weak partial order, i.e. a reflexive, antisymmetric and transitive relation. Sometimes we will speak of a partial order also in the case of a weak partial order, but what is meant will (hopefully) be clear from the context or from the underbarred symbol for the partial order. A *pre-order* $(X, \leq)$ consists of a set X and a transitive relation $\leq$, where for some $x, y \in X$ we may have $x \leq y$ and $y \leq x$ but not $x = y$.

The *downward-closure* of a set $Y \subseteq X$ is $\downarrow Y = \{x \in X \mid \exists y \in Y : x \leq y\}$. Y is *downward-closed* if $Y = \downarrow Y$. *Upward-closed* is defined analogously.

- A *labelled partial order* (*over* some set Z) $p = (X, <, l)$ consists of a partial order $(X, <)$ and a labelling function $l : X \to Z$. As in the case of nets, we index the components of a labelled partial order p with p if necessary, and the components of p_1, p_2 etc. are X_1, X_2, $<_1$, $<_2$ etc. Often the base set of a labelled partial order will be denoted by E instead of X.

  Two labelled partial orders p_1 and p_2 are *isomorphic* if there exists an *isomorphism* $f : X_1 \to X_2$, i.e. a bijection which is *order-preserving* ($x <_1 y \Leftrightarrow f(x) <_2 f(y)$ for all $x, y \in X_1$) and *label-preserving* ($l_2(f(x)) = l_1(x)$ for all $x \in X_1$). As already mentioned, in general we do not distinguish isomorphic labelled or unlabelled partial orders.

- If for some labelled partial order p elements x, y of X are unordered, i.e. we have neither $x < y$ nor $y < x$ (but possibly $x = y$), then we call x and y *concurrent*, denoted by x *co* y. A maximal set of pairwise concurrent elements is called a *cut*. For a set Y of pairwise concurrent elements we let $l^*(Y)$ denote the multiset $\sum_{y \in Y} l(y)$. (Here we assume that λ does not appear as a label.)

  The set of *minimal* (*maximal*) *elements* of some set $Y \subseteq X$ is denoted by $\min Y$ ($\max Y$), or by $\min p$ ($\max p$) if $Y = X$.

- There are two basic relations on labelled partial orders. The first is a generalization from the string case, the second is indigenous to partial orders. Let p_1, p_2 be partial orders labelled over the same alphabet Z.

  p_1 is a *prefix* of p_2 if $X_1 \subseteq X_2$, X_1 is downward-closed in X_2, $<_1 = <_2 \mid_{X_1 \times X_1}$ and $l_1 = l_2 \mid_{X_1}$. A set L of partial orders labelled over some alphabet Z is called *prefix-closed* if it contains all prefixes of all $p \in L$.

  p_1 is *less sequential* ($\preceq$) than p_2 if $X_1 = X_2$, $<_1 \subseteq <_2$ and $l_1 = l_2$. If p_2 is to-tally ordered, then it is called a *linearization* of p_1. In this case, let $x_1, \ldots, x_n$ be the elements of X_2 ordered according to $<_2$; then p_2 can be identified with $l_2(x_1) \ldots l_2(x_n) \in Z^*$. If p_2 has a transitive co-relation, then it is called a *step linearization* of p_1. In this case, let $Y_1, \ldots, Y_n$ be the co-equivalence classes or-dered according to $<_2$; then p_2 can be identified with the sequence of multisets $l_2^*(Y_1) \ldots l_2^*(Y_n) \in \mathcal{M}(Z)^*$.

  A set L of partial orders labelled over some alphabet Z is called *seq-closed* if it contains all sequentializations of every $p \in L$.

Figure 2.4 shows three labelled partial orders; the first is a prefix of the second and less sequential than the third.

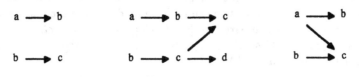

Figure 2.4

- If p is a labelled partial order over Y and $f : Y \to Z \dot\cup \{\lambda\}$ is some function then $f^*(p) = (X', <|_{X' \times X'}, f \circ l|_{X'})$ where $X' = \{x \in X \mid f(l(x)) \neq \lambda\}$.

  This definition of f^* agrees with the definition given in the previous section for those labelled partial orders which can be regarded as sequences, finite multisets or sequences of finite multisets.

Processes of nets

Most often a partial order semantics of nets is based on so-called processes, which in turn are based on a special, very simple net class, the class of causal nets.

- A *causal net* is a finite net (B, E, F) of arc weight 1 without marking and labelling, whose places are called *conditions*, whose transitions are called *events*, and whose arcs are given as $F \subseteq B \times E \cup E \times B$, such that

  - the net is acyclic, i.e. for all $x, y \in B \cup E$ $(x, y) \in F^+$ implies $(y, x) \notin F^+$ (where F^+ is the transitive closure of F)

  - the conditions are not branched, i.e. $|{}^\bullet b|, |b^\bullet| \leq 1$ for all $b \in B$.

- If a causal net is labelled, then both its conditions and events are labelled. A *process* $\pi = (B, E, F, l)$ of a net N *enabled under a marking* M of N consists of a causal net (B, E, F) and a labelling $l : B \cup E \to S_N \cup T_N$ such that

  - conditions correspond to places (or tokens), events to transitions, i.e. $l(B) \subseteq S_N$, $l(E) \subseteq T_N$

  - minimal conditions correspond to the marking M, i.e. for all $s \in S_N$ we have $M(s) = |\{b \in B \mid {}^\bullet b = \emptyset,\ \text{and}\ l(b) = s\}|$

  - the neighbourhood of transitions is respected, i.e. $W_N(s, l(e)) = |l^{-1}(s) \cap {}^\bullet e|$, $W_N(l(e), s) = |l^{-1}(s) \cap e^\bullet|$ for all $s \in S_N$ and $e \in E$.

We write $M[\pi\rangle$ in this situation. The marking M' reached after π is defined by $M'(s) = |\{b \in B \mid b^\bullet = \emptyset\ \text{and}\ l(b) = s\}|$ for $s \in S_N$, and we denote this by $M[\pi\rangle M'$.

A process of N enabled under M_N is simply called a *process of N*. The *initial process* $\pi_0(N)$ is the unique (up to isomorphism) process of N with empty set of events.

- Processes treat places and transitions on an equal footing. Since we are mostly interested in the actions a system can perform we define the *event structure* $ev(\pi)$ of a process π as the labelled partial order $(E, F^+|_{E \times E}, l|_E)$; the *action structure* $ac(\pi)$ of π is $lab_N^*(ev(\pi))$. If $M[\pi\rangle$ or $M[\pi\rangle M'$ we also write $M[ev(\pi)\rangle$ or $M[ev(\pi)\rangle M'$ and $M[ac(\pi)\rangle\rangle$ or $M[ac(\pi)\rangle\rangle M'$.

  The set of action structures of processes of N is denoted by $Proc(N)$. Two nets N_1 and N_2 are *process-equivalent*, if $Proc(N_1) = Proc(N_2)$.

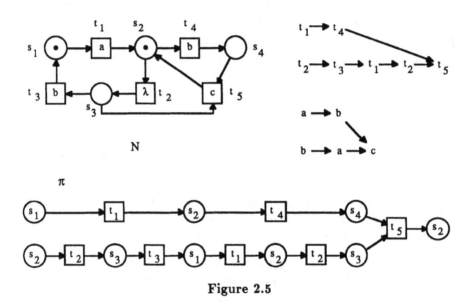

Figure 2.5

A process is intended to model the causality of a system run. Figure 2.5 shows a net N, a process π, and the event and the action structure of π. We can see in this example that the firing of t_4 is directly caused by a firing of t_1, while it is independent of both firings of t_2. This is also represented in the event structure of π. The process also shows that we may have e.g. a token on s_4 and a token on s_3 concurrently.

By definition of a process, $(B \cup E, F^+, l)$ is a labelled partial order. Next, we will see: concurrent events represent transitions that can fire concurrently; concurrent conditions represent tokens that can be on the respective places in the same marking; linearizing or step linearizing the events yields a firing or a step firing sequence of N. Usually, these facts are shown for T-restricted nets only. We state a slightly more general theorem; since its formulation is somewhat awkward we give a corollary afterwards for the case of T-restricted nets.

- The set $Lin(\pi)$ of *linearizations* of the process π is the set of linearizations of $ev(\pi)$. The set $Step(\pi)$ of *step linearizations* of π is the set of step linearizations of $ev(\pi)$.

One can show inductively:

Theorem 2.2.1 *Let $\pi = (B, E, F, l)$ be a process of a net N.*

 i) *Any set of pairwise concurrent conditions can be extended to a cut of $(B \cup E, F^+, l)$*
 by conditions from B and events from $\{e \in E \mid {}^\bullet e = \emptyset \text{ or } e^\bullet = \emptyset\}$. Let $B_1 \subseteq B$
 and $E' \subseteq \{e \in E \mid {}^\bullet e = \emptyset \text{ or } e^\bullet = \emptyset\}$ such that $B_1 \cup E'$ is a cut of $(B \cup E, F^+, l)$,
 and let $M = l^*(B_1)$. Let X be the downward-closure of $B_1 \cup \{e \in E' \mid e^\bullet = \emptyset\}$, let
 $E_1 = X \cap E$, and let w be a (step) linearization of $(E_1, F^+|_{E_1}, l|_{E_1})$. Then $M_N[w\rangle M$.

 ii) *Let E_1 be a set of pairwise concurrent events and $\mu = l^*(E_1)$. Then the conditions of*
 ${}^\bullet E_1$ are pairwise concurrent, and we can extend ${}^\bullet E_1$ to a cut $B_1 \cup E'$ of $(B \cup E, F^+, l)$
 with $B_1 \subseteq B$ and $E' \subseteq \{e \in E \mid {}^\bullet e = \emptyset \text{ or } e^\bullet = \emptyset\}$; let $M = l^*(B_1)$. Then $M[\mu\rangle$.

 iii) *$Lin(\pi) \subseteq FS(N)$, $Step(\pi) \subseteq FSS(N)$, and for M defined by $M_N[\pi\rangle M$ and each*
 $w \in Lin(\pi)$ or $w \in Step(\pi)$ we have $M_N[w\rangle M$.

Corollary 2.2.2 *Let $\pi = (B, E, F, l)$ be a process of a T-restricted net N.*

 i) *Any set of pairwise concurrent conditions can be extended to a cut of $(B \cup E, F^+, l)$*
 consisting of conditions only, a so-called slice. Let $B_1 \subseteq B$ be a slice, $M = l^*(B_1)$,
 X be the downward-closure of B_1, $E_1 = X \cap E$, and let w be a (step) linearization
 of $(E_1, F^+|_{E_1}, l|_{E_1})$; then $M_N[w\rangle M$.

 ii) *Let E_1 be a set of pairwise concurrent events and $\mu = l^*(E_1)$. Then the conditions of*
 ${}^\bullet E_1$ are pairwise concurrent, and we can extend ${}^\bullet E_1$ to a slice B_1. Let $M = l^*(B_1)$;
 then $M[\mu\rangle$.

 iii) *$Lin(\pi) \subseteq FS(N)$, $Step(\pi) \subseteq FSS(N)$, and for M defined by $M_N[\pi\rangle M$ and each*
 $w \in Lin(\pi)$ or $w \in Step(\pi)$ we have $M_N[w\rangle M$.

 One can also show by induction that for each (step) firing sequence of N there exists
a corresponding process.

Theorem 2.2.3 *Let N be a net.*

 i) *For each $w \in FS(N)$ ($w \in FSS(N)$) there exists a process π of N with $w \in Lin(\pi)$*
 ($w \in Step(\pi)$).

 ii) *This process is uniquely determined by w if N is safe.*

 To see that the second part of the theorem does not hold for nets in general, consider
the example of Figure 2.5 and a linearization $t_1 t_4 t_2 \ldots$. Constructing a process inductively,
we start with two conditions representing the initial marking of N, add an event labelled
t_1 and another condition labelled s_2. When we now add an event labelled t_4, we have the
choice whether it 'takes' the token produced by t_1 or the other one. Depending on this
choice, we can construct two different processes.

 Thus, processes individualize tokens while it is a generally made assumption that
tokens on the same place cannot be distinguished. A discussion of these different views
of the token game can be found in [Bra84]. To obtain a similar result as Theorem 2.2.3
ii) it is suggested in [BD87] to form equivalence classes of processes using the so-called
swap-operation. In [Vog90], it is argued that these equivalence classes are in general too

large to describe one system run. In [Vog91a], a different generalization of processes of safe nets to a new partial order semantics of general nets is developed.

Partial words

Processes are meant to model causality, hence they view the concurrent execution of some actions as totally different from their arbitrary interleavings. Thus, e.g. the *Proc*-semantics of the two nets in Figure 2.2 are incomparable. Another view is that concurrency is more than arbitrary interleaving but includes it. From this point of view, the semantics of the first net of Figure 2.2 should include the semantics of the second net. This idea is formalized in the partial-word semantics of [Gra81], also studied e.g. in [Kie88]. A partial word of a net N is a partial order labelled with transitions of N; any set of pairwise concurrent elements of this partial order represents a step that can be fired provided the precedences prescribed by the partial order are observed.

- Let N be a net, M be a marking of N, and $p = (E, <, l)$ be a partial order labelled over T. We call p a *partial word* of N *enabled* under M if for all disjoint subsets B and C of E we have: if all elements of C are pairwise concurrent and B and $B \cup C$ are downward-closed, then the step $l^*(C)$ is enabled under $M + \sum_{e \in B} W(l(e), .) - W(., l(e))$, i.e.

$$\sum_{e \in C} W(., l(e)) \leq M + \sum_{e \in B} W(l(e), .) - W(., l(e)).$$

  We write $M[p\rangle$. The elements of E are called *events*. The marking *reached after* p is $M' = M + \sum_{e \in E} W(l(e), .) - W(., l(e))$, and we write $M[p\rangle M'$. If p is enabled under M_N we simply say that p is a partial word of N.

- A partial word is a behaviour description on the level of concrete transitions. To get a corresponding higher level description in terms of actions, we define the image of a partial word. The *image* q of a partial order p labelled over T_N is $lab_N^*(p)$. If $M[p\rangle$ or $M[p\rangle M'$, then we say that q is *image-enabled* under M (with *follower marking* M') and write $M[q\rangle\rangle$ or $M[q\rangle\rangle M'$; p is called a partial word *underlying* q. If p is a partial word of N, then q is an *image partial word* of N and we denote the set of image partial words of N by $PW(N)$. Two nets N_1 and N_2 are *partial-word-equivalent* if $PW(N_1) = PW(N_2)$.

Figure 2.6 shows a net and one of its partial words; some sets B and C are indicated, and indeed, we can fire the step $2t_4$ if we fire t_1 and t_2 first.

If we want to check whether some labelled partial order is a partial word, we can restrict our considerations to sets B and C where C is a cut.

Lemma 2.2.4 *A partial order labelled over T_N is enabled under some marking of a net N, if the requirements of the definition are satisfied whenever C is a cut.*

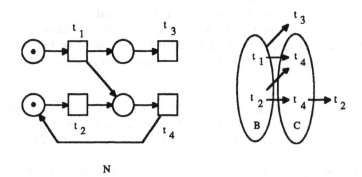

N

Figure 2.6

Proof: Let p, M, B and C be given as in the definition. We can extend C to a cut $C \cup C' \cup B'$ such that $C' \subseteq \min(E-(B \cup C))$ and $B' \subseteq \max B$. Now $B-B'$ and $C \cup C' \cup B'$ are disjoint, and $B-B'$ and $C \cup C' \cup B$ are downward-closed. Thus by assumption of the lemma we have

$$\sum_{e \in C \cup C' \cup B'} W(., l(e)) \leq M + \sum_{e \in B-B'} W(l(e),.) - W(., l(e)).$$

This can be transformed to

$$\sum_{e \in C} W(., l(e)) + \sum_{e \in C'} W(., l(e)) \leq M + \sum_{e \in B}(W(l(e),.) - W(., l(e))) - \sum_{e \in B'} W(l(e),.),$$

which implies the desired inequality, namely

$$\sum_{e \in C} W(., l(e)) \leq M + \sum_{e \in B} W(l(e),.) - W(., l(e)).$$

$\square$

An elegant reformulation of the definition of partial words can be found in [Kie88]:

Proposition 2.2.5 *A labelled partial order p over T_N is enabled under some marking M of a net N if and only if for every step linearization $w \in \mathcal{M}(T)^*$ of p we have $M[w\rangle$.*

It is easy to see that the set of partial words of a net is closed under taking prefixes or making partial words more sequential.

Proposition 2.2.6 *Let p be a partial word of a net N, q a labelled partial order over T.*

i) If q is a prefix of p, then q is a partial word of N.

ii) If p is less sequential than q, then q is a partial word of N and the marking reached after q is the same as the one reached after p.

This proposition shows that concurrency in partial words is just seen as a possibility, and that concurrent actions can also be performed sequentially.

We have used the notation $M[.)$ for sequences of transitions and of steps, for (event structures of) processes and for partial words. We have to show that the respective definitions agree. For firing (step) sequences and partial words this follows directly from the definitions.

Proposition 2.2.7 *Let N be a net, M and M' be markings, and p be a labelled partial order over T such that its co-relation is transitive or, as a special case, total. Let w be the sequence of steps or transitions corresponding to p. Then $M[p\rangle M'$ if and only if $M[w\rangle M'$.*

For processes and partial words the agreement follows from Theorem 2.2.1 and the definition of partial words.

Proposition 2.2.8 *Let π be a process of some net N enabled under some marking M such that $M[\pi\rangle M'$. Then $ev(\pi)$ is a partial word enabled under M such that $M[ev(\pi)\rangle M'$.*

Since processes are intended to model causality which in turn is thought to describe just the necessary precedences, one would expect that the labelled partial orders $ev(\pi)$ with π a process are the least sequential partial words. This is only partly true.

Theorem 2.2.9 *Let N be a net. Then $\{p \mid p$ is a least sequential partial word of $N\} \subseteq \{ev(\pi) \mid \pi$ is a process of $N\}$. If N is safe, then equality holds.*

This is an important result that relates the two approaches to a semantics with 'true concurrency'. This result is obtained in [Kie88]; since the proof given there is quite complicated, we give a short proof here, which is based on the following version of the well-known marriage theorem (see e.g. [Bol90, III Corollary 9]).

Theorem 2.2.10 *Let G be a (finite) bipartite graph with vertex classes V_1 and V_2, and $f : V_2 \to \mathbb{N}_0$ be a function. Then G contains a subgraph H such that every $x \in V_2$ has degree $f(x)$ and every $y \in V_1$ has degree at most 1 if and only if for every subset U of V_2 there are at least $\sum_{x \in U} f(x)$ vertices in V_1 adjacent to a vertex in U.*

Proof of 2.2.9:   Let p be a partial word of N. In view of Proposition 2.2.8, we have to construct a process π such that $ev(\pi) \preceq p$. Let $E_\pi = E_p$ and let l_π equal l_p on this set E. We will construct the conditions with label s for each place s separately.

Let $B_s = \{(s, i) \mid 1 \le i \le M_N(s)\} \cup \{(e, i) \mid e \in E, 1 \le i \le W(l(e), s)\}$ be the set of conditions with label s and let $e(e, i) \in F_s$. Thus, we have the right number of conditions to represent the initial marking and the postsets of the events are correct, too.

Define a bipartite graph G with vertex classes B_s and E and edges $(s, i)e$ for $e \in E$, $(s, i) \in B_s$ and $(e', i)e$ for $e \in E$, $(e', i) \in B_s$ such that $e' <_p e$; let $f(e) = W(s, l(e))$. If we find a subgraph H as in 2.2.10 and let $ce \in F_s$ whenever ce is an edge in H, then the presets of the events have the right size, and there is a directed path of length 2 in π from some e' to some e only if $e' <_p e$. Thus, letting B_π be the union of the B_s, F_π the union of the F_s, the resulting process π satisfies $ev(\pi) \preceq p$ (which also ensures that π is in fact acyclic, since p is a partial order).

It remains to check the condition of 2.2.10. Let $U \subseteq E$; we have to compare $|U|$ with the number of neighbours of U in B_s. Let U' be the downward-closure of U in p and put $C = \max U'$, $B = U' - C$. Applying the definition of a partial word gives

$$\sum_{e \in C} W(s, l(e)) \leq M_N(s) + \sum_{e' \in B} (W(l(e'), s) - W(s, l(e'))),$$

or equivalently

$$\sum_{e \in B \cup C} W(s, l(e)) \leq M_N(s) + \sum_{e' \in B} W(l(e'), s).$$

Now we have

$$\begin{aligned}
\sum_{e \in U} W(s, l(e)) &\leq \sum_{e \in U'} W(s, l(e)) \leq M_N(s) + \sum_{e' \in B} W(l(e'), s) \\
&= M_N(s) + \sum_{e' < e \in U} W(l(e'), s),
\end{aligned}$$

and the latter is just the number of neighbours of events in U.

If for processes π, π' of a safe net we have $ev(\pi) \preceq ev(\pi')$, then every linearization of π' is also one of π. Since for safe nets every linearization determines the process, we have $\pi = \pi'$. Now the second part of the theorem follows from the first.   □

In view of Theorem 2.2.9, one can say that partial words describe the causality of a system run, but include some additional ordering information. For example, due to some (partial) observation we know that event e really happend after e', although it was not caused by e'; this ordering information can be incorporated in a partial word.

Corollary 2.2.11   *i)* If two nets have the same event structures of processes, then they have the same partial words. For safe nets the reverse implication holds, but not for nets in general.

   ii) Process equivalence implies partial-word equivalence, but not vice versa – not even for safe nets.

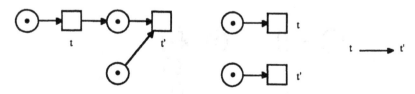

Figure 2.7

Proof: i) The partial words of a net are all labelled partial orders that are more sequential than the event structures of the processes. For safe nets the event structures of processes are the least sequential partial words. The nets in Figure 2.7 have the same partial words, and a partial word is shown which is an event structure of a process for the first net only.

ii) Apply i); Figure 2.8 shows two partial-word-equivalent safe nets and an image of a partial word that is the action structure of a process only for the first net.   □

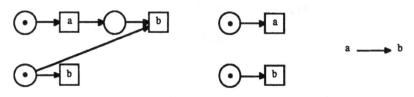

Figure 2.8

For some purposes the self-concurrent firing of a transition creates problems, as we will see in Chapter 5. Therefore, semiwords are introduced in [Sta81].

- A partial word is a *semiword* if x *co* y and $x \neq y$ implies $l(x) \neq l(y)$. Two nets are *semiword-equivalent*, if they have the same images of semiwords. We denote the set of image semiwords of a net N by $SW(N)$.

Proposition 2.2.12   *i) For safe nets without isolated transitions all partial words are semiwords, and partial-word equivalence and semiword equivalence coincide.*

  ii) In general, partial-word equivalence does not imply semiword equivalence, and the reverse implication does not hold either.

Proof: i) In a safe net, a transition with empty preset must have an empty postset. If $s \in {}^\bullet t$, then the self-concurrent firing of t requires two tokens on s.

ii) Figure 2.9 shows two nets with the same image semiwords and a distinguishing image of a partial word. Figure 2.10 shows two nets with the same image partial words and a distinguishing image of a semiword.   □

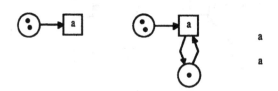

Figure 2.9

Observe that the image of a semiword might contain autoconcurrent actions, i.e. x *co* y does not imply $lab(l(x)) \neq lab(l(y))$.

Let us note that all the semantics we have defined in this section coincide for sequential nets.

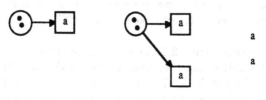

Figure 2.10

Proposition 2.2.13 *For sequential nets language-, step-language-, process-, partial-word- and semiword-equivalence coincide.*

Proof: This follows from the definition of step sequences, Theorem 2.2.1, Theorem 2.2.9, and the definition of semiwords.   □

A third possibility to give a partial order semantics to nets, especially to safe nets, is given by traces in the sense of Mazurkiewicz, see e.g. [Maz87,AR88,Die90]. The partial order of a trace is based on a static dependence relation. This approach allows a rich mathematical theory, but only on the level of unlabelled nets, and we will not deal with traces in this book.

2.3   Branching-time semantics of nets

In the last two sections we have described the behaviour of a net by its language, its step language, its processes or its partial language. All these behaviour notions define so-called linear-time semantics; i.e. an image firing sequence or a process etc. describe one run of the system without giving information on conflicts that occurred and were resolved during this run.

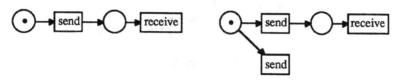

Figure 2.11

Consider the two nets in Figure 2.11. They model a channel that can be used once, and they can both perform a unique 'send' followed possibly by a unique 'receive'. Thus, they have the same language, and since they are sequential, they also have the same action structures of processes etc. Nevertheless, they are quite different: only the second net can choose to perform 'send' in such a way that 'receive' is not possible any more, i.e. this channel can accept a message and forget it without delivery; in a respective environment, this can create a deadlock which would not be possible with the first net. A branching-time semantics takes such possible choices into account.

In this section we will present some branching-time semantics, which were mostly developed in a setting of process algebras like CCS or TCSP. All of them are interleaving semantics.

To distinguish the nets of Figure 2.11 we can use the failure semantics. It was developed for TCSP [BHR84,BR84] and translated to nets in the following form in [Pom86]. The *failure semantics* of a net N is $\mathcal{F}(N) = \{(w, X) \mid w \in \Sigma^*, X \subseteq \Sigma$, and there is a marking M such that $M_N[w\rangle M$ and $\forall a \in X : \neg M[a\rangle)\}$. In this semantics we add to an image firing sequence w a negative look-ahead of one action, i.e. a set of actions that can be *refused* after w. An element (w, X) of $\mathcal{F}(N)$ is called a *failure pair*, w is sometimes called a *history*, and X a *refusal set*. Two nets N_1 and N_2 are *failure-equivalent* ($\mathcal{F}$-*equivalent*) if $\mathcal{F}(N_1) = \mathcal{F}(N_2)$.

Whenever we introduce some semantics $Sem(N)$ of nets N in the following, nets are called *Sem-equivalent* if they have the same *Sem*-semantics; this is the corresponding semantic equivalence.

Note that an image firing sequence does not necessarily determine a unique follower marking. The second net N_2 of Figure 2.11 can perform a in such way that b is possible afterwards, hence $(ab, \emptyset) \in \mathcal{F}(N_2)$, but it can also perform a such that b is impossible, thus $(a, \{b\}) \in \mathcal{F}(N_2)$. Due to the latter failure pair the two nets of Figure 2.11 are not failure-equivalent.

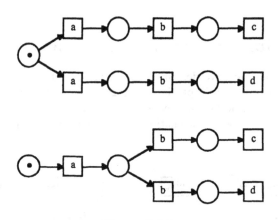

Figure 2.12

Figure 2.12 shows a typical example of failure-equivalent nets. In the first net the choice between c and d is already made before performing a, in the second net only afterwards. This is ignored in the failure semantics, since in both cases the choice has been made after ab, i.e. it has already been made when either c or d can occur.

Another example of failure-equivalent nets is shown in Figure 2.13. The refusal set is just some set of actions that can be refused, not necessarily the complete set of such actions; hence e.g. $(a, \{a, b\}) \in \mathcal{F}(N)$ implies $(a, \{a\}) \in \mathcal{F}(N)$, and both these failure pairs are contained in the failure semantics of both nets in Figure 2.13. In other words, we are just interested in the 'worst' situations while situations that are not so bad are 'covered up'.

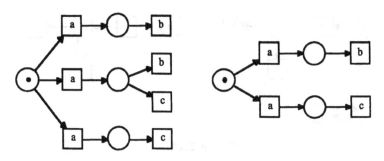

Figure 2.13

An important aspect of systems behaviour is divergence, infinite internal looping. This is taken into account in the failure semantics as defined in [BR84]. Translating this definition to nets gives the *failure/divergence semantics* (*$\mathcal{FD}$-semantics*) of a net as $\mathcal{FD}(N) = (F(N), D(N))$ where $D(N) = \{w \in \Sigma^* \mid \exists v \in L^\omega(N) : v \text{ is a prefix of } w\}$, $F(N) = \{(w, X) \mid w \in D(N), X \subseteq \Sigma\} \cup \mathcal{F}(N)$.

The idea of this semantics is, that divergence is catastrophic since it absorbs all processing resources of the system. Instead of declaring every behaviour after divergence impossible, this catastrophe is modelled by considering every behaviour after divergence as possible. If $v \in L^\omega(N)$ is a prefix of $w \in \Sigma^*$, then v is a finite sequence, thus its relevant underlying firing sequence, which is infinite, ends in an infinite suffix of λ-labelled transitions. Such a sequence v is called a *divergence string*, and $D(N)$ contains all divergence strings and their continuations. $F(N)$ is similar to $\mathcal{F}(N)$, but since every behaviour is possible after divergence, $F(N)$ also contains the elements of $D(N)$ paired with arbitrary refusal sets.

Figure 2.14

Figure 2.14 shows two nets, which are $\mathcal{F}$-equivalent, since $\mathcal{F}$-semantics ignores divergence; but they are not $\mathcal{FD}$-equivalent, since the first net N_1 can diverge immediately, i.e. $\lambda \in D(N_1)$, whereas the second net cannot.

On the other hand, the nets in Figure 2.15 are $\mathcal{FD}$-equivalent, since both can perform a and diverge afterwards, so that all further behaviour is ignored. But they are not $\mathcal{F}$-equivalent, since the first net can 'really' perform b after a, while the second net can 'really' perform c.

The 'covering up' mentioned above is formulated in the following proposition, which is easily proven from the definitions.

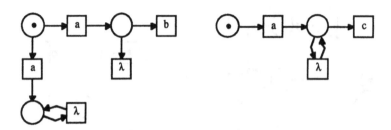

Figure 2.15

Proposition 2.3.1 *For all nets N and $X \subseteq Y \subseteq \Sigma$ we have that $(w, Y) \in \mathcal{F}(N)$ implies $(w, X) \in \mathcal{F}(N)$, and $(w, Y) \in F(N)$ implies $(w, X) \in F(N)$. On the other hand, for all $X, Y \subseteq \Sigma$ we have that $(w, Y) \in \mathcal{F}(N)$ and $(wa, \emptyset) \notin \mathcal{F}(N)$ for all $a \in X$ implies $(w, X \cup Y) \in \mathcal{F}(N)$, and analogously for $F(N)$.*

Next we come to an equivalence that is often studied in the literature (especially for process algebras), and that distinguishes much finer than $\mathcal{F}$-equivalence. In a way, it takes the full branching behaviour of a net into account while abstracting from internal moves (and divergence). Two systems are called bisimilar, if there is a relation between the system states with the following properties:

- The initial states are related.

- If the systems are in related states, each of them can simulate the other; i.e., if one system performs some actions, the other can perform the same actions in such a way that the systems end up in related states again.

Formally, two nets N_1 and N_2 are *(interleaving) bisimilar* [Mil83,Par81], if there exists a *bisimulation* between them, i.e. a relation $\mathcal{B} \subseteq [M_{N_1}\rangle \times [M_{N_2}\rangle$ such that

i) $(M_{N_1}, M_{N_2}) \in \mathcal{B}$

ii) If $(M_1, M_2) \in \mathcal{B}$ and $M_1[w\rangle\rangle M_1'$, $w \in \Sigma^*$, then there is $M_2' \in [M_{N_2}\rangle$ such that $M_2[w\rangle\rangle M_2'$ and $(M_1', M_2') \in \mathcal{B}$.

iii) vice versa, i.e. if $(M_1, M_2) \in \mathcal{B}$ and $M_2[w\rangle\rangle M_2'$, $w \in \Sigma^*$, then there is $M_1' \in [M_{N_1}\rangle$ such that $M_1[w\rangle\rangle M_1'$ and $(M_1', M_2') \in \mathcal{B}$.

Bisimulation (more precisely: bisimilarity) can easily be checked to be an equivalence. It distinguishes e.g. the nets in Figure 2.12 or in Figure 2.13. It ignores divergence, therefore it identifies the nets of Figure 2.14 and distinguishes those of Figure 2.15. It is an interleaving equivalence, i.e. it identifies the nets of Figure 2.2.

Thus, bisimulation ignores divergence and 'true concurrency', but otherwise it is a very fine equivalence. Figure 2.16 shows a typical example of bisimilar nets; the second net is just a partial unwinding of the first one and has two identical copies of the 'c-branch' instead of just one.

Bisimulation preserves a lot of system properties. For example, one sees easily:

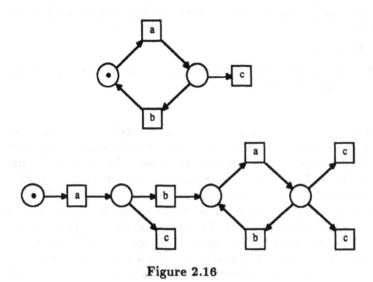

Figure 2.16

Proposition 2.3.2 *If two nets N_1 and N_2 are bisimilar, then some $a \in \Sigma$ is image-live in N_1 if and only if a is image-live in N_2.*

Bisimulation abstracts from internal moves, but does not ignore them. The first two nets of Figure 2.17 are not bisimilar, since only the second can internally decide not to make an a-move.

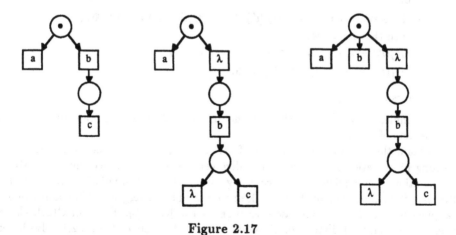

Figure 2.17

It is well known that in order to check a bisimulation it is enough to consider moves that consist of at most one visible action – as we have just seen, exactly one visible action would not be correct here.

Proposition 2.3.3 *For nets N_1 and N_2, a relation $\mathcal{B} \subseteq [M_{N_1}\rangle \times [M_{N_2}\rangle$ is a bisimulation if and only if:*

i) $(M_{N_1}, M_{N_2}) \in \mathcal{B}$

ii) *If $(M_1, M_2) \in \mathcal{B}$ and $M_1[a\rangle\rangle M_1'$, $a \in \Sigma \cup \{\lambda\}$, then there is $M_2' \in [M_{N_2}\rangle$ such that $M_2[a\rangle\rangle M_2'$ and $(M_1', M_2') \in \mathcal{B}$.*

iii) vice versa

In [GW89b], the authors argue that the way bisimulation abstracts from internal actions is not totally natural: a property that bisimilar nets without internal actions have may be violated for nets with internal actions, namely that 'in two bisimilar processes every computation in the one process corresponds to a computation in the other, in such a way that all intermediate states of these computations correspond as well'. (Here 'process' is the system model used in [GW89b], hence in our context one should read 'net' instead of 'process'.) Indeed, the second and the third net of Figure 2.17 are bisimilar, where the additional b of the third net corresponds to a sequence in the second net with an internal action before and after b. The state after the first and the state before the second internal action have no correspondence to states in the run that consists of the additional b. Therefore the following refined version of bisimulation is suggested in [GW89b]:

Two nets N_1 and N_2 are *branching bisimilar* if there exists a *branching bisimulation* between them, i.e. a relation $\mathcal{B} \subseteq [M_{N_1}\rangle \times [M_{N_2}\rangle$ such that

i) $(M_{N_1}, M_{N_2}) \in \mathcal{B}$

ii) If $(M_1, M_2) \in \mathcal{B}$ and $M_1[t_1\rangle M_1'$ then

  - $lab_1(t_1) = \lambda$ and $(M_1', M_2) \in \mathcal{B}$

    or

  - there exists a sequence $M_{20}[t_1'\rangle M_{21} \ldots [t_n'\rangle M_{2n}[t_2\rangle M_2'$, $n \in \mathbb{N}_0$,
    such that $M_2 = M_{20}$,
    $lab_2(t_i') = \lambda$, $(M_1, M_{2i}) \in \mathcal{B}$, $i = 1, \ldots, n$, and
    $lab_1(t_1) = lab_2(t_2)$, $(M_1', M_2') \in \mathcal{B}$.

iii) vice versa

For example, the second and the third net of Figure 2.17 are not branching bisimilar, while the nets of Figure 2.16 are.

To obtain the last two equivalences of this section, we will refine bisimulation and branching bisimulation further such that divergence is taken into account. In [BKO87] a bisimulation with explicit divergence is suggested, where a bisimulation can relate two states only if both or none of them can lead to immediate divergence. This idea can also be applied to branching bisimulation, as suggested in [Gla90b]. Rob van Glabbeek has developed an even finer distinction [Gla90a]; consider the nets of Figure 2.18. Both nets can diverge after a, but only the second can diverge in such a way that b remains possible all the time.

Two nets N_1 and N_2 are *(branching) bisimilar with bisimilar divergence* if there exists a (branching) bisimulation $\mathcal{B}$ such that

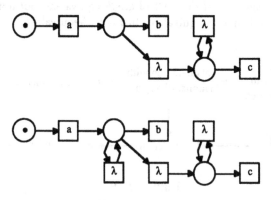

Figure 2.18

i) Assume $(M_{10}, M_{20}) \in \mathcal{B}$ such that there exists an infinite sequence $M_{10}[t_{10})M_{11}$ $[t_{11})M_{12}\ldots$ with $lab_1(t_{1i}) = \lambda$, $i \in \mathbb{N}_0$. Then we can find an infinite sequence $M_{20}[t_{20})M_{21}[t_{21})M_{22}\ldots$ with $lab_2(t_{2i}) = \lambda$, $i \in \mathbb{N}_0$, and we can partition the two infinite sequences of states into finite, non-empty groups $P_{10} = M_{10}\ldots M_{1i_0}$, $P_{11} = M_{1i_0+1}\ldots M_{1i_1}$, $P_{12} = M_{1i_1+1}\ldots M_{1i_2}$, $\ldots$ and $P_{20} = M_{20}\ldots M_{2j_0}$, $P_{21} = M_{2j_0+1}\ldots M_{2j_1}$, $P_{22} = M_{2j_1+1}\ldots M_{1j_2}$, $\ldots$ such that for all $k \in \mathbb{N}_0$ each state in P_{1k} is bisimilar to each state in P_{2k}.

ii) vice versa

Condition i) describes that, if the first net diverges, the second net can do so in such a way that it passes through related states.

I do not want to argue that branching bisimulation with bisimilar divergence is just the right equivalence for some purpose. We will just need a very strong equivalence to make one of our results below as general as possible, and this equivalence is the best I could find.

It is not difficult to see the following relations between the equivalences defined in this section:

Proposition 2.3.4   *i) Branching bisimulation with bisimilar divergence implies both, branching bisimulation and bisimulation with bisimilar divergence. Each of these implies bisimulation, this implies $\mathcal{F}$-equivalence, which in turn implies language equivalence. For conflict-free nets all these equivalences coincide.*

ii) Bisimulation with bisimilar divergence implies $\mathcal{FD}$-equivalence.

No other implications hold in general: The nets of Figure 2.15 are $\mathcal{FD}$-equivalent, but not language-equivalent, hence $\mathcal{FD}$-equivalence does not imply any of the other equivalences. The nets of Figure 2.14 are branching bisimilar, but not $\mathcal{FD}$-equivalent, hence branching bisimulation or any weaker equivalence does not imply $\mathcal{FD}$-equivalence or (branching) bisimulation with bisimilar divergence. Furthermore, Figure 2.17 shows nets which are bisimilar (with bisimilar divergence), but not branching bisimilar (with bisimilar divergence); the nets of Figure 2.13 are $\mathcal{F}$-equivalent, but not bisimilar; and the nets of Figure 2.11 are language-equivalent, but not $\mathcal{F}$-equivalent.

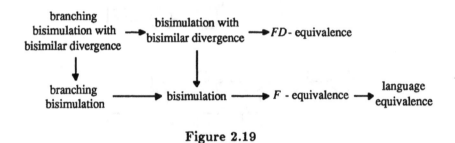

Figure 2.19

Figure 2.19 shows the implications presented in Proposition 2.3.4.

Chapter 3

Parallel Composition and Deadlocking

For the modular construction of Petri nets, it is desirable to have operations on nets that are intuitively meaningful, but also natural with respect to the graphical presentation of nets. In this chapter, we will mainly study a parallel composition operator indexed with a set A of actions; when composed with this operator, the component nets run in parallel, but have to synchronize when performing actions from A, i.e. they have to perform these together. Graphically, this operator stands for merging transitions.

The operator is introduced in the first section, and it is shown how the system runs of the composed system can be determined from the system runs of the components; here we consider the system runs introduced in Sections 2.1 and 2.2, in particular the firing sequences and processes. In the second section, we explain the ideas of external equivalence and full abstractness, and we show that the two versions of failure semantics introduced in Section 2.3 are just what we need to build deadlock-free systems and deadlock- and divergence-free systems with the parallel composition operator. Several variations of this approach are discussed in the third section, most notably the following two. For applications, the class of safe nets is particularly important, and we develop the necessary modifications for the construction of safe deadlock-free nets. Also, we show how to construct live nets; the corresponding variation of failure semantics will turn up again in the next chapter in a different context. Finally, some other operators on nets are introduced in the fourth section; these are well known from process algebra, but are only of secondary interest in our considerations.

3.1 Parallel composition with synchronization

Parallel systems often consist of several components that work in parallel and communicate with each other from time to time. This communication may be synchronous or asynchronous; in the synchronous case, two components communicate by performing some actions together. To describe how a system is constructed from synchronously communicating components, we introduce the operator $\parallel_A$, where A is the set of actions that have to be performed together; it corresponds to the parallel composition with synchronization of $TCSP$ [BHR84], and similar operators have also been studied widely in net theory, at least for unlabelled nets, see e.g. [Hac76b] as an early reference. Graphically, this operator stands for merging transitions, namely those transitions with label in A.

A straightforward merging of transitions is not always possible. It would require that each transition with label in A has a unique equally-labelled 'partner' in the other component such that we can merge each transition with its partner. This requirement is met in a natural way if the labellings of the components are injective for A (e.g. in the case of unlabelled nets) and each $a \in A$ occurs as a label in both components; then $\|_A$ takes the union of the components and identifies equally labelled transitions with label in A. In general, it must be possible to synchronize every a-labelled transition of one component with every a-labelled transition from the other component for $a \in A$. To do this, every such transition is split up into an appropriate number of copies before merging takes place. Figure 3.1 shows an example of a net N_1 modelling two measuring devices and a net N_2 modelling a channel. Figure 3.2 shows $N_1 \|_{\{send\ data\}} N_2$, where the two measuring devices send their data on the same channel. Of course, N_1 can be constructed from a net N modelling one measuring device as $N \|_\emptyset N$.

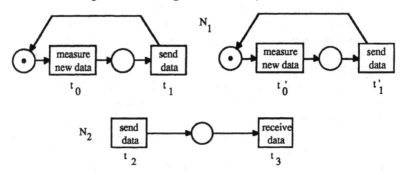

Figure 3.1

Formally, the places and transitions of the composed net are pairs, and we use λ as a dummy component where necessary; this ensures that we take the disjoint union of the place sets and the disjoint union over the unsynchronized transitions, and it allows a uniform treatment of synchronized and unsynchronized transitions.

Definition 3.1.1 Let N_1, N_2 be nets, $A \subseteq \Sigma$. Then the *parallel composition* $N = N_1 \|_A N_2$ of N_1 and N_2 *with synchronization over* A is defined by

$$S = S_1 \times \{\lambda\} \cup \{\lambda\} \times S_2$$

$$
\begin{aligned}
T = \ & \{(t_1, t_2) \mid t_1 \in T_1, t_2 \in T_2, lab_1(t_1) = lab_2(t_2) \in A\} \\
& \cup \{(t_1, t_2) \mid t_1 \in T_1, t_2 = \lambda, lab_1(t_1) \notin A \text{ or } t_1 = \lambda, t_2 \in T_2, lab_2(t_2) \notin A\}
\end{aligned}
$$

$$
W((s_1, s_2), (t_1, t_2)) = \left\{
\begin{array}{ll}
W_1(s_1, t_1) & \text{if } s_1 \in S_1, t_1 \in T_1 \\
W_2(s_2, t_2) & \text{if } s_2 \in S_2, t_2 \in T_2 \\
0 & \text{otherwise}
\end{array}
\right.
$$

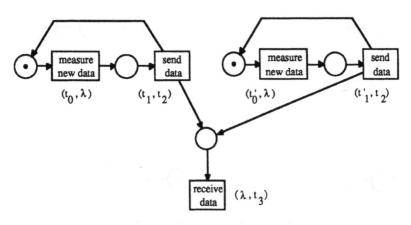

Figure 3.2

$$W((t_1,t_2),(s_1,s_2)) = \begin{cases} W_1(t_1,s_1) & \text{if } s_1 \in S_1, t_1 \in T_1 \\ W_2(t_2,s_2) & \text{if } s_2 \in S_2, t_2 \in T_2 \\ 0 & \text{otherwise} \end{cases}$$

$$M_N = M_{N_1} \dot{\cup} M_{N_2}, \text{ i.e. } M_N(s_1,s_2) = \begin{cases} M_{N_1}(s_1) & \text{if } s_1 \in S_1 \\ M_{N_2}(s_2) & \text{if } s_2 \in S_2 \end{cases}$$

$$lab(t_1,t_2) = \begin{cases} lab_1(t_1) & \text{if } t_1 \in T_1 \\ lab_2(t_2) & \text{if } t_2 \in T_2. \end{cases}$$

A special parallel composition is defined by $N_1 \parallel N_2 = N_1 \parallel_{\alpha(N_1) \cap \alpha(N_2)} N_2$. We consider N_1 and N_2 with some transitions multiplied (or deleted) as subnets of N; this justifies the notation $M_{N_1} \dot{\cup} M_{N_2}$.

It is easy to obtain the following result.

Proposition 3.1.2 *The operators $\parallel$ and $\parallel_A$ for every $A \subseteq \Sigma$ are associative and commutative. For all nets N_1 and N_2 we have $\alpha(N_1 \parallel N_2) = \alpha(N_1) \cup \alpha(N_2)$. For every $A \subseteq \Sigma$ the net consisting of the transition set A, no places and the identity labelling is the unit for $\parallel_A$; the empty net is the unit for $\parallel$.*

This result allows us to write $\parallel_{i \in I} N_i$ for a finite family $(N_i)_{i \in I}$ of nets. Note that in general we do not have $(N_1 \parallel_A N_2) \parallel_B N_3 = N_1 \parallel_A (N_2 \parallel_B N_3)$. Figure 3.3 shows nets N_1, N_2, N_3 and the different nets $(N_1 \parallel_{\{a\}} N_2) \parallel_{\{b\}} N_3$ and $N_1 \parallel_{\{a\}} (N_2 \parallel_{\{b\}} N_3)$. Also we have in this example $\alpha(N_2 \parallel_{\{b\}} N_3) = \{a,c\} \neq \alpha(N_2) \cup \alpha(N_3)$.

The system runs of the composed net can be determined from the system runs of the components, and we describe next how this can be done for $\parallel_A$, first on the level of transitions, then on the level of actions. As system runs we consider firing sequences, firing step sequences, processes, partial words, and semiwords.

Let pr_i denote the projection of a tuple to the i-th component. Let (B,E,F,l) be a causal net where l labels $B \cup E$ with tuples. Analogously to the definition of f^* on

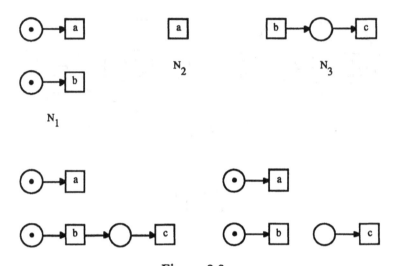

Figure 3.3

labelled partial orders we define $pr_i^*(B, E, F, l) = (B', E', F', pr_i \circ l \mid_{B' \cup E'})$, where $B' = \{b \in B \mid pr_i(l(b)) \neq \lambda\}$, $E' = \{e \in E \mid pr_i(l(e)) \neq \lambda\}$, and $F' = F \cap (B' \times E' \cup E' \times B')$. In the following, we write $pr_1^*(M)$ and $pr_2^*(M)$ for a marking M of a net $N_1 \parallel_A N_2$; in this notation, we regard M as a labelled partial order over $S_1 \times \{\lambda\} \cup \{\lambda\} \times S_2$ with empty ordering. Thus, $pr_1^*(M)$ is a marking of N_1, namely that defined by $(pr_1^*(M))(s_1) = M(s_1, \lambda)$, and similarly for $pr_2^*(M)$.

Note that we might have $ev(pr_i^*(\pi)) \neq pr_i^*(ev(\pi))$ for some process π of some net $N_1 \parallel_A N_2$, since applying pr_i^* to π might delete a condition that 'carries the causal precedence' between two events that are not deleted. Figure 3.4 shows a process π of the net in Figure 3.2, $ev(pr_2^*(\pi))$ – which is the event structure of a process of N_2 in Figure 3.1 – and $pr_2^*(ev(\pi))$ – which is not such an event structure.

First, we prove a lemma that shows that the projections of a system run of the composed net are system runs of the components.

Lemma 3.1.3 *Let N_1, N_2 be nets, $A \subseteq \Sigma$, and $N = N_1 \parallel_A N_2$.*

- *i)* *If π is a process of N with $M[\pi\rangle M'$, then for $i = 1, 2$ we have that $pr_i^*(\pi)$ is a process of N_i with $pr_i^*(M)[pr_i^*(\pi)\rangle pr_i^*(M')$.*

- *ii)* *If p is a partial word of N with $M[p\rangle M'$, then $pr_i^*(p)$ is a partial word of N_i with $pr_i^*(M)[pr_i^*(p)\rangle pr_i^*(M')$ for $i = 1, 2$.*

- *iii)* *If p is a semiword of N and lab_1, lab_2 are injective for A, then $pr_i^*(p)$ is a semiword of N_i with $pr_i^*(M)[pr_i^*(p)\rangle pr_i^*(M')$ for $i = 1, 2$.*

- *iv)* *If $w \in FSS^\infty(N)$ ($w \in FS^\infty(N)$), then $pr_i^*(w) \in FSS^\infty(N_i)$ ($pr_i^*(w) \in FS^\infty(N_i)$). If for a sequence w of steps or transitions we have $M[w\rangle$ or $M[w\rangle M'$ in N then $pr_i^*(M)[pr_i^*(w)\rangle$ or $pr_i^*(M)[pr_i^*(w)\rangle pr_i^*(M')$ in N_i for $i = 1, 2$.*

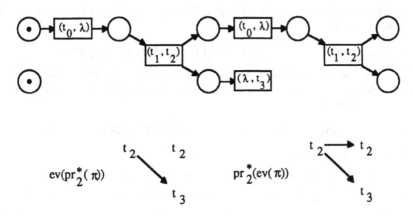

Figure 3.4

Proof: i) The requirements for $pr_i^*(\pi)$ being a process are easily checked. Observe that a condition of π with label (s_1, λ) can only be adjacent to events (t_1, t_2) with $t_1 \neq \lambda$. Thus the minimal (maximal) conditions in $pr_1^*(\pi)$ are also minimal (maximal) in π, and analogously for $pr_2^*(\pi)$. Furthermore each marking M of N can be seen as the disjoint union of a marking of N_1 and a marking of N_2, namely $pr_1^*(M)$ and $pr_2^*(M)$.

ii) By Theorem 2.2.10 we can find a process π with $M[\pi\rangle M'$ and $ev(\pi) \preceq p$. Since $ev(pr_i^*(\pi)) \preceq pr_i^*(ev(\pi)) \preceq pr_i^*(p)$ we conclude from i) that $pr_i^*(p)$ is a partial word as desired.

iii) Take two events e_1, e_2 of $pr_1^*(p)$ with the same label t_1. Either $lab_1(t_1) \notin A$, then $l(e_1) = l(e_2) = (t_1, \lambda)$, or $lab_1(t_1) \in A$, then there is a unique $t_2 \in T_2$ with $lab_2(t_2) = lab_1(t_1)$ and $l(e_1) = l(e_2) = (t_1, t_2)$. In either case e_1 and e_2 are ordered in p and thus in $pr_1^*(p)$. Hence the claim follows from ii).

iv) Finite firing (step) sequences are special partial words and applying pr_i^* preserves the transitivity of co and totality of the ordering. Thus we can apply ii). The extension to infinite sequences is obvious.   □

The following theorem describes how the processes and partial words etc. of a composed net can be determined from the processes and partial words etc. of the components.

Theorem 3.1.4 *Let N_1, N_2 be nets, $A \subseteq \Sigma$ and $N = N_1 \|_A N_2$.*

i) *The processes π of N with $M[\pi\rangle M'$ are those labelled causal nets that can be constructed as $\pi = (B_1 \cup B_2, E_1 \cup E_2, F_1 \cup F_2, l)$ from processes π_i of N_i with $pr_i^*(M)[\pi_i\rangle$ $pr_i^*(M')$, $i = 1, 2$, such that*

- $B_1 \cap (B_2 \cup E_2) = \emptyset = B_2 \cap (B_1 \cup E_1)$
- $\{e \in E_1 \mid lab_1(l_1(e)) \in A\} = E_1 \cap E_2 = \{e \in E_2 \mid lab_2(l_2(e)) \in A\}$
- $lab_1(l_1(e)) = lab_2(l_2(e))$ *for all* $e \in E_1 \cap E_2$
- $(F_1 \cup F_2)^+$ *is a partial order*

- $l(x) = (l_1(x), l_2(x))$ for all $x \in B_1 \cup B_2 \cup E_1 \cup E_2$, where $l_i(x) = \lambda$ if $x \notin B_i \cup E_i$ for $i = 1, 2$.

If π is constructed from π_1 and π_2, then $pr_1^*(\pi) = \pi_1$ and $pr_2^*(\pi) = \pi_2$.

ii) The partial words p of N with $M[p\rangle M'$ are those labelled partial orders that can be constructed as $p = (E_1 \cup E_2, <, l)$ from partial words p_i of N_i with $pr_i^*(M)[p_i\rangle pr_i^*(M')$, $i = 1, 2$, such that

- $\{e \in E_1 \mid lab_1(l_1(e)) \in A\} = E_1 \cap E_2 = \{e \in E_2 \mid lab_2(l_2(e)) \in A\}$
- $lab_1(l_1(e)) = lab_2(l_2(e))$ for all $e \in E_1 \cap E_2$
- $< \supseteq (<_1 \cup <_2)^+$ and $(<_1 \cup <_2)^+$ is a partial order.
- $l(e) = (l_1(e), l_2(e))$, for all $e \in E_1 \cup E_2$, where $l_i(e) = \lambda$ if $e \notin E_i$ for $i = 1, 2$.

For a labelled partial order p over T we have $M[p\rangle M'$ if and only if for $i = 1, 2$ we have that $pr_i^*(M)[pr_i^*(p)\rangle pr_i^*(M')$.

iii) If lab_1 and lab_2 are injective for A, then the semiwords of N can be constructed from the semiwords of N_1 and N_2 as in ii).

iv) The finite firing (step) sequences of N can be determined as partial words of N with total order or transitive co-relation from the firing (step) sequences of N_1 and N_2 as in ii).

We have for $w \in T^\infty$ or $w \in \mathcal{M}(T)^\infty$ that $M[w\rangle$ if and only if $pr_i^*(M)[pr_i^*(w)\rangle$ for $i = 1, 2$. The (step) sequence w is infinite if and only if at least one of $pr_1^*(w)$, $pr_2^*(w)$ is; both, $pr_1^*(w)$ and $pr_2^*(w)$, are (step) sequences just as w.

Proof: i) It is easy to see that the construction gives a process of N. (Note that the conditions refer to specific representatives of π_1 and π_2, and applying the construction to isomorphic representatives π_1' and π_2' might yield a process π' of N which is not isomorphic to π.) A given process π of N can be obtained by this construction from $pr_1^*(\pi)$ and $pr_2^*(\pi)$, which are processes of N_1 and N_2 by the preceding lemma.

Note that for a causal net π which is properly labelled over $S \cup T$ it would not be enough to require that $pr_1^*(\pi)$ and $pr_2^*(\pi)$ are processes. In such a π a transition from $T_1 \setminus T_2$ could be adjacent to a 'token' from S_2, since this would not be visible in $pr_1^*(\pi)$ or $pr_2^*(\pi)$; but such an adjacency is impossible for a process of N.

ii) If p is constructed from p_1 and p_2, then the construction of i) applied to processes π_1 and π_2 with $ev(\pi_1) \preceq p_1$, $ev(\pi_2) \preceq p$ yields a process π with $ev(\pi) \preceq p$, hence p is a partial word by Theorem 2.2.10. Every partial word p can be constructed from $pr_1^*(p)$ and $pr_2^*(p)$, which are partial words by the preceding lemma. Note that it would not suffice to construct partial words with $< = (<_1 \cup <_2)^+$: if p is a partial word of N, $pr_i^*(p) = p_i$, $i = 1, 2$, then we might have in p some immediate predecessor $e_1 \in E_1 \setminus E_2$ of some $e_2 \in E_2 \setminus E_1$, but we would not have $e_1(<_1 \cup <_2)^+ e_2$.

For the second part of ii) observe that a labelled partial order p over T is constructed from $pr_1^*(p)$ and $pr_2^*(p)$ as described in the first part. Also a proof directly from the definition of a partial word can be given: checking the enabledness condition for a place of the i-th component it is obviously enough to consider $pr_i^*(M)$ and $pr_i^*(p)$.

Note that in the case of partial words it is no problem that p might contain some ordering that is not visible in any of the projections, since partial words are closed under making them more sequential, see Proposition 2.2.6.

iii), iv) analogous; for firing (step) sequences a proof can also easily be obtained by induction. □

This theorem describes a composition of processes and partial words etc. which is similar to the operator $\|_A$ on nets. But here, each event e_1 of one system run with $lab_1(l_1(e)) \in A$ is combined with exactly one event e_2 of the second system which satisfies $lab_2(l_2(e_2)) = lab_1(l_1(e_1))$. Similarly to the construction in the previous theorem, we define a parallel composition with synchronization on action-labelled partial orders.

Definition 3.1.5 Let p_1, p_2 be labelled partial orders over Σ, $A \subseteq \Sigma$. If $p_1 = (X_1, <_1, l_1)$ and $p_2 = (X_2, <_2, l_2)$ are representatives such that

- $\{x \in X_1 \mid l_1(x) \in A\} = X_1 \cap X_2 = \{x \in X_2 \mid l_2(x) = A\}$
- $l_1(x) = l_2(x)$ for all $x \in X_1 \cap X_2$
- $(<_1 \cup <_2)^+$ is a partial order on $X_1 \cup X_2$,

then $p = (X_1 \cup X_2, (<_1 \cup <_2)^+, l_1 \cup l_2) \in p_1 \|_A p_2$, where $l_1 \cup l_2$ is the labelling that coincides with l_1 on X_1 and with l_2 on X_2. We call p a *parallel composition* of p_1 and p_2 *with synchronization* over A.

For the special case that the order of p_1 and p_2 is total, i.e. that we have sequences $v_1, v_2 \in \Sigma^*$, we write $v_1 \mathbb{I}_A v_2$ for $\{v \in \Sigma^* \mid \exists p \in v_1 \|_A v_2 : p \preceq v\}$. This notation is also used for $v_1, v_2 \in \Sigma^\infty$, where we transfer the above construction to possibly infinite labelled partial orders. Alternatively we can define (since in this book partial orders are always finite)

$$v_1 \mathbb{I}_A v_2 = \{v \in \Sigma^\infty \mid v = a_1 a_2 \ldots, \ v_1 = b_1 b_2 \ldots, \ v_2 = c_1 c_2 \ldots, \text{ such that for all } i$$
$$\text{either } a_i = b_i c_i \in (\Sigma - A) \cup \{\lambda\} \text{ or } a_i = b_i = c_i \in A\}.$$

(Note that $b_i c_i \in (\Sigma - A) \cup \{\lambda\}$ implies that $b_i = \lambda$ or $c_i = \lambda$.)

Definition 3.1.5 refers to specific representatives of p_1 and p_2. This requires some care since for some representatives composition might be impossible (i.e. $(<_1 \cup <_2)^+$ is not a partial order), while different results are possible for other representatives. For example, Figure 3.5 shows a labelled partial order p and the two elements of $p \|_{\{a\}} p$. Nevertheless, we have preferred not to distinguish always between isomorphism classes and representatives since this gets very tedious. Also the operator $\|$ on labelled partial orders requires some special care, and we postpone its definition to Chapter 7, where we will need it.

Note that $\mathbb{I}_\emptyset$ is the well-known shuffle operator. One easily sees that a criterion for the synchronizability of partial words can be given with the following projection function: for a partial order p labelled over Σ and $A \subseteq \Sigma$ the projection of p onto A is $id_A^*(p)$, where id_A is the identity on A and maps $\Sigma - A$ to $\{\lambda\}$.

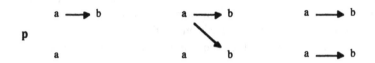

p

Figure 3.5

Proposition 3.1.6 *Let p_1 and p_2 be partial orders labelled over Σ and $A \subseteq \Sigma$. Then $p_1 \parallel_A p_2 \neq \emptyset$ if and only if $id_A^*(p_1) \parallel_A id_A^*(p_2) \neq \emptyset$ if and only if there are $p_1' \succeq id_A^*(p_1)$ and $p_2' \succeq id_A^*(p_2)$ with $p_1' = p_2'$. If $p_1, p_2 \in \Sigma^*$, then $p_1 \parallel_A p_2 \neq \emptyset$ if and only if $p_1 \parallel\!\parallel_A p_2 \neq \emptyset$ if and only if $id_A^*(p_1) = id_A^*(p_2)$.*

Remark: For traces in the sense of Mazurkiewicz and the operator $\parallel$ on labelled partial orders (to be defined in Chapter 7) a corresponding result does not hold in general, but only under a certain condition which is determined in [DV89].

Now we can lift Theorem 3.1.4 to the action level. For example a similar result as part ii) of the following theorem, but for unlabelled nets, can be found in [Gra81].

Theorem 3.1.7 *Let N_1, N_2 be nets, $A \subseteq \Sigma$, and $N = N_1 \parallel_A N_2$. Let M, M' be markings of N.*

i)

$$\{ac(\pi) \mid \pi \text{ is a process of } N \text{ and } M[ac(\pi)\rangle\rangle M'\} = \bigcup_{p_1 \in \Pi_1, p_2 \in \Pi_2} p_1 \parallel_A p_2,$$

where $\Pi_i = \{ac(\pi_i) \mid \pi_i \text{ is a process of } N_i \text{ and } pr_i^*(M)[ac(\pi_i)\rangle\rangle pr_i^*(M')\}, i = 1, 2.$

$$Proc(N) = \bigcup_{p_1 \in Proc(N_1), p_2 \in Proc(N_2)} p_1 \parallel_A p_2$$

ii) *$\{p \mid p$ is an image partial word of N and $M[p\rangle\rangle M'\} = \{p \mid$ there are image partial words p_i of N_i with $pr_i^*(M)[p_i\rangle\rangle pr_i^*(M')$, $i = 1, 2$, and q such that: $p \succeq q \in p_1 \parallel_A p_2\}$.*

$$PW(N) = \{p \mid \exists p_1 \in PW(N_1),\ p_2 \in PW(N_2),\ q : p \succeq q \in p_1 \parallel_A p_2\}$$

iii) *If lab_1 and lab_2 are injective for A, then the same equalities as in ii) hold for image semiwords and $SW(N)$.*

iv) *The same equalities as in ii) hold for image firing (step) sequences, $L(N)$ and $SL(N)$, especially:*
For all $a \in A : M[a\rangle\rangle M'$ in N if and only if $pr_i^*(M)[a\rangle\rangle pr_i^*(M')$ for $i = 1, 2$.
For all $a \in (\Sigma - A) \cup \{\lambda\} : M[a\rangle\rangle M'$ in N if and only if $pr_1^*(M)[a\rangle\rangle pr_1^*(M')$ and $pr_2^*(M)[\lambda\rangle\rangle pr_2^*(M')$ or $pr_1^*(M)[\lambda\rangle\rangle pr_1^*(M')$ and $pr_2^*(M)[a\rangle\rangle pr_2^*(M')$.

$$L(N) = \{w \in \Sigma^* \mid \exists w_1 \in L(N_1), w_2 \in L(N_2) : w \in w_1 \parallel\!\parallel_A w_2\}$$

$$L^\omega(N) = \{w \in \Sigma^\infty \mid \ \exists w_1 \in L^\infty(N_1), w_2 \in L^\infty(N_2) :$$
$$w_1 \in L^\omega(N_1) \text{ or } w_2 \in L^\omega(N_2), w \in w_1 \parallel\!\parallel_A w_2\}$$

Proof: i) Assume we are given processes π_1 of N_1 and π_2 of N_2. Applying the construction in Theorem 3.1.4 i) and taking the action structures of the resulting processes gives the same result as applying $\parallel_A$ to $ac(\pi_1)$ and $ac(\pi_2)$. Thus the result follows from 3.1.4.

ii), iii), iv) are analogous.   □

In Part iv) of the previous theorem we have shown a characterization of $M\langle a\rangle\rangle M'$. This characterization corresponds to the definition of an operational semantics for parallel composition in a process algebra like TCSP; the composed system can perform some $a \in A$ if both components can, it can perform $a \in \Sigma - A$ if one component can while the other stays idle. Due to this result, it is not much of a surprise that we will be able to show in the following, that failure semantics and bisimulation are congruences with respect to $\parallel_A$, results that are known from process algebras.

Corollary 3.1.8 *Process-, partial-word-, step-language- and language-equivalence – and for nets whose labelling is injective for A also semiword equivalence – are congruences with respect to $\parallel_A$, $A \subseteq \Sigma$. For nets without dead transitions they are also congruences with respect to $\parallel$, where in the case of semiword equivalence we apply $\parallel$ only to nets N_1, N_2 whose labelling is injective for $\alpha(N_1) \cap \alpha(N_2)$.*

If nets N_1, N_2 are language-equivalent, then for all $A \subseteq \Sigma$ and nets N the nets $N \parallel_A N_1$ and $N \parallel_A N_2$ are marking-equivalent on S_N.

Proof: For $\parallel$ observe that nets without dead transitions can only be equivalent if they have the same alphabet. Thus if N_1, N_2 are equivalent to N'_1, N'_2, then in $N_1 \parallel N_2$ and $N'_1 \parallel N'_2$ the operator $\parallel$ coincides in both cases with $\parallel_{\alpha(N_1) \cap \alpha(N_2)}$.   □

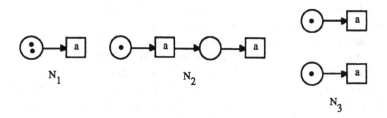

Figure 3.6

We close this section with an example showing that semiword equivalence is not a congruence with respect to $\parallel_A$ in general. Figure 3.6 shows the semiword-equivalent nets N_1 and N_2, and furthermore a net N_3 such that $N_1 \parallel_{\{a\}} N_3$ and $N_2 \parallel_{\{a\}} N_3$ are not semiword-equivalent; only $N_1 \parallel_{\{a\}} N_3$ has an image semiword consisting of two concurrent a's.

3.2   External equivalences based on deadlocking and divergence

When composing nets in a modular fashion, it is important that we can calculate the behaviour of the composed net from the behaviour of the components. In other words, we should use a semantics which gives as semantic equivalence a congruence with respect to the composition operators we use.

The last result of the previous section indicates that in the case of congruences for $\|_A$ we have quite a number of semantics to choose from, and in fact we will show some more congruence results in this chapter. Thus, it is to some degree a matter of taste which congruence one chooses. One way to reach an agreement in this situation would be to agree upon a very rough equivalence – more precisely, to agree that nets which are not in this rough equivalence should definitely be distinguished. Then, one can determine the coarsest congruence with respect to the operators in question that refines the rough equivalence; such a congruence is called *fully abstract* [Mil77] with respect to the operators and the rough equivalence. This fully abstract equivalence is just the right one for a modular construction – in the sense that it distinguishes exactly those nets that have to be distinguished according to our agreement and to our desire to have a congruence.

If we change the rough equivalence to a finer equivalence that lies between the rough and the fully abstract equivalence, then we end up with the same fully abstract equivalence for this finer equivalence. Therefore, it is desirable to find a very rough equivalence such that the corresponding fully abstract equivalence is nevertheless very distinctive; then this fully abstract equivalence will be agreeable for many people. In the third part of Section 3.3, we will present an example of an interesting equivalence that lies strictly between a rough equivalence considered earlier and the corresponding fully abstract equivalence.

A slightly different approach to modular net construction is to exchange subnets in such a way that the behaviour of the whole net remains unchanged. Regarding the operators $\|_A$ we call nets N_1 and N_2 externally equivalent if for all nets N and all $A \subseteq \Sigma$ the nets $N \|_A N_1$ and $N \|_A N_2$ have the same behaviour in some sense. The aim is to characterize this external equivalence by some internal equivalence that refers only to the nets N_1 and N_2, but not to all possible environments N and synchronization sets A. (The notions external and internal equivalence were introduced in [Bau88].)

Thus, we start with some basic behaviour or – what amounts to the same – with some rough equivalence and end up with the more elaborate external (or internal) equivalence. That it is indeed more elaborate follows from the fact that at least one of our operators has a unit: if $N_\emptyset$ is the empty net, and N_1 and N_2 are externally equivalent, then $N_1 = N_\emptyset \|_\emptyset N_1$ and $N_2 = N_\emptyset \|_\emptyset N_2$ have the same basic behaviour.

Again we can argue that the external equivalence is just the right equivalence for this kind of modular approach and the rough equivalence we started from. In general, this external equivalence does not have to be a congruence; it takes into account the exchange of nets in contexts built with only one application of one of our operators, while full abstractness considers all possible contexts that can be built with any number of applications of these operators. But in the natural cases we consider here, external equivalence turns out to be a congruence, and both approaches to modular construction agree. At the end of this section, we will present an artificial rough equivalence for which the related external equivalence and the related fully abstract congruence differ.

To define the first rough equivalence we consider the ability or inability to come to a standstill, i.e. to deadlock. This ability is certainly important for systems that should run 'forever' like operating systems or manufacturing processes: they should not deadlock. But also in general, it should be possible to agree that a system that can deadlock is not equivalent to a system that cannot deadlock.

The second equivalence is concerned with divergence. Divergence, i.e. an infinite firing sequence of internal transitions, might be acceptable if we consider for example a busy-wait loop, but in general a system design that allows an infinite internal looping is most likely incorrect. At some stage such a system would not show any visible behaviour just as in the case of deadlock, but we could not even be sure that it will not show some visible action in the future.

Definition 3.2.1 A net can *deadlock* if it has a reachable marking M such that $\neg M[a\rangle)$ for all $a \in \Sigma$; otherwise it is called *deadlock-free*. Two nets are *deadlock-similar* if both can deadlock or both are deadlock-free. Nets N_1 and N_2 are *deadlock-equivalent*, if for all nets N and all $A \subseteq \Sigma$ the nets $N \parallel_A N_1$ and $N \parallel_A N_2$ are deadlock-similar.

A net can *diverge* if it has a reachable marking under which an infinite sequence of internal transitions is enabled; otherwise it is called *divergence-free*. Two nets are *deadlock/divergence-similar*, if either both are deadlock- and divergence-free or both can deadlock or diverge. Nets N_1 and N_2 are *deadlock/divergence-equivalent*, if for all nets N and all $A \subseteq \Sigma$ the nets $N \parallel_A N_1$ and $N \parallel_A N_2$ are deadlock/divergence-similar.

The results on deadlock equivalence and its internal characterization we present in the following we take from [Vog89]. The approach taken here has some similarities with the testing equivalences defined in [DNH84] – mainly in a setting of CCS processes. (Similar ideas as in [DNH84] have also been expressed in [Dar82]). For example, by the definition of deadlock equivalence we can test some nets N_1 and N_2 for equivalence by choosing appropriate nets N and sets A. In [DNH84] testing refers to some observers and their abilities, and varying the abilities of the observers one can argue not only for the testing equivalences of [DNH84], but for example also for bisimulation; see [Abr87]. Here, we refer to principles of modular system design and some basic features of systems, and we would not argue that for example divergence is observable. Still, deadlock/divergence equivalence coincides with the must-equivalence of [DNH84]; but deadlock equivalence does not have a counterpart.

To compare systems, often pre-orders $\sqsubseteq$ are considered instead of equivalences, where $N_1 \sqsubseteq N_2$ can be read as 'N_2 implements N_1', see e.g. [DNH84]. For practical applications this is very important: think of N_1 as a specification (that possibly allows some freedom) and of N_2 as an implementation, where the decisions left open in the specification have been made; $N_1 \sqsubseteq N_2$ means that N_2 is a correct implementation of N_1. An example will be discussed below.

Most often the considerations for equivalences are very similar as for the related pre-orders, but slightly less tedious. Therefore, in general, we only consider equivalences in this book. Just as an example, we will introduce the pre-orders related to deadlock- and deadlock/divergence-equivalence.

Definition 3.2.2 A net N_2 is a *deadlock-free implementation* of a net N_1, denoted $N_1 \sqsubseteq_D N_2$, if for all nets N and all $A \subseteq \Sigma$ deadlock-freeness of $N \parallel_A N_1$ implies that of $N \parallel_A N_2$.

A net N_2 is a *deadlock/divergence-free implementation* of N_1, denoted $N_1 \sqsubseteq_{DD} N_2$, if for all nets N and all $A \subseteq \Sigma$ deadlock- and divergence-freeness of $N \parallel_A N_1$ implies that of $N \parallel_A N_2$.

In the following we will compare deadlock- and deadlock/divergence-equivalence with $\mathcal{F}$- and $\mathcal{FD}$-equivalence. First, we note some obvious connections between deadlock- and deadlock/divergence-similarity on the one hand and $\mathcal{F}$- and $\mathcal{FD}$-equivalence on the other.

Lemma 3.2.3   *i) A net N can deadlock if and only if for some $w \in \Sigma^*$ we have $(w, \Sigma) \in \mathcal{F}(N)$. $\mathcal{F}$-equivalence implies deadlock similarity.*

ii) A net N is divergence-free if only if $D(N) = \emptyset$. If N is divergence-free, then $\mathcal{F}(N) = F(N)$. $\mathcal{FD}$-equivalence implies deadlock/divergence similarity.

Proof: We only show the last part of ii). Assume $\mathcal{FD}(N_1) = \mathcal{FD}(N_2)$ for some nets N_1, N_2. Now N_1 can diverge iff $D(N_1) \neq \emptyset$ iff N_2 can diverge. If N_1 and N_2 are divergence-free, then they are $\mathcal{F}$-equivalent, thus deadlock-similar by i). $\square$

Now we show that $\mathcal{F}$- and $\mathcal{FD}$-equivalence are congruences with respect to $\parallel_A$.

Theorem 3.2.4 $\mathcal{F}$-*equivalence is a congruence with respect to* $\parallel_A$ *for all $A \subseteq \Sigma$; for nets N_1, N_2, and $A \subseteq \Sigma$ we have:*

$$\mathcal{F}(N_1 \parallel_A N_2) = \{(w, X) \mid w \in \Sigma^*, \ X \subseteq \Sigma, \ and \ \exists (w_i, X_i) \in \mathcal{F}(N_i), i = 1, 2 :$$
$$w \in w_1 \parallel\!\!\parallel_A w_2 \ and \ X - A \subseteq X_1 \cap X_2 \ and \ X \cap A \subseteq X_1 \cup X_2\}$$

Proof: The congruence result follows, once we have shown the above equality. Let $N = N_1 \parallel_A N_2$.

'$\subseteq$' Let $(w, X) \in \mathcal{F}(N)$; then there exists M with $M_N[w\rangle\rangle M$ and $\neg M[a\rangle\rangle$ for all $a \in X$. By Theorem 3.1.7 iv) there exist w_1, w_2 such that $M_{N_1}[w_1\rangle\rangle pr_1^*(M)$, $M_{N_2}[w_2\rangle\rangle pr_2^*(M)$, and $w \in w_1 \parallel\!\!\parallel_A w_2$. Put $X_1 = \{a \in \Sigma \mid \neg pr_1^*(M)[a\rangle\rangle\}$ and $X_2 = \{a \in \Sigma \mid \neg pr_2^*(M)[a\rangle\rangle\}$. Now Theorem 3.1.7 iv) shows that for $a \in X \cap A$ we must have $a \in X_1$ or $a \in X_2$, and for $a \in X - A$ we must have $a \in X_1$ and $a \in X_2$.

'$\supseteq$' Let an element (w, X) of the right-hand side be given, and let $M_{N_1}[w_1\rangle\rangle M_1$, $M_{N_2}[w_2\rangle\rangle M_2$ such that $\neg M_1[a\rangle\rangle$ for all $a \in X_1$ and $\neg M_2[a\rangle\rangle$ for all $a \in X_2$. Again by Theorem 3.1.7 iv), we have $M_N[w\rangle\rangle M_1 \dot\cup M_2$. For $a \in A$, $M_1 \dot\cup M_2[a\rangle\rangle$ implies $M_1[a\rangle\rangle$ and $M_2[a\rangle\rangle$, hence $\neg M_1 \dot\cup M_2[a\rangle\rangle$ for $a \in X \cap A$. For $a \in \Sigma - A$, $M_1 \dot\cup M_2[a\rangle\rangle$ implies $M_1[a\rangle\rangle$ or $M_2[a\rangle\rangle$, hence we have $\neg M_1 \dot\cup M_2[a\rangle\rangle$ for $a \in X - A$. $\square$

Theorem 3.2.5 $\mathcal{FD}$-*equivalence is a congruence with respect to* $\parallel_A$ *for all $A \subseteq \Sigma$; for nets N_1, N_2, and $A \subseteq \Sigma$ we have*

$$F(N_1 \parallel_A N_2) = \{(w, X) \mid w \in \Sigma^*, X \subseteq \Sigma, \ and \ \exists (w_i, X_i) \in F(N_i), \ i = 1, 2 :$$
$$w \in w_1 \parallel\!\!\parallel_A w_2 \ and \ X - A \subseteq X_1 \cap X_2 \ and \ X \cap A \subseteq X_1 \cup X_2\}$$
$$\cup \ \{(w, X) \mid w \in D(N_1 \parallel_A N_2), X \subseteq \Sigma\}$$

$$D(N_1 \|_A N_2) \;=\; \{w \in \Sigma^* \mid \exists (w_1, \emptyset) \in F(N_1), (w_2, \emptyset) \in F(N_2), v \in \Sigma^* :$$
$$(w_1 \in D(N_1) \text{ or } w_2 \in D(N_2)) \text{ and } v \in w_1 \|_A w_2 \text{ is a prefix of } w\}$$

Proof: For the equality of the D-part we will apply the last part of Theorem 3.1.7 iv).

'$\subseteq$' Obviously, it is enough to consider some $w \in D(N_1 \|_A N_2)$ such that no proper prefix of w is in $D(N_1 \|_A N_2)$. Such a w is a divergence string, thus we can find $w_1 \in L^\infty(N_1)$, $w_2 \in L^\infty(N_2)$, such that $w \in w_1 \|_A w_2$ and $w_1 \in L^\omega(N_1)$ or $w_2 \in L^\omega(N_2)$. Since for finite w_i we have $w_i \in L^\infty(N_i)$ iff $(w_i, \emptyset) \in F(N_i)$, and $w_i \in L^\omega(N_i)$ iff $w_i \in D(N_i)$, $i = 1, 2$, we are done.

'$\supseteq$' Again we only have to consider $w \in \Sigma^*$ such that no proper prefix of w is in the right-hand side. If, say, w_1 has a proper prefix in $D(N_1)$, then we can construct from this prefix and an appropriate prefix of w_2 a proper prefix of w belonging to the right-hand side. By choice of w this is not possible. Therefore one of w_1 and w_2 is a divergence string, the other an image firing sequence; we conclude that w is a divergence string of $N_1 \|_A N_2$.

For the F-part we have that (w, X) is in both sets if $w \in D(N_1 \|_A N_2)$. Otherwise we conclude from the D-part that the $(w, X), (w_i, X_i)$ we have to consider belong in fact to $\mathcal{F}(N), \mathcal{F}(N_i)$, and the result follows from the previous theorem. $\square$

Now we are ready to show that $\mathcal{F}$- and $\mathcal{FD}$-equivalence are the internal equivalences belonging to deadlock- and deadlock/divergence-similarity. Therefore, by the above theorems, they are also the coarsest congruences refining the similarities. We can also give an internal characterization of the pre-orders $\sqsubseteq_D$ and $\sqsubseteq_{DD}$, and they turn out to be pre-congruences (i.e. $N_1 \sqsubseteq_D N_1'$ and $N_2 \sqsubseteq_D N_2'$ implies $N_1 \|_A N_2 \sqsubseteq_D N_1' \|_A N_2'$, and the same implication holds for $\sqsubseteq_{DD}$).

Theorem 3.2.6   *i) Let N_1, N_2 be nets with $\alpha(N_1) \cup \alpha(N_2) \neq \Sigma$. Then*

- *$N_1 \sqsubseteq_D N_2$ if and only if $\mathcal{F}(N_1) \supseteq \mathcal{F}(N_2)$,*
- *N_1 and N_2 are deadlock-equivalent if and only if they are $\mathcal{F}$-equivalent,*
- *$N_1 \sqsubseteq_{DD} N_2$ if and only if $\mathcal{FD}(N_1) \supseteq \mathcal{FD}(N_2)$ (componentwise),*
- *N_1 and N_2 are deadlock/divergence-equivalent if and only if they are $\mathcal{FD}$-equivalent.*

ii) Let Σ' be an infinite subset of Σ, let $\mathcal{N}$ be the class of nets N for which $\alpha(N) \cap \Sigma'$ is finite. Then $\mathcal{N}$ is closed under $\|_A$ for all $A \subseteq \Sigma$, and on $\mathcal{N}$

- *$\sqsubseteq_D (\sqsubseteq_{DD})$ is the coarsest pre-congruence with respect to $\|_A$, $A \subseteq \Sigma$, such that: if $N_1 \sqsubseteq_D N_2$ ($N_1 \sqsubseteq_{DD} N_2$) and N_1 is deadlock- (deadlock/divergence-) free, then N_2 is deadlock- (deadlock/divergence-) free.*
- *$\mathcal{F}$- ($\mathcal{FD}$-) equivalence is fully abstract with respect to $\|_A$, $A \subseteq \Sigma$, and deadlock- (deadlock/divergence-) similarity.*

All these results also hold if in Definition 3.2.1 and 3.2.2 only finite nets are considered.

Proof: i) We only have to show the assertions concerning $\sqsubseteq_D$ and $\sqsubseteq_{DD}$. If $\mathcal{F}(N_1) \supseteq \mathcal{F}(N_2)$, then by Theorem 3.2.4 $\mathcal{F}(N \parallel_A N_1) \supseteq \mathcal{F}(N \parallel_A N_2)$ for all nets N and $A \subseteq \Sigma$; thus, by Lemma 3.2.3 i), $N \parallel_A N_2$ can deadlock only if $N \parallel_A N_1$ can; therefore $N_1 \sqsubseteq_D N_2$. If $\mathcal{FD}(N_1) \supseteq \mathcal{FD}(N_2)$, then similarly $N \parallel_A N_2$ can diverge only if $N \parallel_A N_1$ can; if $N \parallel_A N_2$ is divergence-free but can deadlock, then $N \parallel_A N_1$ can diverge or deadlock; therefore $N_1 \sqsubseteq_{DD} N_2$.

Now assume $N_1 \sqsubseteq_D N_2$. Let $A = \alpha(N_1) \cup \alpha(N_2)$, $(w, X) \in \mathcal{F}(N_2)$, $w = w_1 \ldots w_m$ with $w_i \in \Sigma$. Consider the net N (see Figure 3.7) defined by

$$\begin{aligned}
S &= \{s_a, s'_a, s_j, s \mid a \in A, \ j = 1, \ldots, m+2\} \\
T &= \{t_a, t'_b, t_j, t \mid a \in A, \ b \in A \cap X, \ j = 1, \ldots, m+1\}
\end{aligned}$$

W takes values in $\{0, 1\}$ and is 1 for the pairs

- $(s_j, t_j), (t_j, s_{j+1})$ for $j = 1, \ldots, m+1$

- $(t_j, s_a), (s'_a, t_{j+1})$ for $j = 1, \ldots, m$, $a = w_j$

- $(s_a, t_a), (t_a, s'_a)$, for $a \in A$

- $(s, t), (t, s), (s, t_{m+1})$

- $(t_{m+1}, s_b), (s_{m+2}, t'_b), (s'_b, t'_b), (t'_b, s)$ for $b \in A \cap X$

M is 1 on s_1 and s, and 0 everywhere else.
$lab(t_a) = a$ for $a \in A$,
$lab(t), lab(t_j), lab(t'_b) \in \Sigma - A$ for $j = 1, \ldots, m+1$ and $b \in X \cap A$.

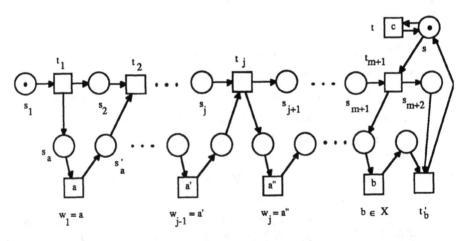

Figure 3.7

The only way that $N \parallel_A N_i$ can deadlock is to block t by firing $t_1 \ldots t_{m+1}$ interspersed with actions that give the sequence w. This interspersed sequence must correspond to a firing sequence of N_i that yields a marking M_i with $\neg M_i[b)$ for all $b \in X \cap A$; otherwise we could mark s again. Therefore $N \parallel_A N_2$ can deadlock, thus $N \parallel_A N_1$ can, too. Hence $(w, X \cap A) \in \mathcal{F}(N_1)$, and also $(w, X) \in \mathcal{F}(N_1)$, since N_1 does not have a transition with label in $X - A$. Therefore $\mathcal{F}(N_2) \subseteq \mathcal{F}(N_1)$. If N_1 and N_2 are finite, then A and thus N are finite, too. Therefore $\mathcal{F}(N_2) \subseteq \mathcal{F}(N_1)$ also in the case that in Definition 3.2.2 only finite nets are considered.

Finally assume $N_1 \sqsubseteq_{DD} N_2$. For $w \in D(N_2)$ consider the net N of Figure 3.7, but without the transitions t_{m+1} and t'_b. $N \parallel_A N_i$ cannot deadlock, and it can diverge if and only if a prefix of w is a divergence string of N_i. Thus, $N \parallel_A N_2$ can diverge, hence also $N \parallel_A N_1$, and we conclude that $D(N_2) \subseteq D(N_1)$. For $(w, X) \in F(N_2)$ with $w \notin D(N_2)$ we can use the same construction as above. Since $N \parallel_A N_2$ cannot diverge, we get that $w \in D(N_1)$ or $(w, X) \in \mathcal{F}(N_1)$, hence $(w, X) \in F(N_1)$ and $F(N_2) \subseteq F(N_1)$.

ii) It is obvious that $\mathcal{N}$ is closed under $\parallel_A$, and for nets from $\mathcal{N}$ we can apply i) with test nets N taken from $\mathcal{N}$, too. That the relations are pre-congruences, congruences respectively, follows from i) and the previous theorems; that they are the coarsest pre-congruences, fully abstract respectively, as desired follows directly from the definitions of $\sqsubseteq_D$, $\sqsubseteq_{DD}$, deadlock- and deadlock/divergence-equivalence. □

The preconditions of this theorem, that $\alpha(N_1) \cup \alpha(N_2) \neq \Sigma$ and that nets are taken from the class $\mathcal{N}$, are somewhat technical; often all nets under consideration will have a finite alphabet, e.g. if they are finite themselves, and then these preconditions are trivially satisfied.

Of course, the definition of deadlock- and deadlock/divergence-equivalence depends crucially on the context nets N that are considered. Thus, it should be stressed that the context nets in the above proof are safe and without λ-labels, they are finite for nets with finite alphabet, and their labelling can be chosen to be injective for such nets. For the parallel composition of safe nets, see also the next section.

Exchanging a net N_1 by a deadlock- (or deadlock/divergence-) free implementation N_2 of N_1 preserves by definition only a very simple feature of behaviour. By Theorem 3.2.6, in fact in a much stronger sense behaviour that is guaranteed by the first net will be preserved. In N_2 some nondeterministic choices of N_1 are resolved, but otherwise it can exhibit the same actions. Consider N_1, the first net of Figure 2.13. It can perform a, and then it is guaranteed that it can perform b or c – since $(a, \{b, c\})$ is not in its failure semantics. Thus, a net N_2 that can just perform a and nothing else is not a deadlock-free implementation of N_1, since $(a, \{b, c\}) \in \mathcal{F}(N_2)$. On the other hand, if N_2 can just perform a followed by b, then we have $N_1 \sqsubseteq_D N_2$. When N_1 performs a it has the nondeterministic choice, which of the three a-transitions it fires. It can always choose the upper a-transition, and this choice is implemented in N_2.

If $N_1 \sqsubseteq_D N_2$, then by Theorem 3.2.6 N_2 is a more deterministic version of N_1, but only with respect to interleaving behaviour. In this approach to implementation, we cannot specify that some concurrency must be possible or must not be possible. This might be a drawback, but on the other hand both of the following views are reasonable:

- If the specification allows some concurrency, the implementation may interleave the respective actions, e.g. if only one processor is available.

– If the specification does not prescribe concurrency, one may still be able to detect some possible concurrency, and implementing it can speed up the system (see e.g. [Ace89]).

Therefore, it seems to be reasonable to leave open, how concurrency of the specification is treated in the implementation.

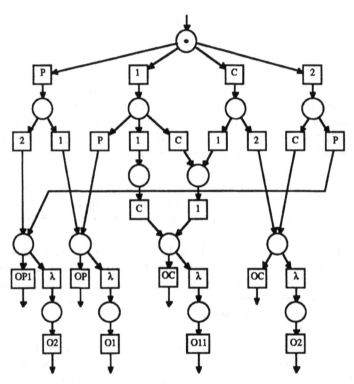

C	customer chooses chocolate	OC	customer gets chocolate
		$O2$	customer gets a 2-piece
P	customer chooses peanuts	OP	customer gets peanuts
		$OP1$	customer gets peanuts and a 1-piece
2	customer inserts a 2-piece	$O1$	customer gets a 1-piece
1	customer inserts a 1-piece	$O11$	customer gets two 1-pieces

Figure 3.8

An example

Let us look at a more elaborate example. Figure 3.8 shows the specification SV of a vending machine (in the vain of [Hoa85]) that sells peanuts for one (monetary unit) or chocolate for two. If the customer chooses chocolate, the machine accepts payment of two either as one 2-piece or as two 1-pieces; then it outputs a chocolate or it internally decides to return the inserted money, e.g. because there is no chocolate available at the moment or because the machine finds the inserted coins suspicious. If the customer chooses peanuts, the machine accepts one coin and outputs the peanuts and the change if necessary – or again it returns the inserted money. The choice between chocolate and peanuts and the insertion of money can be done in any order. In Figure 3.8 we have used the following abbreviation: each arc going out from the drawing at the bottom goes to the top place with the ingoing arc.

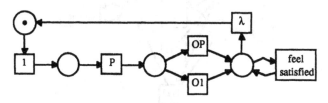

Figure 3.9

This specification is purely sequential, possibly because the designer found it easier to think sequentially. Admittedly, the specification is not completely satisfactory, since it allows the machine to return the money in every case. But it certainly specifies some minimal expectations of both, the owner of the machine and the customers. Indeed, we can convince ourselves that the machine as specified works to the (repeated) satisfaction for the customers shown in Figures 3.9 – 3.11.

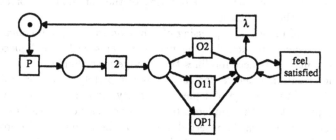

Figure 3.10

The first customer inserts a 1-piece first and and presses the peanuts-button after-wards. The second customer presses the peanuts-button first, inserts a 2-piece afterwards (maybe he is just interested in the change), and would also accept two 1-pieces – which the machine will never output. The third customer is quite in a hurry to get some choco-late; so she presses the chocolate-button while at the same time inserting 1-pieces until something is output by the machine; hopefully it is chocolate, but grudgingly she would also accept some number of 1-pieces. If *Cust*, any of these customers, interacts with the machine as specified, then satisfaction will invariably be possible, i.e. 'feel satisfaction' is image-live in $Cust \parallel_A SV$ where $A = \alpha(SV)$, and especially $Cust \parallel_A SV$ cannot deadlock. To see this, it is advisable not to draw the net $Cust \parallel_A SV$, but to use Theorem 3.1.7 iv) which says that $Cust \parallel_A SV$ can perform e.g. P if both $Cust$ and SV can perform P under the corresponding markings.

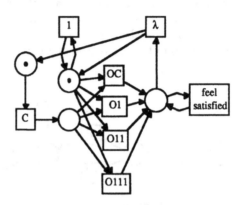

Figure 3.11

Next, we will give a deadlock-free implementation IV of SV, hence we will have that $Cust \parallel_A IV$ is deadlock-free, too; thus, due to the special form of the customers, 'feel satisfaction' will be image-live in any of these nets.

Figure 3.12 shows IV, and we have used the following abbreviation: each arc going out from the drawing at the bottom represents three arcs going to the three top places with the ingoing arcs. The triples labelling the outgoing arcs give the arc weights of the corresponding three arcs, where the ordering of the components corresponds to the graphical ordering of the top places.

In IV, insertion of money and pressing of a button may happen concurrently. Also, a storage s of 1-pieces is modelled. The machine starts with five 1-pieces, and every insertion and output of a 1-piece modifies this storage accordingly.

There is one situation where a choice that is completely nondeterministic in SV is determined in IV: if the storage of 1-pieces is empty, and the customer wants peanuts and inserts a 2-piece, then the implemented machine has no choice but to return the money. The decision to return the inserted money can be made by IV immediately after a button is pressed and independently of the insertion of money. This is a difference to SV, but a difference which is immaterial for deadlocks and ignored by the failure semantics.

To convince ourselves that indeed $SV \sqsubseteq_D IV$, we need the following lemma.

Lemma 3.2.7 *We say that an action $a \in \Sigma$ is possible after $w \in \Sigma^*$ in a net N, if $(wa, \emptyset) \in \mathcal{F}(N)$. Let N_1 and N_2 be nets such that $L(N_1) \supseteq L(N_2)$, and for all $w \in L(N_2)$ and sets X of actions possible after w in N_1 we have $(w, X) \in \mathcal{F}(N_2) \Rightarrow (w, X) \in \mathcal{F}(N_1)$. Then $\mathcal{F}(N_1) \supseteq \mathcal{F}(N_2)$.*

Proof: Let $(w, X) \in \mathcal{F}(N_2)$, $Y := \{a \in X \mid (wa, \emptyset) \in \mathcal{F}(N_1)\}$, and $Y' = X - Y$. By Proposition 2.3.1 we have $(w, Y) \in \mathcal{F}(N_2)$, and by assumption $(w, Y) \in \mathcal{F}(N_1)$. By

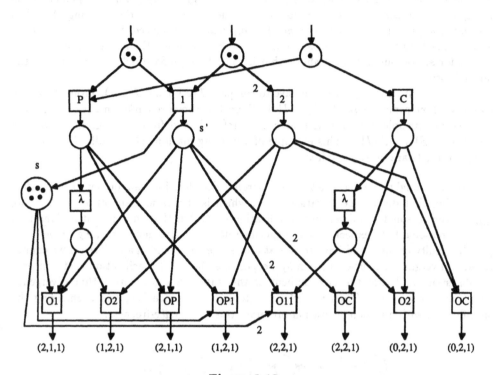

Figure 3.12

definition of Y' we have $\neg M[a\rangle)$ for all $a \in Y'$ and M such that $M_{N_1}[w\rangle) M$. Hence $(w, X) \in \mathcal{F}(N_1)$. □

Before we consider the failure semantics of SV and IV in order to show $SV \sqsubseteq_D IV$, we have a look at the languages of SV and IV, and we make the following definitions.

$$
\begin{array}{llll}
L_1 & = & \{1P, P1\}, & a_{11} = OP, \quad a_{12} = O1 \\
L_2 & = & \{2P, P2\}, & a_{21} = OP1, \quad a_{22} = O2 \\
L_3 & = & \{11C, 1C1, C11\}, & a_{31} = OC, \quad a_{32} = O11 \\
L_4 & = & \{2C, C2\}, & a_{41} = OC, \quad a_{42} = O2
\end{array}
$$

Let $L = (\bigcup_{i=1}^{4} L_i\{a_{i1}, a_{i2}\})^*$; it is not too difficult to see that $L(SV)$ is the set of prefixes of strings from L. We claim that any $w \in L(IV)$ is a prefix of some $v \in L$; furthermore, that if $w \in L(IV) \cap L$ then the marking reached after w equals the initial marking except on s.

We show this by induction, and the claim is obvious for $w = \lambda$. Now let $w \in L(IV)$ such that the claim holds for any shorter string; let w' be a maximal proper prefix of w with $w' \in L$, i.e. the marking reached after w' is as described above.

After w' any sequence from $\bigcup L_i$ can occur, and no action that is possible can be refused. Say, $v' \in L_i$ occurs and observe that s now carries at least as many tokens as s'.

The next action can only be a_{i1} or a_{i2}. Whether a_{i1} is actually possible depends on the firing of the corresponding λ-transition, and in the case $i = 2$ on the marking of s; thus a_{i1} may always be refused at this stage. But a_{i2} cannot be refused; for $i = 1, 3$ recall the above remark on the marking of s'. After a_{ij} we have reached the initial marking again, except for s; also our argument shows that w must be a prefix of $w'v'a_{ij}$ and thus of the desired form.

In order to show $SV \sqsubseteq_D IV$, we have to compare $\mathcal{F}(SV)$ and $\mathcal{F}(IV)$. From the above claim, we first conclude $L(SV) \supseteq L(IV)$. From the above remarks on actions that are possible in principle and actions that can be refused, and from Lemma 3.2.7, we now arrive at $\mathcal{F}(SV) \supseteq \mathcal{F}(IV)$. This example of a vending machine also fulfills $SV \sqsubseteq_{DD} IV$, since both nets are divergence-free.

Now one might wonder whether a diverging specification can be sensible at all – considering that divergence is catastrophic, which is the basic assumption when using $\sqsubseteq_{DD}$. One reason for a diverging specification could be, that in some situations it is just not reasonable to expect that the specified device is divergence-free, e.g. if we use a possibly faulty channel and have to repeat a message again and again. Even if we cannot avoid a diverging specification, it may be possible that this specification just concerns a subsystem, and that the environment of this subsystem guarantees that divergence situations are avoided. Then the whole system is divergence-free, and exchanging the specification with an implementation preserves this divergence-freeness.

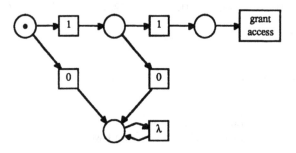

Figure 3.13

Another possibility is that divergence indicates underspecification. Figure 3.13 concerns the input of some code in order to get access to some proctected item; the net shown specifies that input of the correct code guarantees access. (For simplicity we have assumed a binary two digit code, namely '11'). It is left open – indicated by divergence – what happens on input of a wrong digit. Figure 3.14 shows a deadlock/divergence-free implementation of this specification, where the divergence is replaced by some nondiverging behaviour.

In the above example of the vending machine we have shown that the given internal characterization of $\sqsubseteq_D$ makes it possible to prove this relation. Of course, it is desirable to automate such a proof – at least for finite nets. This is in general not possible, as the following theorem shows. For bounded nets it can be done; but the proof below suggests

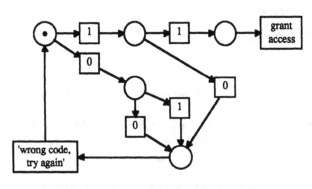

Figure 3.14

to use $\mathcal{R}(N)$, and it should be remarked that already the construction of $\mathcal{R}(N)$ might require exponential space.

Theorem 3.2.8 $\mathcal{F}$- *and* $\mathcal{FD}$-*equivalence are undecidable for finite nets. For finite and bounded nets* $\sqsubseteq_D$ *and* $\sqsubseteq_{DD}$ *are decidable.*

Proof: Let $L_0(N) = \{w \in L(N) \mid M_N[w\rangle M$ and $M(s) = 0$ for all $s \in S\}$. It is undecidable whether $L_0(N_1) = L_0(N_2)$ for λ-free labelled finite nets N_1 and N_2 by a result of [Hac76a].

Let λ-free labelled finite nets N_1 and N_2 be given, and let $A = \alpha(N_1) \cup \alpha(N_2)$. We may assume that N_1 and N_2 carry some token initially, otherwise we add to both nets a marked place and a transition with a new label that empties this place. The equality or inequality of the L_0-languages stays the same.

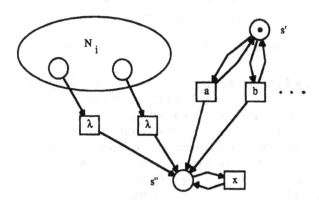

Figure 3.15

Now we modify each net N_i, $i = 1, 2$, to a net N_i' as follows. We add a place s' with one token, an empty place s'', transitions t_a, $a \in A$, and t_s, $s \in S_i$, and one transition t_x with $x \in \Sigma - A$, furthermore arcs of weight 1 specified by

$- (s', t_a), (t_a, s'), (t_a, s'')$ for $a \in A$,

$- (s, t_s), (t_s, s'')$ for $s \in S_i$,

$- (s'', t_x), (t_x, s'')$.

Each transition t_a, $a \in A \cup \{x\}$, is labelled a, each transition t_s, $s \in S_i$, is labelled λ. See Figure 3.15.

Since N_i' is divergence-free, N_1' and N_2' are $\mathcal{FD}$-equivalent if and only if they are $\mathcal{F}$-equivalent. $\mathcal{F}(N_i)$ is essentially $\{(w, \emptyset) \mid w \in (A \cup \{x\})^*\} \cup \{(w, \{x\}) \mid w \in L_0(N_i)\}$. Thus N_1' and N_2' are $\mathcal{F}$-equivalent if and only if $L_0(N_1) = L_0(N_2)$, and thus the undecidability follows.

Now assume we are given finite bounded nets N_1 and N_2 and $A = \alpha(N_1) \cup \alpha(N_2)$. We will turn each net N_i, $i = 1, 2$, into a finite automaton $\mathcal{AF}_i$ and into a finite automaton $\mathcal{AFD}_i$.

We have already mentioned that we can view $\mathcal{R}(N_i)$ as a finite automaton with initial state M_{N_i}. We obtain $\mathcal{AF}_i$ from $\mathcal{R}(N_i)$ by adding a state *fail*, which is the only final state, and arcs from each $M \in [M_{N_i}\rangle$ to *fail* with label X for all $X \subseteq A$ such that $\neg M[a\rangle$ for all $a \in X$. Obviously $\mathcal{F}(N_1) \supseteq \mathcal{F}(N_2)$ if and only if the language accepted by $\mathcal{AF}_1$ contains the language accepted by $\mathcal{AF}_2$, and this is decidable.

To obtain $\mathcal{AFD}_i$ from $\mathcal{AF}_i$ we add another state *div*. If from $M \in [M_{N_i}\rangle$ we can reach a cycle of λ-arcs via a path of λ-arcs, then we add a λ-arc from M to *div*. For all $a \in A$ we add an a-arc from *div* to *div*, and for all $X \subseteq A$ we add an X-labelled arc from *div* to *fail*. Now $F(N_1) \supseteq F(N_2)$ if and only if the language of $\mathcal{AFD}_1$ contains the language of $\mathcal{AFD}_2$; furthermore $D(N_1) \supseteq D(N_2)$ if and only if the language of $\mathcal{AFD}_1$ with final state *div* contains the language of $\mathcal{AFD}_2$ with final state *div*. Again the decidability follows.   □

At least we can decide whether a finite net is divergence-free, even if it is unbounded – a result which is also useful for the next chapter.

Theorem 3.2.9 *It is decidable whether a finite net is divergence-free.*

Proof: Let N be a finite net, and let N_1 be obtained from N by deleting all transitions that are not λ-labelled. For each $t \in T_1$ we define $continual(t) = \{M : S \to \mathbb{N}_0 \mid \exists w \in T_1^\omega : M[w\rangle$ in N_1 and w contains t infinitely often$\}$.

The sets $continual(t)$ are upward-closed with respect to componentwise $\leq$. By [VJ85, Theorem 3.13] we can effectively determine the (finitely many) minimal elements of $continual(t)$ for all $t \in T_1$. By inspection of the coverability graph of N, see e.g. [Pet81,Rei85], we can decide whether any of these elements can be covered by a reachable marking of N, i.e. whether some element of some $continual(t)$ is reachable in N. Now the result follows, since any infinite sequence of λ-transitions must contain some $t \in T_1$ infinitely often.   □

We conclude this section with an example of a rough equivalence, for which the related external equivalence and the related coarsest congruence differ. The rough equivalence in itself is quite uninteresting, and the purpose of this example is just to show that external equivalence and coarsest congruence do not have to coincide.

Let nets N_1 and N_2 be 1-a-related if they are isomorphic or if both do not have exactly one a-transition. Let N_1 and N_2 be nets without places, such that N_1 consists of two a-transitions and N_2 consist of two a-transitions and a b-transition. N_1 and N_2 are externally equivalent for the following reasons:

- If $a \notin A$, then $N \parallel_A N_1$ and $N \parallel_A N_2$ both have at least two a-transitions, thus they are 1-a-equivalent.

- If $a \in A$, then $N \parallel_A N_1$ and $N \parallel_A N_2$ have an even number of a-transitions (possibly zero), thus they are 1-a-equivalent.

But the external equivalence is not a congruence. Let $N_\emptyset$ be the empty net, N_a the net consisting of one a-transition. Then $N_\emptyset \parallel_{\{a\}} N_1 = N_\emptyset$ and $N_\emptyset \parallel_{\{a\}} N_2$ consists of one b-transition; these nets are not externally equivalent, since $N_a \parallel_\emptyset N_\emptyset = N_a$ and $N_a \parallel_\emptyset (N_\emptyset \parallel_{\{a\}} N_2)$ are not 1-a-related.

Observe that, for associative and commutative operators like $\parallel$, external equivalence and coarsest congruence always coincide. If N_1 and N_2 are externally equivalent, then for all nets N, N' the nets $(N \parallel N') \parallel N_1 = N \parallel (N' \parallel N_1)$ and $(N \parallel N') \parallel N_2 = N \parallel (N' \parallel N_2)$ are related by the rough equivalence. Thus $N' \parallel N_1$ and $N' \parallel N_2$ are externally equivalent, and by commutativity external equivalence is a congruence, hence it is by definition necessarily the coarsest congruence contained in the rough equivalence.

3.3 Modifications of failure semantics

The approach presented in the previous section can be modified in many ways in order to meet related or additional requirements. In the first part of this section, we deal with the important class of safe nets; the problem is that, even though we are ultimately interested in constructing safe deadlock-free nets, we have to work with components that might not be safe if considered in isolation. In the second part, we study wich subnets can be exchanged such that image-liveness of actions or certain reachability properties are preserved; it turns out that the usual failure semantics is too weak to give an internal characterization of this external equivalence, and that we have to consider some stronger versions that take into account the refusal of sequences of actions. The third part is a collection of various ideas; for example, divergence is based on an infinite firing sequence, and for this reason questions of fairness are of interest; also, one can consider the infinite visible behaviour of a net (as opposed to divergence).

3.3.1 Modifications for safe nets

For both, applications and theoretical considerations, the class of safe nets is of special interest. One way to adapt the results of Section 3.2 to this class would be to require that all nets we work with are safe. In general, changing the class of nets under consideration can change the fully abstract and the external equivalence, even if we keep the rough equivalence and the operators. But in this case, no modification would be necessary for the following reason. Failure equivalence implies deadlock similarity and it is a congruence on the class of all nets, hence this also holds for safe nets. Therefore, also in the class of safe nets, failure equivalence implies the fully abstract congruence corresponding to deadlock similarity, and this, in turn, always implies the corresponding external equivalence. The crucial point is that, as already remarked, the context nets N used in the proof of Theorem 3.2.6 are safe. Therefore, the external equivalence implies failure equivalence, just as in the class of all nets. The same considerations go through for the case of deadlock/divergence similarity.

Unfortunately, this way of adaption is much too restrictive, since the components of a safe net may be unsafe, if considered in isolation; Figure 3.16 shows a safe net $N = N' \|_{\{a,b\}} N''$, which is composed of the two unsafe nets N' and N''. Therefore, we consider safeness as an additional property that is equally interesting as deadlock- and divergence-freeness, and we modify the approach of the previous section as follows.

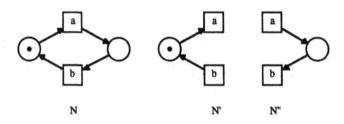

N N' N''

Figure 3.16

Definition 3.3.1 Nets N_1 and N_2 are *safe/deadlock- (safe/deadlock/divergence-) similar,* if: N_1 is safe and deadlock- and divergence-free if and only if N_2 is. They are *safe/deadlock- (safe/deadlock/divergence-) equivalent,* if for all $A \subseteq \Sigma$ and nets N the nets $N \|_A N_1$ and $N \|_A N_2$ are safe/deadlock- (safe/deadlock/divergence-) similar.

With this definition, not only a divergence string has to be considered as a catastrophic behaviour; also an image firing sequence that leads to a marking with more than one token on some place gives rise to a catastrophe that – like divergence – cannot be remedied. Accordingly, we modify the failure semantics.

Definition 3.3.2 The $\mathcal{FU}$- and $\mathcal{FDU}$-*semantics* of a net N ($\mathcal{U}$ stands for unsafe) are $\mathcal{FU}(N) = (F_u(N), U(N))$ and $\mathcal{FDU}(N) = (F_{du}(N), D_u(N))$, where
$U(N) = \{w \in \Sigma^* \mid$ for some prefix v of w we have
$\qquad\qquad M_N[v\rangle\rangle M$ and $M(s) \geq 2$ for some $s \in S\}$,
$F_u(N) = \mathcal{F}(N) \cup \{(w, X) \mid w \in U(N), X \subseteq \Sigma\}$,
$D_u(N) = U(N) \cup D(N)$,
$F_{du}(N) = F_u(N) \cup F(N)$.

Theorem 3.3.3 $\mathcal{FU}$- *and* $\mathcal{FDU}$-*equivalence are congruences with respect to* $\|_A$, $A \subseteq \Sigma$. *For* $A \subseteq \Sigma$ *and nets* N_1, N_2 *we have*

$$F_u(N_1 \|_A N_2) = \{(w, X) \mid \exists(w_i, X_i) \in F_u(N_i),\ i = 1,2 :$$
$$w \in w_1 \, \mathbb{I}_A \, w_2,\ X - A \subseteq X_1 \cap X_2,\ and\ X \cap A = X_1 \cup X_2\}$$
$$\cup\ \{(w, X) \mid w \in U(N_1 \|_A N_2),\ X \subseteq \Sigma\}$$

$$U(N_1 \|_A N_2) = \{w \mid \exists(w_i, \emptyset) \in F_u(N_i),\ i = 1,2,\ v \in \Sigma^* :$$
$$(w_1 \in U(N_1)\ or\ w_2 \in U(N_2))\ and\ v \in w_1 \, \mathbb{I}_A \, w_2\ is\ a\ prefix\ of\ w\}$$

The same holds for F_u *and* U *replaced by* F_{du} *and* D_u *throughout.*

Proof: analogous to the proof of Theorem 3.2.5.   □

Theorem 3.3.4 *i) Let* N_1, N_2 *be nets with* $\alpha(N_1) \cup \alpha(N_2) \neq \Sigma$. N_1 *and* N_2 *are safe/deadlock- (safe/deadlock/divergence-) equivalent if and only if they are* $\mathcal{FU}$- ($\mathcal{FDU}$-) *equivalent.*

ii) Let $\mathcal{N}$ *be defined as in Theorem 3.2.6. In* $\mathcal{N}$ $\mathcal{FU}$- ($\mathcal{FDU}$-) *equivalence is fully abstract with respect to* $\|_A$, $A \subseteq \Sigma$, *and safe/deadlock- (safe/deadlock/divergence-) similarity.*

These results also hold if in Definition 3.3.1 only finite nets are considered.

Proof: In the proof of Theorem 3.2.6, we have shown how we can put deadlock/divergence-equivalent nets into test environments in order to show that they are $\mathcal{FD}$-equivalent. Similarly, we can show that safe/deadlock- or safe/deadlock/divergence-equivalent nets must be $\mathcal{FU}$- or $\mathcal{FDU}$-equivalent. In the same way as we tested $D(N_1) = D(N_2)$ in the proof of Theorem 3.2.6, we can test whether $U(N_1) = U(N_2)$ or $D_u(N_1) = D_u(N_2)$. In the same way as we tested $(w, X) \in F(N_i)$ for $w \notin D(N_i)$, we can test whether $(w, X) \in F_u(N_i)$ for $w \notin U(N_i)$ and $(w, X) \in F_{du}(N_i)$ for $w \notin D_u(N_i)$. Thus, the external equivalences are at least as fine as $\mathcal{FU}$- and $\mathcal{FDU}$-equivalence.

On the other hand, if nets are $\mathcal{FU}$-equivalent, then either their U-semantics is non-empty and they are both not safe, or their U-semantics is empty, in which case both nets are safe, the F_u-semantics coincides with the $\mathcal{F}$-semantics and the nets are deadlock-similar. Similarly, if nets are $\mathcal{FDU}$-equivalent, then either their D_u-semantics is nonempty and they are both unsafe or divergent, or their D_u-semantics is empty, in which case both nets are safe and divergence-free, the F_{du}-semantics coincides with the $\mathcal{F}$-semantics and the nets are deadlock-similar. Hence, $\mathcal{FU}$-equivalence implies safe/deadlock-similarity, and $\mathcal{FDU}$-equivalence implies safe/deadlock/divergence-similarity. Furthermore, $\mathcal{FU}$- and $\mathcal{FDU}$-equivalence are congruences by Theorem 3.3.3. Thus, $\mathcal{FU}$- and $\mathcal{FDU}$-equivalence are at least as fine as the fully abstract congruences, which are always at least as fine as the external equivalences. Thus, the theorem follows.   □

$\mathcal{FU}$- and $\mathcal{FDU}$-equivalence can be decided for finite nets, even if they are unbounded. This is not very surprising, since the unsafe markings indicate a catastrophe where further behaviour is of no interest.

Theorem 3.3.5 *For finite nets $\mathcal{FU}$- and $\mathcal{FDU}$-equivalence are decidable.*

Proof: For a finite net N we construct finite automata $A(U)$, $A(F_u)$, $A(D_u)$, $A(F_{du})$ that recognize $U(N)$, $F_u(N)$, $D_u(N)$, $F_{du}(N)$.

For $A(U)$ we consider $\mathcal{R}(N)$ restricted to the safe markings as a finite automaton with initial state M_N. (The case that M_N is unsafe is easily treated, thus we may ignore it.) We add a new state *unsafe*, which will be the only final state. Finally, for a state $M \neq unsafe$ of $A(U)$ and some unsafe M' with $M[t\rangle M'$ we add an arc labelled $lab(t)$ from M to *unsafe*, and we add arcs from *unsafe* to *unsafe* for every action under consideration, i.e. from $\alpha(N) \cup \alpha(N')$ if we compare N and N'.

To obtain $A(D_u)$ from $A(U)$ we add a λ-labelled arc from state $M \neq unsafe$ to *unsafe*, whenever there is a λ-labelled path from M to a λ-labelled cycle. $A(U)$ and $A(D_u)$ can be modified to $A(F_u)$ and $A(F_{du})$ by adding a state *fail* in the same way as in the proof of Theorem 3.2.8.   □

3.3.2   Modifications for reachability and liveness

In many circumstances, the net properties of interest refer to the reachable markings instead of the actions a net can perform. When exchanging N_1 and N_2, we cannot expect that the reachable markings are the same, since it would already be much too restrictive to require that N_1 and N_2 have the same set of places. Instead, we can require that $N \parallel_A N_1$ and $N \parallel_A N_2$ are marking-equivalent on S_N.

Definition 3.3.6 *Nets N_1 and N_2 are* externally marking-equivalent *if for all $A \subseteq \Sigma$ and nets N the nets $N \parallel_A N_1$ and $N \parallel_A N_2$ are marking-equivalent on S_N.*

It should be remarked that this external equivalence does not fit into the same scheme as those we have considered above; it is not based on a rough equivalence that is defined for all nets, but it depends on the special form of the nets $N \parallel_A N_i$. To check marking- of the composed net as required, one has to know which way the nets have been composed in order to identify the set S_N in $S_{N\parallel_A N_i}$. For this reason, it does not make sense in this situation to look for the coarsest congruence contained in some rough equivalence. To make the coarsest congruence approach feasible, one could refine the nets we use by adding a set of observable places; nets with observable places (but without a transition labelling) are considered e.g. in [DCDMPS88], [PS90].

An internal characterization of external marking equivalence is easy to obtain.

Theorem 3.3.7 *Nets N_1 and N_2 are externally marking-equivalent if and only if they are language-equivalent. External marking equivalence is a congruence with respect to $\parallel_A$, $A \subseteq \Sigma$.*

Proof: That N_1 and N_2 must be language-equivalent can be tested with the same nets that we have used in the proof of Theorem 3.2.6 to test $D(N_1) = D(N_2)$ – but no transition with label not in $\alpha(N_1) \cup \alpha(N_2)$ is needed. Now the result follows from Corollary 3.1.8.

$\square$

Since language equivalence is not really satisfactory, it would be nice to find an external equivalence based on markings that justifies to take into account the branching behaviour in some sense. One idea one could come up with is the following (where again we cannot speak of a coarsest congruence contained in some rough equivalence).

Definition 3.3.8 A place s of a net N is *invariantly reachable*, if for every $M \in [M_N)$ there is some $M' \in [M)$ such that $M'(s) \geq 1$.

Nets N_1 and N_2 are *IR-equivalent*, if for all $A \subseteq \Sigma$ and nets N every place $s \in S_N$ is invariantly reachable in $N \parallel_A N_1$ if and only if it is in $N \parallel_A N_2$.

To motivate this definition, one may think of a place s that is marked if the net has successfully terminated; of course, one would like to know whether successful termination is always possible.

Definition 3.3.8 has the same format as the definition of liveness. Therefore we consider in this context also the following external equivalence.

Definition 3.3.9 Nets N_1, N_2 are *image-live-similar* if the same actions from Σ are image-live in N_1 and in N_2. The nets are *image-live-equivalent*, if for all $A \subseteq \Sigma$ and nets N the nets $N \parallel_A N_1$ and $N \parallel_A N_2$ are image-live-similar.

From our considerations in the previous section we conclude the following implication.

Proposition 3.3.10 *If nets N_1, N_2 are IR-equivalent or we have $\alpha(N_1) \cup \alpha(N_2) \neq \Sigma$ and N_1, N_2 are image-live-equivalent, then they are $\mathcal{F}$-equivalent. This also holds if in the definition of IR-equivalence and image-live equivalence only finite nets are considered.*

Proof: Consider the nets shown in Figure 3.7. For image-live equivalence the image-liveness of $lab(t)$ has to be considered; for IR-equivalence, t can be deleted and the invariant reachability of s has to be considered. $\square$

The following example shows that $\mathcal{F}$-equivalence is not strong enough to imply IR- or image-live-equivalence. The nets N_1, N_2 in Figure 3.17 are $\mathcal{F}$-equivalent, but if we consider a net N with just one a-transition that puts a token on an empty place s, then s is invariantly reachable in $N \parallel_{\{a\}} N_1$, but not in $N \parallel_{\{a\}} N_2$. If we add to N a c-labelled transition on a loop with s, then c is image-live in $N \parallel_{\{a\}} N_1$, but not in $N \parallel_{\{a\}} N_2$.

We see from this example that the negative look-ahead of one action, which we have with $\mathcal{F}$-semantics, is not enough. In the net N_1 we can always fire a, either immediately or after firing b first, in N_2 it might be impossible to fire a eventually.

This motivates our next semantics, which one could call impossible-future semantics; it corresponds to the possible-future semantics defined in [RB81] as failure semantics corresponds to readiness semantics. We denote it by $\mathcal{F}^+$, since the refusal sets are subsets of Σ^+ and not necessarily of Σ as in the $\mathcal{F}$-semantics.

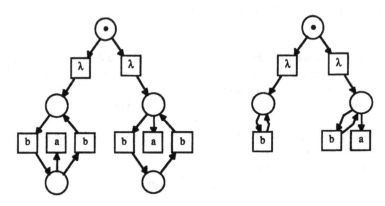

Figure 3.17

Definition 3.3.11 The $\mathcal{F}^+$-*semantics* of a net N is

$$\mathcal{F}^+(N) = \{(w, X) \mid w \in \Sigma^*,\ X \subseteq \Sigma^+ \text{ and there exists } M \text{ such that}$$
$$M_N[w\rangle\rangle M \text{ and } \neg M[v\rangle\rangle \text{ for all } v \in X\}.$$

In fact, this equivalence is strictly stronger than *IR*- and image-live-equivalence, as we will see immediately. The internal characterization of *IR*- and image-live-equivalence we give next is more elaborate – and less intuitive one might feel.

Definition 3.3.12 The $\mathcal{F}^{++}$-*semantics* of a net N is

$$\mathcal{F}^{++}(N) = \{(w, X) \mid w \in L(N),\ X \subseteq \Sigma^+, \text{ and } X = \emptyset \text{ or}$$
$$\text{there are some } w' \in X, \text{ a prefix } v \text{ of } w', \text{ and } M \text{ such that}$$
$$M_N[wv\rangle\rangle M \text{ and } \neg M[v'\rangle\rangle \text{ for all } vv' \in X\}.$$

Note that we do not require anything about sequences $u \in X$ that do not have v as a prefix. Thus, we have the following relationship between $\mathcal{F}^{++}$-semantics and $\mathcal{F}^+$-semantics. Let $(wv, Y) \in \mathcal{F}^+(N)$ with $Y \neq \emptyset$ and $X' = \{vv' \mid v' \in Y\}$; now construct X from X' by adding any action sequences that do not have v as a prefix. Then $(w, X) \in \mathcal{F}^{++}(N)$. In this way, the element (wv, Y) covers up the element (w, X), which lies in its past.

Consider the nets in Figure 3.18 and choose $w = \lambda$, $v = a$, and $Y = \{a\}$. We have $(wv, Y) = (a, \{a\}) \in \mathcal{F}^+(N_1) \cap \mathcal{F}^+(N_2)$. From $X' = \{aa\}$ we can obtain $X = \{aa, b\}$ as described above; thus $(w, X) = (\lambda, \{aa, b\}) \in \mathcal{F}^{++}(N_1) \cap \mathcal{F}^{++}(N_2)$. This element $(\lambda, \{b, aa\})$ is in fact also in $\mathcal{F}^+(N_1)$; but if we only know $\mathcal{F}^{++}(N_1)$, we cannot see this any more; and indeed, $(\lambda, \{b, aa\})$ is in $\mathcal{F}^{++}(N_2)$ as well, but not in $\mathcal{F}^+(N_2)$. In this sense, the element $(a, \{a\}) \in \mathcal{F}^+(N_1)$ covers up the element $(\lambda, \{b, aa\})$, which is in the past of $(a, \{a\})$.

In fact, all $(w, X) \in \mathcal{F}^{++}(N)$ with nonempty X can be constructed in the way described above. Thus, we have $(w, X) \in \mathcal{F}^{++}(N)$ if and only if either $X = \emptyset$ and $(w, \emptyset) \in$

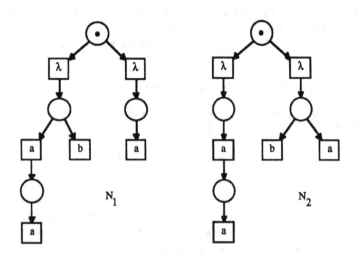

Figure 3.18

$\mathcal{F}^+(N)$ or for some prefix v of some $w' \in X$ we have $(wv, Y) \in \mathcal{F}^+(N)$, where $Y = \{v' \mid vv' \in X\}$. Therefore, $\mathcal{F}^{++}$-semantics can be obtained from $\mathcal{F}^+$-semantics by some closure operation; in particular, $\mathcal{F}^+$-equivalence implies $\mathcal{F}^{++}$-equivalence. (Similarly, the usual failure semantics can be obtained from readiness semantics [OH86] by some closure operation.)

On the other hand, if $(w, X) \in \mathcal{F}^{++}(N)$ and X is a nonempty subset of Σ, then we must have $v = \lambda$ and thus $(w, X) \in \mathcal{F}(N)$. This way it is easy to see that $\mathcal{F}(N) = \{(w, X) \in \mathcal{F}^{++}(N) \mid X \subseteq \Sigma\}$. In particular, $\mathcal{F}^{++}$-equivalence implies $\mathcal{F}$-equivalence.

We state how bisimulation, $\mathcal{F}^+$-, $\mathcal{F}^{++}$- and $\mathcal{F}$-equivalence are related.

Proposition 3.3.13 *Bisimulation implies $\mathcal{F}^+$-equivalence, $\mathcal{F}^+$-equivalence implies $\mathcal{F}^{++}$-equivalence, and $\mathcal{F}^{++}$-equivalence implies $\mathcal{F}$-equivalence, but in general not vice versa.*

Proof: The first implication is obvious, the other two implications we have already seen above.

Counterexamples for the converse implications are shown in Figures 2.13, 3.17, and 3.18. The nets in Figure 3.17 are $\mathcal{F}$-equivalent, but not $\mathcal{F}^{++}$-equivalent, since $(\lambda, \{b^n a \mid n \in \mathbb{N}_0\})$ only belongs to the $\mathcal{F}^{++}$-semantics of the right-hand-side net. The nets in Figure 2.13 are $\mathcal{F}^+$-equivalent, but not bisimilar, as we have seen. The nets in Figure 3.18 are not $\mathcal{F}^+$-equivalent, since $(\lambda, \{aa, b\}) \in \mathcal{F}^+(N_1) - \mathcal{F}^+(N_2)$. In order to show $\mathcal{F}^{++}(N_1) = \mathcal{F}^{++}(N_2)$, we first observe that in both nets every $X \subseteq \Sigma^+$ can be refused after a, b and aa. Thus, we only have to consider failure pairs $(\lambda, X) \in \mathcal{F}^{++}(N_i)$.

If $a \in X$, then v cannot be λ, since a cannot be refused after λ in both nets. Hence we must find some suitable $v \neq \lambda$ that satisfies the definition of the $\mathcal{F}^{++}$-semantics. In both nets, this cannot be a, since $M[\lambda\rangle\rangle$ for all markings M. Thus, v is either b and we have $b \notin X$, but some $bv' \in X$ with $v' \in \Sigma^+$; or it is aa and we have $aa \notin X$, but some

$aav' \in X$ with $v' \in \Sigma^+$. In this two cases we have indeed $(\lambda, X) \in \mathcal{F}^{++}(N_1) \cap \mathcal{F}^{++}(N_2)$. Hence, the sets $\mathcal{F}^{++}(N_1)$ and $\mathcal{F}^{++}(N_2)$ coincide on pairs (λ, X) for which $a \in X$.

If $a \notin X$, then $(\lambda, X) \in \mathcal{F}^+(N_1) \subseteq \mathcal{F}^{++}(N_1)$. To see that $(\lambda, X) \in \mathcal{F}^{++}(N_2)$ we have to consider two subcases:

- If $aa \notin X$, then $(\lambda, X) \in \mathcal{F}^+(N_2) \subseteq \mathcal{F}^{++}(N_2)$.

- If $aa \in X$, then choosing $v = a$ we get $(\lambda, X) \in \mathcal{F}^{++}(N_2)$, since we can refuse every $v' \in \Sigma^+$ after firing the right-hand-side a.

Therefore, we have that $\mathcal{F}^{++}(N_1)$ and $\mathcal{F}^{++}(N_2)$ coincide on pairs (λ, X), and hence N_1 and N_2 are $\mathcal{F}^{++}$-equivalent. $\square$

In our next result, we show that IR- and image-live-equivalence coincide, and that both can be internally characterized by $\mathcal{F}^{++}$-equivalence. This result has the drawback, that I can only state it for the case that infinite context nets are allowed. It is an open problem whether IR- and image-live-equivalence imply $\mathcal{F}^{++}$-equivalence, if the nets N in Definitions 3.3.8 and 3.3.9 are restricted to be finite.

Theorem 3.3.14 *IR-equivalence, $\mathcal{F}^{++}$-equivalence and – for nets N_1, N_2 with $\alpha(N_1) \cup \alpha(N_2) \neq \Sigma$ – image-live equivalence coincide.*

Proof: a) Assume that N_1 and N_2 are $\mathcal{F}^{++}$-equivalent, $A \subseteq \Sigma$ and N is a net.

i) First, we show that N_1 and N_2 are IR-equivalent. Assume that there exists some $s \in S_N$ which is not invariantly reachable in $N \parallel_A N_1$; let $M_{N\parallel_A N_1}[w\rangle\rangle M$ such that no marking that is reachable from M in $N \parallel_A N_1$ marks s.

By Theorem 3.1.7 we can find w_0 and w_1 such that $M_N[w_0\rangle\rangle pr_1^*(M)$, $M_{N_1}[w_1\rangle\rangle pr_2^*(M)$, and $w \in w_0 \parallel\!\parallel_A w_1$. Let Y be the set of all $w_0' \in \Sigma^*$ such that $pr_1^*(M)[w_0'\rangle\rangle M_0'$ and $M_0'(s) \geq 1$. Let X be the set of all $w_1' \in \Sigma^*$ with $id_A^*(w_1') = id_A^*(w_0')$ for some $w_0' \in Y$.

For $pr_2^*(M)[w_1'\rangle\rangle$, $id_A^*(w_1') = id_A^*(w_0')$, and $w_0' \in Y$ we can find $w' \in w_0' \parallel\!\parallel_A w_1'$ by Proposition 3.1.6, and then we have $M [w'\rangle\rangle M_0' \dot\cup M_1'$ in $N \parallel_A N_1$ with $(M_0' \dot\cup M_1')(s) = M_0'(s) \geq 1$. Since this is impossible by choice of M, we conclude that $\neg pr_2^*(M)[w_1'\rangle\rangle$ for all $w_1' \in X$.

Thus we have $(w_1, X) \in \mathcal{F}^{++}(N_1) = \mathcal{F}^{++}(N_2)$. If $X = \emptyset$, then $Y = \emptyset$; this implies $M_{N\parallel_A N_2} [w\rangle\rangle pr_1^*(M) \dot\cup M_2$ for some M_2, where by choice of Y it is impossible to mark s from $pr_1^*(M) \dot\cup M_2$. Hence s is not invariantly reachable in $N \parallel_A N_2$.

Therefore we can assume $X \neq \emptyset$. Let v_2 be the prefix of some string from X with $M_{N_2}[w_1 v_2\rangle\rangle M_2$ that exists according to the definition of $\mathcal{F}^{++}(N_2)$. By choice of X we find some v_0 that is a prefix of some string of Y such that $id_A^*(v_2) = id_A^*(v_0)$. This v_0 is enabled under $pr_1^*(M)$, therefore we find M_0, M_2' and $v \in v_0 \parallel\!\parallel_A v_2$ such that $M_{N\parallel_A N_2} [w\rangle\rangle pr_1^*(M) \dot\cup M_2' [v\rangle\rangle M_0 \dot\cup M_2$. If $M_0 \dot\cup M_2 [v'\rangle\rangle M'$ with $M'(s) \geq 1$, then $v' \in v_0' \parallel\!\parallel_A v_2'$ with $M_0[v_0'\rangle\rangle pr_1^*(M')$ and $M_2[v_2'\rangle\rangle pr_2^*(M')$. Thus $v_0 v_0' \in Y$, $id_A^*(v_0) = id_A^*(v_2)$, $id_A^*(v_0') = id_A^*(v_2')$, and $v_2 v_2' \in X$. By definition of $\mathcal{F}^{++}(N_2)$ this implies $\neg M_2[v_2'\rangle\rangle$, a contradiction. We conclude that s is not marked under any marking reachable from $M_0 \dot\cup M_2$, and that s is not invariantly reachable in $N \parallel_A N_2$.

ii) The case of image-liveness is similar to i); some differences arise since the place s in i) is internal to N, whereas some image-live action a may belong to the synchronization set A or a-labelled transitions may be internal to N, N_1 and N_2.

Assume that some $a \in \Sigma$ is not image-live in $N \|_A N_1$; let $M_{N\|_A N_1}[w)\rangle M$ such that no w' enabled under M contains a.

By Theorem 3.1.7 we can find w_0 and w_1 such that $M_N[w_0)\rangle pr_1^*(M)$, $M_{N_1}[w_1)\rangle pr_2^*(M)$ and $w \in w_0 \, \|\|_A \, w_1$. Let X be the set of all $w_1' \in \Sigma^*$ such that $id_A^*(w_1') = id_A^*(w_0')$ for some w_0' with $pr_1^*(M)[w_0')\rangle$ and w_1' or w_0' contains a.

For each $w_1' \in X$ with $pr_2^*(M)[w_1')\rangle$ and the corresponding w_0' we can find $w' \in w_0' \, \|\|_A \, w_1'$ with $M[w')\rangle$; by choice of M this is impossible, hence $\neg pr_2^*(M)[w_1')\rangle$ for all $w_1' \in X$.

Thus we have $(w_1, X) \in \mathcal{F}^{++}(N_1) = \mathcal{F}^{++}(N_2)$. If $X = \emptyset$, then $a \in A$ and no image firing sequence enabled under $pr_1^*(M)$ contains a; thus $M_{N\|_A N_2}[w)\rangle \, pr_1^*(M)\dot\cup M_2$ for some M_2, and no image firing sequence enabled under $pr_1^*(M)\dot\cup M_2$ contains a. Hence a is not image-live in this case.

Therefore we can assume $X \neq \emptyset$. Let v_2 be the prefix of some string w_1' from X with $M_{N_2}[w_1 v_2)\rangle M_2$ that exists according to the definition of $\mathcal{F}^{++}(N_2)$. We can find v_0 such that v_0 is a prefix of the corresponding string w_0' that exists by choice of X, and v_0 satisfies $id_A^*(v_0) = id_A^*(v_2)$. This v_0 is enabled under $pr_1^*(M)$, therefore we find M_0, M_2' and $v \in v_0 \, \|\|_A \, v_2$ such that $M_{N\|_A N_2}[w)\rangle \, pr_1^*(M)\dot\cup M_2' [v)\rangle M_0 \dot\cup M_2$. If $M_0\dot\cup M_2[v')\rangle$ such that v' contains a, then $v' \in v_0' \, \|\|_A \, v_2'$ with $M_0[v_0')\rangle$ and $M_2[v_2')\rangle$. Thus by considering $v_0 v_0'$ we see that $v_2 v_2' \in X$. By definition of $\mathcal{F}^{++}(N_2)$ this implies $\neg M_2[v_2')\rangle$, a contradiction. We conclude that such a v' cannot exist, and that a is not image-live in $N \|_A N_2$.

b) Let N_1 and N_2 be *IR*-equivalent or image-live-equivalent, where $\alpha(N_1)\cup\alpha(N_2) \neq \Sigma$ in the latter case. The test nets that show $\mathcal{F}^{++}(N_1) = \mathcal{F}^{++}(N_2)$ are constructed in a similar fashion as before. But this time I was not able to restrict the proof to the use of finite test nets only. The test nets can be chosen to be injectively labelled and safe, but to ease understanding we will use safe nets with arbitrary labels.

Let $(w, X) \in \mathcal{F}^{++}(N_1)$ and $A = \alpha(N_1) \cup \alpha(N_2)$. We construct a net N that starts with a path labelled w; at the end of the path the single token from a place s is removed, where (in the case of image-liveness) s is on a loop with a transition with label not in $\alpha(N_1)\cup\alpha(N_2)$. In the next part of N the strings of X are represented as a tree – common prefixes correspond to the same path of the tree. At the end of a string from X a token can be put onto s again. Compare Figure 3.19.

Let $w = w_1 \ldots w_n$; then N is formally defined as follows:

$S = \{s, s_i, s_v \mid i = 0, \ldots, n, \ v \text{ is a non-empty prefix of some } v' \in X\}$

$T = \{t, t_i, t_v, t_u' \mid i = 1, \ldots, n, \ v \text{ is a non-empty prefix of some } v' \in X, \ u \in X\}$

$$(t \in T \text{ only in the case of image-liveness})$$

All arc weights are 1 and the arcs are

- $(s_{i-1}, t_i), (t_i, s_i)$ for $i = 1, \ldots, n$

- $(s, t), (t, s), (s, t_n)$

- $(s_n, t_a), (t_v, s_v), (s_v, t_{va})$ for $a \in \Sigma, v \in \Sigma^*$

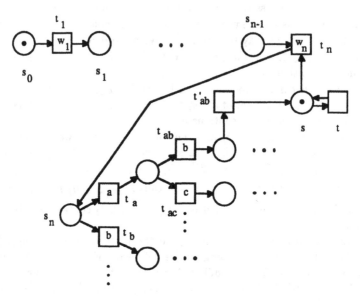

Figure 3.19

– $(s_u, t'_u), (t'_u, s)$

All places are unmarked except for one token on s_0 and, if $w \neq \lambda$, on s. (For $w = \lambda$ take $n = 0$.)

The labelling is

- $lab(t_i) = w_i$

- $lab(t_{va}) = a$ for $a \in \Sigma$

- $lab(t'_u) = \lambda$

- $lab(t) \in \Sigma - A$ in the case of image-liveness

In $N \parallel_A N_i$ the place s is not invariantly reachable and – in the case of image-liveness – $lab(t)$ is not image-live if and only if: we can fire w in N_i to remove the token from s; if $X \neq \emptyset$, we furthermore fire in N_i some suitable prefix v of some $v' \in X$. Now only the place s_v is marked, and no v' can be fired in N_i such that $vv' \in X$ – otherwise we could fire $t'_{vv'}$ in N in order to mark s again.

This situation can arise in $N \parallel_A N_1$, thus it must be possible in $N \parallel_A N_2$, and we conclude that $(w, X) \in \mathcal{F}^{++}(N_2)$.   □

Finally, we show that $\mathcal{F}^+$-equivalence is a congruence. The case of $\mathcal{F}^{++}$-equivalence has not been settled yet, and seems to be much more intricate. The congruence result for $\mathcal{F}^+$ and its proof are analogous to Theorem 3.2.4, but the formulation is different. In order to compare the formulation below with the one of Theorem 3.2.4, consider the case that $X \subseteq \Sigma$; then in the equality below we have $v \in \Sigma$ and

   – either $v = v_1 = v_2 \in A$ and $v \in X_1 \cup X_2$

   – or $v \notin A$, hence either $v_1 = v$, $v_2 = \lambda$, thus $v \in X_1$ or $v_2 = v$, $v_1 = \lambda$, thus $v \in X_2$; therefore $v \in X_1 \cap X_2$.

Theorem 3.3.15 $\mathcal{F}^+$-*equivalence is a congruence with respect to* $\|_A$, $A \subseteq \Sigma$; *for* $A \subseteq \Sigma$ *and nets* N_1 *and* N_2 *we have*

$$\mathcal{F}^+(N_1 \|_A N_2) = \{(w, X) \mid \exists (w_i, X_i) \in \mathcal{F}^+(N_i), \ i = 1, 2, \ such \ that \ w \in w_1 \|\!\|_A w_2$$
$$and \ v \in v_1 \|\!\|_A v_2 \ implies \ v_1 \in X_1 \ or \ v_2 \in X_2 \ for \ all \ v \in X\}.$$

3.3.3 Further modifications

Many people would define the behaviour of a net in the first place by its language. Examples like the nets in Figure 2.11 show that the language ignores deadlocks; when faced with such example, a natural idea would be to consider what is often called the completed-trace semantics, namely the set of those image firing sequences that (possibly) lead to a deadlock.

Put $L_{mf}(N) = \{w \in \Sigma^* \mid M_N[w\rangle M, \ \neg M[a\rangle \text{ for all } a \in \Sigma\}$, and call nets N_1, N_2 *mf*-*equivalent* (see e.g. [Bau88]) if $L_{mf}(N_1) = L_{mf}(N_2)$; *mf* stands for maximal finite. But, considering $L_{mf}(N)$ alone is not really satisfactory; for example, a net that can just fire an infinite a-sequence would have $L_{mf}(N) = \emptyset$, and thus would have the same behaviour as the empty net. As an alternative, we can combine the language and the completed traces, and call nets N_1, N_2 *mf'*-*equivalent* if $(L(N_1), L_{mf}(N_1)) = (L(N_2), L_{mf}(N_2))$.

These equivalences lie (strictly) between the rough equivalence we have studied in Section 3.2 and the corresponding external equivalence.

Proposition 3.3.16 $\mathcal{F}$-*equivalence implies* mf'-*equivalence, mf'-equivalence implies mf-equivalence, which in turn implies deadlock similarity.*

Proof: For all nets N, we have $L(N) = \{w \mid (w, \emptyset) \in \mathcal{F}(N)\}$ and $L_{mf}(N) = \{w \mid (w, \Sigma) \in \mathcal{F}(N)\}$, and N can deadlock if and only $L_{mf}(N) \neq \emptyset$. $\qquad\square$

Therefore, the external equivalences based on mf-equivalence and on mf'-equivalence are failure equivalence again, and no modification is necessary if we start our considerations with mf-equivalence or mf'-equivalence instead of deadlock similarity.

One way to incorporate $L(N)$ and $L_{mf}(N)$ in one set would be to consider all maximal image firing sequences, no matter whether the underlying firing sequence is finite or infinite, see again e.g. [Bau88]. Let $L_{mc}(N) = L^\omega(N) \cup L_{mf}(N)$, and call nets N_1 and N_2 *mc*-*equivalent* if $L_{mc}(N_1) = L_{mc}(N_2)$; *mc* stands for maximal countable.

Figure 3.20 shows two nets that are language and *mc*-equivalent, but not *mf*-equivalent. But for divergence-free nets we have that $L(N)$ is the set of finite prefixes of strings from $L_{mc}(N)$ and $L_{mf}(N) = L_{mc}(N) \cap \Sigma^*$, hence that *mc*-equivalence implies *mf*-equivalence. Neither does *mf*-equivalence imply *mc*-equivalence in general; Figure 3.21

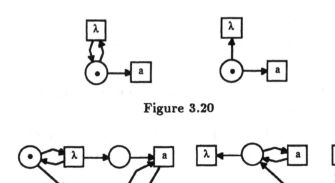

Figure 3.20

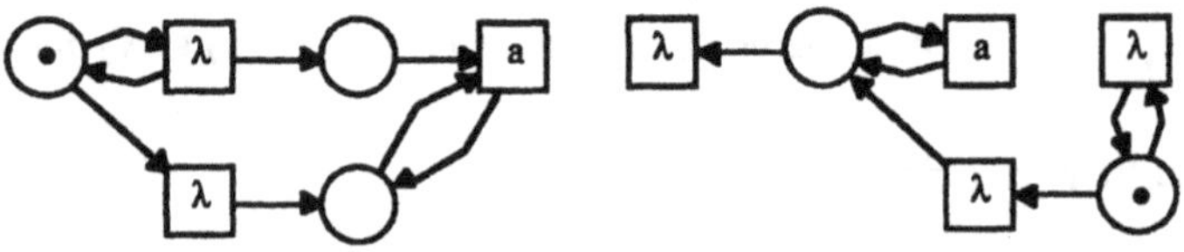

Figure 3.21

shows two nets that are even $\mathcal{F}$-equivalent, and which additionally can only diverge as long as no visible action has occurred; but only the second net can fire an infinite a-sequence.

For a finite net this situation can only occur if the net can diverge and is unbounded. This is established in the following theorem, which shows that for an image-finite net N the set $L^\omega(N)$ can be determined from $L(N)$. Also a static condition is given that is sufficient for enabling-finiteness, and enabling-finiteness is related to image-finiteness.

This relation shows that image-finiteness is a very plausible assumption. First of all, it is realistic to assume that for any given state a system has only finitely many states it can evolve into by one – internal or visible – action, i.e. enabling-finiteness is a realistic assumption. This assumption gives image-finiteness for nets without internal transitions, and for nets with internal transitions image-finiteness can only be violated if divergence is possible; but divergence, i.e. an infinite internal evolution, is in general considered to be undesirable. Note that, in general, for a reachable marking of an image-finite net there may be infinitely many markings reachable by at most one visible action, as long as these actions are different.

Theorem 3.3.17 $i)$ *If for a net N only finitely many places are marked initially, ${}^\bullet t \neq \emptyset$ for all $t \in T$, and $x^\bullet$ is finite for all $x \in S \cup T$, then N is enabling-finite.*

$ii)$ *If a net is enabling-finite and divergence-free, then it is image-finite. If a finite net is bounded or divergence-free, then it is image-finite.*

$iii)$ *If a net N is image-finite, then for all $w \in \Sigma^\omega$ we have: $w \in L^\omega(N)$ if and only if every finite prefix of w is in $L(N)$.*

Proof: i) Since $t^\bullet$ is finite for all $t \in T$, every reachable marking marks only finitely many places. This, together with the finiteness of $s^\bullet$ for all $s \in S$, implies that N is enabling-finite.

ii) Let M be a reachable marking. We define a directed graph whose vertices are markings that can be reached from M by firing λ-transitions only. In this graph, there is an arc from M' to M'' if $M'[t\rangle M''$ for some λ-transition t. Since N is enabling-finite, the graph is locally finite. Any vertex can be reached from M via a directed path; thus if this graph were infinite, König's Lemma would imply the existence of an infinite path, i.e. N would be divergent. Therefore, there are only finitely many markings that can be reached from M by firing λ-transitions; from each of these we can reach only finitely many markings by firing an a-transition due to enabling-finiteness; and as above from each of these markings only finitely many markings can be reached via λ-transitions. Therefore for any $M \in [M_N\rangle$ and any $a \in \Sigma$ there are only finitely many M' with $M[a\rangle\rangle M'$.

A finite net is obviously enabling-finite. Therefore, if a finite net is divergence-free, then it is image-finite by the first part; if it is bounded, then it has only finitely many reachable markings, hence it is trivially image-finite.

iii) The first implication is obvious, hence let us assume that every finite prefix of w is in $L(N)$. We define a directed graph with vertices M_N and (M, v), where $v \neq \lambda$ is a finite prefix of w with $M_N[v\rangle\rangle M$; there is an arc from M_N to (M', a) (from (M, v) to (M', va)) if $M_N[a\rangle\rangle M'$ $(M[a\rangle\rangle M')$ for $a \in \Sigma$. The graph is locally finite, but infinite, and every vertex is reachable from M_N. Thus by König's Lemma there is an infinite path starting in M_N, and this path corresponds to the image firing sequence w. □

Reading the proof carefully, we obtain the following sharpened version that we will need later on.

Corollary 3.3.18 *If a net N is enabling-finite and for some $w \in \Sigma^\omega$ no prefix of w is a divergence string of N, then $w \in L^\omega(N)$ if and only if every finite prefix of w is in $L(N)$.*

We call nets N_1 and N_2 *externally mc-equivalent*, if for all $A \subseteq \Sigma$ and nets N the nets $N \|_A N_1$ and $N \|_A N_2$ are *mc*-equivalent. In [Vog89], the following variation of the $\mathcal{FD}$-semantics is defined:

$$
\begin{aligned}
\mathcal{FD}'(N) &= (F'(N), D'(N)), \text{,where} \\
D'(N) &= L^\omega(N) \\
F'(N) &= \mathcal{F}(N) \cup \{(w, X) \mid w \in D(N) \cap \Sigma^*, X \subseteq \Sigma\}.
\end{aligned}
$$

The difference to the $\mathcal{FD}$-semantics is that $D'(N)$ is not closed under elongation and may contain infinite strings, and $F'(N)$ differs from $F(N)$ according to the first of these changes. Besides the mathematical motivation we will quote below from [Vog89], this semantics can be motivated by the following example.

A system uses a faulty channel and thus may be forced to repeat a message infinitely often; if this repetition is an internal action, the system may be unable to avoid divergence. Of course we will expect that the channel will work properly after several repetitions. Then the system recovers from the undesirable divergence situation, i.e. once we observe the reception of the message we know that the 'crisis is over' and we can consider possible actions as before. This is modelled by the $\mathcal{FD}'$-semantics, where the possibility of divergence does not obscure any future behaviour. Especially we can see from the $\mathcal{FD}'$-semantics, which actions the system can actually perform after a divergence situation, thus the nets of Figure 2.15 are distinguished.

The following two results are shown in [Vog89] for finite nets, but the proofs carry over to infinite nets.

Theorem 3.3.19 $\mathcal{FD}'$-equivalence is a congruence with respect to $\|_A$, $A \subseteq \Sigma$. For $A \subseteq \Sigma$ and nets N_1, N_2 we have:

$$
\begin{aligned}
F'(N_1 \|_A N_2) \;=\; & \{(w, X) \mid \exists(w_i, X_i) \in F'(N_i),\ i = 1, 2 : \\
& \quad w \in w_1 \amalg_A w_2,\ X - A \subseteq X_1 \cap X_2 \text{ and } X \cap A \subseteq X_1 \cup X_2\} \\
\cup\; & \{(w, X) \mid w \in D'(N) \cap \Sigma^*,\ X \subseteq \Sigma\}
\end{aligned}
$$

$$
\begin{aligned}
D'(N_1 \|_A N_2) \;=\; & \{w \mid \exists w_1 \in D'(N_1),\ (w_2, \emptyset) \in F'(N_2) : w \in w_1 \amalg_A w_2\} \\
\cup\; & \{w \mid \exists(w_1, \emptyset) \in F'(N_1),\ w_2 \in D'(N_2) : w \in w_1 \amalg_A w_2\} \\
\cup\; & \{w \mid \exists w_1 \in D'(N_1),\ w_2 \in D'(N_2) : w \in w_1 \amalg_A w_2\}
\end{aligned}
$$

Theorem 3.3.20   i) Let N_1, N_2 be nets with $\alpha(N_1) \cup \alpha(N_2) \neq \Sigma$. Then N_1 and N_2 are externally mc-equivalent if and only if they are $\mathcal{FD}'$-equivalent.

    ii) On $\mathcal{N}$ as in Theorem 3.2.6 $\mathcal{FD}'$-equivalence is fully abstract with respect to $\|_A$, $A \subseteq \Sigma$, and mc-equivalence.

Another variation concerns the notion of deadlock; our notion of deadlock includes the case that no visible action can occur, but an infinite sequence of λ-transitions can; L_{mf} is defined accordingly. Would we define a deadlock as a marking where no transition is enabled and modify deadlock similarity accordingly, then the corresponding external equivalence would be induced by $\mathcal{F}'(N) = \{(w, X) \mid w \in \Sigma^*,\ X \subseteq \Sigma$ and there exists M such that $M_N[w\rangle\rangle M,\ \neg M[a\rangle$ for all $a \in X$, and no λ-transition is enabled under $M\}$. $\mathcal{F}'$-equivalence is a congruence with respect to $\|_A$, $A \subseteq \Sigma$, and Theorem 3.2.4 holds with $\mathcal{F}$ replaced by $\mathcal{F}'$. But observe that $\mathcal{F}'$-equivalence does not imply language equivalence.

One can also vary the composition operators. One can fix a set $A \subseteq \Sigma$ and consider only the single operator $\|_A$ in the definition of e.g. deadlock equivalence or external mc-equivalence. The resulting deadlock$_A$- and external mc_A-equivalence have been characterized in [Gol88a].

Instead of composing nets by merging transitions, one can consider the merging of places. We will look into such an approach in the next chapter. Furthermore one can compose two nets by connecting them with new arcs. Such a composition is suggested in [Bau88], and in this approach the labelling of one of the nets is only used to direct the insertion of the new arcs (thus composition is not commutative). After the arc insertion, each of these labels is turned to λ (hiding, see next section). As explained in [Vog89] this composition can also be described by applying $\|_A$ and hiding all actions from A afterwards; thus it resembles the parallel operator in CCS. In [Vog89] the external equivalence based on this operator and mc-equivalence is characterized by giving yet another variation of the $\mathcal{FD}$-semantics. (In [Bau88] a sufficient condition for divergence-free nets being externally equivalent in this sense is given; it is based on a semantics that is somewhat similar to the readiness semantics of [OH86].)

As pointed out above, $\mathcal{F}$- and $\mathcal{FD}$-semantics are both interleaving semantics; the histories, i.e. the first components of the failure pairs, are sequences. Of course, one could also describe the histories by step sequences ([Pom86]; [TV89] gives such a semantics for TCSP) or partial orders, as it is done in [Pom88]. We will discuss this in Chapter 5.

That we are working with interleaving semantics indicates that we implicitly assume that a net 'runs on one processor'. For the deadlocking behaviour this is of little importance, since a deadlock is a deadlock no matter how many processors we have. But for divergence it does make a difference: if we have a net consisting of two parts, each 'run on its own processor', and one part diverges, then the corresponding processor would be kept in an infinite sequence of internal actions, but the other could show some visible behaviour; thus the whole net should not be considered as diverging. This is a question of fairness; an infinite sequence of internal transitions from only one part of the net would violate fairness – more precisely, the finite delay property – if transitions from the other part are enabled, too; see e.g. [Car86] for detailed considerations about fairness. A combination of the *mc*-approach and fairness considerations can be found in [Gol88a], where we find the following definitions and the result below:

- A finite or infinite firing sequence $w = t_1 t_2 \ldots$ of a net N is *fair* with respect to some $t \in T$ if t appears infinitely often in w or there is some $n \in \mathbb{N}$ that is less or equal the length of w such that for all $m \geq n$ t is not enabled under the marking reached after $t_1 t_2 \ldots t_m$. For finite w this means that t is not enabled under the marking reached after w.

- $L_{fair}(N) = \{v \in \Sigma^\infty \mid v = lab^*(w) \text{ for some } w \in FS^\infty(N) \text{ that is fair for all } t \in T\}$.

  Nets N_1 and N_2 are *fair-equivalent*, if $L_{fair}(N_1) = L_{fair}(N_2)$. They are *externally fair-equivalent* if for all $A \subseteq \Sigma$ and all nets N the nets $N \parallel_A N_1$ and $N \parallel_A N_2$ are fair-equivalent.

- $\mathcal{F}_{fair}(N) = \{(v, X) \mid X \subseteq \Sigma \text{ and there is some } w \in FS^\infty(N) \text{ such that}$
  $lab^*(w) = v \text{ and } w \text{ is fair for all } t \in T \text{ with } lab(t) \in X \cup \{\lambda\}\}$

Theorem 3.3.21 *Let N_1 and N_2 be safe, finite nets without loops such that $^\bullet t \neq \emptyset$ for all $t \in T_1 \cup T_2$. N_1 and N_2 are externally fair-equivalent if and only if $\mathcal{F}_{fair}(N_1) = \mathcal{F}_{fair}(N_2)$.*

Observe the close relationship between $\mathcal{F}$- and $\mathcal{F}_{fair}$-semantics. If N is divergence-free, then for all $v \in \Sigma^*$ and $X \subseteq \Sigma$ we have $(v, X) \in \mathcal{F}(N)$ if and only if $(v, X) \in \mathcal{F}_{fair}(N)$.

3.4 Further operators and congruence results

In this section, we will introduce three more operators that are of some importance, but not in the center of our interest. Furthermore, we will show some more congruence results.

The first two operators can change the labelling of a net. In the first case, we change the name of visible actions; this is called relabelling. It is useful, for example, if we have modelled some protocol such that the only visible actions are 'send data' and 'receive data'; if we want to work with several channels that use this protocol, we include the channel name in the action, i.e. a model for channel c is obtained by replacing 'send data' by 'send data on c' etc. Now we can model a sender with two available channels by having actions 'send data on c_1', 'send data on c_2'. Synchronizing the channels c_1 and c_2 and this sender over the set $A = \{$'send data on c_1', 'send data on c_2'$\}$ keeps the channels disjoint.

In the second case, we change a visible action into an internal action; this is called hiding. It allows to abstract from details that were important at some stage, e.g. in order to synchronize some subnets, but are not of interest any more.

Definition 3.4.1 Let f be a function from $\Sigma \cup \{\lambda\}$ to $\Sigma \cup \{\lambda\}$ with $f(\lambda) = \lambda$ and $f^{-1}(\lambda) = \{\lambda\}$. Then the *relabelling* $N[f]$ of a net N is obtained from N by changing the labelling such that

$$lab_{N[f]}(t) = f(lab_N(t))$$

Let $A \subseteq \Sigma$. Then *hiding* A changes a net N to the net $N \setminus A$ obtained by changing the labelling such that

$$lab_{N \setminus A}(t) = \begin{cases} \lambda & \text{if } lab_N(t) \in A \\ lab_N(t) & \text{otherwise} \end{cases}$$

Analogously, for a partial order p labelled over Σ the relabelling $p[f]$ is defined. Hiding A gives $p \setminus A = id^*_{\Sigma-A}(p)$.

Theorem 3.4.2 *Language-, $\mathcal{F}$-, $\mathcal{F}D$-, $\mathcal{F}^+$-, $\mathcal{F}U$-, and $\mathcal{F}DU$-equivalence are congruences with respect to relabelling. Language- and $\mathcal{F}^+$-equivalence, and – for enabling-finite nets – $\mathcal{F}D$- and $\mathcal{F}DU$-equivalence are congruences with respect to hiding. We have:*

i)

$$
\begin{aligned}
L(N[f]) &= \{w[f] \mid w \in L(N)\} \\
\mathcal{F}(N[f]) &= \{(w[f], X) \mid (w, f^{-1}(X)) \in \mathcal{F}(N)\} \\
F(N[f]) &= \{(w[f], X) \mid (w, f^{-1}(X)) \in F(N)\} \\
D(N[f]) &= \{w[f] \mid w \in D(N)\} \\
\mathcal{F}^+(N[f]) &= \{(w[f], X) \mid (w, \{v \mid v[f] \in X\}) \in \mathcal{F}^+(N)\} \\
U(N[f]) &= \{w[f] \mid w \in U(N)\} \\
D_u(N[f]) &= \{w[f] \mid w \in D_u(N)\} \\
F_u(N[f]) &= \{(w[f], X) \mid (w, f^{-1}(X)) \in F_u(N)\} \\
F_{du}(N[f]) &= \{(w[f], X) \mid (w, f^{-1}(X)) \in F_{du}(N)\}
\end{aligned}
$$

ii)

$$
\begin{aligned}
L(N \setminus A) &= \{w \setminus A \mid w \in L(N)\} \\
\mathcal{F}^+(N \setminus A) &= \{(w \setminus A, X) \mid (w, \{v \mid v \setminus A \in X\}) \in \mathcal{F}^+(N)\}
\end{aligned}
$$

For an enabling-finite net N:

$$
\begin{aligned}
F(N \setminus A) &= \{(w \setminus A, X) \mid (w, X \cup A) \in F(N)\} \\
&\quad \cup \ \{(w, X) \mid w \in D(N \setminus A)\} \\
D(N \setminus A) &= \{w \mid \exists v \in D(N) : v \setminus A \text{ is prefix of } w\} \\
&\quad \cup \ \{w \mid \exists v \in \Sigma^* : v \setminus A \text{ is a prefix of } w \text{ and} \\
&\qquad \forall n \in \mathbb{N} \ \exists v' \in A^n : (vv', \emptyset) \in F(N)\}
\end{aligned}
$$

$$U(N \setminus A) = \{w \mid \exists v \in U(N) : v \setminus A \text{ is prefix of } w\}$$
$$F_{du}(N \setminus A) = \{(w \setminus A, X) \mid (w, X \cup A) \in F_{du}(N)\}$$
$$\cup \ \{(w, X) \mid w \in D_u(N \setminus A)$$
$$D_u(N \setminus A) = \{w \mid \exists v \in D_u(N) : v \setminus A \text{ is prefix of } w\}$$
$$\cup \ \{w \mid \exists v \in \Sigma^* : v \setminus A \text{ is a prefix of } w \text{ and}$$
$$\forall n \in \mathbb{N} \ \exists v' \in A^n : (vv', \emptyset) \in F_{du}(N)\}\}$$

Proof: i) obvious

ii) The equations for $L(N \setminus A)$ and $\mathcal{F}^+(N \setminus A)$ are obvious. For the other equations there are some less obvious subcases; we demonstrate them for the $\mathcal{F}D$-case, the $\mathcal{F}D\mathcal{U}$-case is analogous.

a) Let $v \in \Sigma^*$ such that for all $n \in \mathbb{N}$ there is some $v_n \in A^n$ with $(vv_n, \emptyset) \in F(N)$. Let $u_n \in FS(N)$ be a firing sequence underlying vv_n. We define a directed graph with vertex set $\{u \mid u \text{ is a prefix of some } u_n\}$ and arcs uu' if $u' = ut$ for some $t \in T$. Since N is enabling-finite, this is an infinite, locally finite tree. By König's Lemma we find an infinite firing sequence whose image under $lab_{N \setminus A}$ is a prefix of $v \setminus A$. Hence, $v \setminus A$ and any elongations are in $D(N \setminus A)$.

b) Let $(w, X) \in F(N \setminus A)$ such that $w \notin D(N \setminus A)$. We can elongate any firing sequence of $N \setminus A$ that gives rise to (w, X) by finitely many λ-transitions to a sequence u such that, under the marking M reached after u, no λ-transition is enabled (otherwise w would be a divergence string). Thus, in N we have $\neg M[a\rangle$ for all $a \in X \cup A$, hence $(lab_N(u), X \cup A) \in F(N)$ and $lab_N(u) \setminus A = w$. □

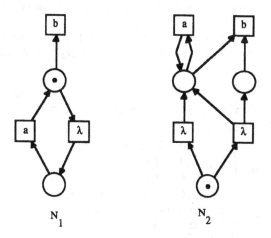

Figure 3.22

$\mathcal{F}$-equivalence is not a congruence with respect to hiding. Figure 3.22 shows two $\mathcal{F}$-equivalent nets – the essential elements of their $\mathcal{F}$-semantics are $(a^n, \{b\}), (a^n b, \{a, b\})$ for

$n \in \mathbb{N}_0$. But we have $(\lambda, \{b\}) \in \mathcal{F}(N_2 \setminus \{a\}) - \mathcal{F}(N_1 \setminus \{a\})$. Observe that, despite the similarities between $\mathcal{F}$- and $\mathcal{F}^+$-semantics, $\mathcal{F}^+$-equivalence *is* a congruence with respect to hiding.

Theorem 3.4.3 *Bisimulation, branching bisimulation and branching bisimulation with bisimilar divergence are congruences with respect to* $\|_A$, $A \subseteq \Sigma$, *relabelling and hiding. If* $\mathcal{B}$ *is a bisimulation of any of the above types for nets* N_1 *and* N_2, *then it is also one for* $N_1[f]$ *and* $N_2[f]$, $N_1 \setminus A$ *and* $N_2 \setminus A$ *resp.*

Proof: The detailed proofs are not difficult, but tedious. For relabelling and hiding the appropriate bisimulations are as stated in the theorem. For parallel composition let $\mathcal{B}_1$, $\mathcal{B}_2$ be bisimulations of any type for N_1 and N_1', N_2 and N_2'. The reachable markings of $N_1 \|_A N_2$ can be seen as pairs of reachable markings of N_1 and N_2 (and analogously for $N_1' \|_A N_2'$). Thus, we can define $\mathcal{B}$ as the restriction of $\{(M_1 \dot\cup M_2, M_1' \dot\cup M_2') \mid (M_1, M_1') \in \mathcal{B}_1, (M_2, M_2') \in \mathcal{B}_2\}$ to the reachable markings of $N_1 \|_A N_2$ and $N_1 \|_A N_2$; $\mathcal{B}$ is a bisimulation for these nets.   □

The third operator of this section is the choice operator +. It gives a net which behaves like either one of its two components. The choice is made when the first transition fires. In situations where initially both, visible and internal transitions, are enabled this can have surprising effects. One effect is that nearly no equivalence we have considered is a congruence with respect to +. This is not very important for us since we introduce + only to describe some example nets more easily. In general, one can either refine all the equivalence notions such that special care is taken of initially possible internal actions, see e.g. [BKO87], or one can restrict the comparison, for example, to nets for which under the initial marking no internal transitions are enabled. We will restrict the choice operator to root unwound nets; for this notion and a more general treatment of choice see e.g. [GV87].

Definition 3.4.4 A net is *root unwound* if the initial marking is safe and no initially marked place has an ingoing arc. The *choice* $N = N_1 + N_2$ of root unwound nets N_1 and N_2 is defined by

$$
\begin{aligned}
S &= (S_1 - \{s \mid M_{N_1}(s) = 1\}) \,\dot\cup\, (S_2 - \{s \mid M_{N_2}(s) = 1\}) \\
  &\quad \dot\cup\, \{(s_1, s_2) \mid s_i \in S_i, \; M_{N_i}(s_i) = 1, \; i = 1, 2\} \\
T &= T_1 \dot\cup T_2 \\
W(x, y) &= \begin{cases}
W_1(x, y) & \text{if } x, y \in S_1 \cup T_1 \\
W_2(x, y) & \text{if } x, y \in S_2 \cup T_2 \\
W_1(s_1, t) & \text{if } x = (s_1, s_2), \; t \in T_1 \\
W_2(s_2, t) & \text{if } x = (s_1, s_2), \; t \in T_2
\end{cases} \\
M_N(s) &= \begin{cases}
1 & \text{if } s = (s_1, s_2) \in S_1 \times S_2 \\
0 & \text{otherwise}
\end{cases} \\
lab(t) &= \begin{cases}
lab_1(t) & \text{if } t \in T_1 \\
lab_2(t) & \text{if } t \in T_2.
\end{cases}
\end{aligned}
$$

Figure 3.23 shows an example for the choice operator.

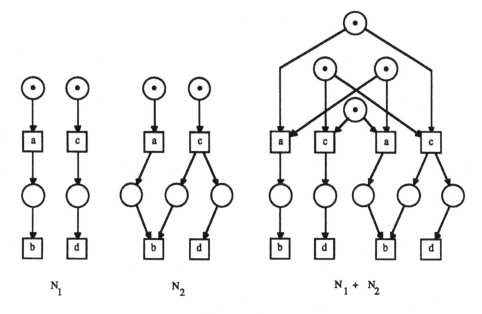

Figure 3.23

Chapter 4

Behaviour Preserving Refinement of Places and Transitions

In Chapter 3, we have considered parallel composition $\|_A$ with synchronous communication and the exchange of nets that are connected to some environment by application of $\|_A$, i.e. by merging transitions. Our interest has been to determine when such an exchange preserves behaviour in some sense. In this chapter, we regard the refinement of places as a special case of such an exchange; to get a suitable framework for this special case, we restrict the general approach of Chapter 3 to the exchange of nets that are in some sense deterministic. For these $\mathcal{F}$-deterministic nets, we show favourable decidability and more powerful behaviour preservation results.

In the second half of this chapter, we develop an analogous approach for the refinement of transitions. We define a parallel composition $\|_P$ with asynchronous communication and study the corresponding exchange of nets. We characterize for which nets such an exchange always preserves deadlock- and deadlock/divergence-freeness; in the first case, we can again use the $\mathcal{F}^{++}$-semantics of Section 3.3.2. Then we look for a suitable notion of deterministic nets and come up with what we call I, O-nets. This restricted class of nets gives a framework for the refinement of transitions with improved behaviour preservation and decidability results.

The first section of this chapter explains our refinement technique, i.e. how we can regard the refinement of places and transitions as a special case of exchanging nets that are connected with their environment via $\|_A$ or $\|_P$. In Section 4.2, we study $\mathcal{F}$-deterministic nets and their exchange; the results of this section apply especially to the refinement of places, and we put special emphasis on the place refinement of safe nets. The general approach for the operator $\|_P$ is developed in Section 4.3, i.e. we study the exchange of nets that are connected to their environment by merging places; we give characterization results, but also show that the external equivalences of interest are undecidable. Finally, in Section 4.4, we restrict the general approach of Section 4.3 to I, O-nets; we show that it is decidable whether a net is an I, O-net and whether I, O-nets are externally equivalent, and we prove strong results regarding behaviour preservation; we apply our results to transition refinement and also study transition refinement of safe nets.

4.1   Refinement techniques

In the previous chapter, we have studied the parallel composition of nets. In the first place, this corresponds to a bottom-up design of a system: subsystems are designed first, then they are assembled to larger subsystems, and finally these are composed to give the entire system. Such a bottom-up design is supported by the congruence results of Chapter 3.

When studying external equivalences we are more concerned with a top-down design: a coarse system model is designed first, and then it is refined stepwise by replacing parts of the design by more detailed models. The coarse model is relatively small; thus it is not too hard to verify that it meets our requirements. When we refine parts by more detailed models, we also have to study relatively small models in order to ensure that this refinement preserves the relevant behaviour. This way we know that the system we end up with meets our requirements without studying the entire system as a whole.

Exchanging nets N_1 and N_2 that are composed via $\|_A$ with a common context net N can be seen as such a refinement step. But talking of Petri nets and refinement, it is most natural to think of the refinement of the basic units of nets: the places and transitions. Therefore, we will henceforth understand by a refinement always the refinement of places or transitions.

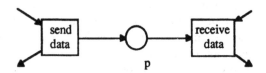

Figure 4.1

A refinement of a place may look like shown in Figures 4.1 to 4.3. Figure 4.1 shows a subnet modelling a communication via some channel p with unbounded buffering capacity.

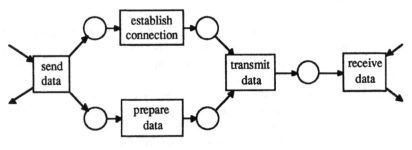

Figure 4.2

Figure 4.2 shows a refinement of the place p. We can see some of the subactivities that are necessary for the communication. Graphically, we have replaced the circle p by some subnet – except that, instead of one arc from 'send data' to p, we have two arcs from 'send data' to the subnet. These two arcs indicate that some concurrent subactions are initiated.

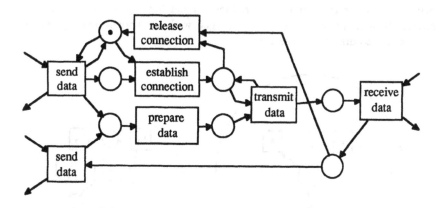

Figure 4.3

Figure 4.3 shows a more intricate model of the channel. Here the link is disconnected only if the message has been acknowledged. Furthermore there is the possibility to send data immediately without disconnecting and reestablishing the link first. Observe that this refinement can only be used safely under suitable circumstances; after 'send data' and 'establish connection' the next 'send data' is only possible if 'receive data' has occurred in between. This is no problem if, for example, the place p is safe in the coarse model.

Graphically, this refinement is not such a simple replacement. Not only is the place p empty while the refining net carries a token initially; we also have an arc from 'receive data' to the subnet, although there is no arc from 'receive data' to p; and even more surprisingly, we suddenly have two copies of 'send data'.

This last feature might seem somewhat exotic; but the example shows its usefulness, and it fits easily into our general approach, which is as follows. It is clear that, in order to refine a place, we have to delete the place and to insert a new subnet. It is not so clear how we should connect this refinement net to its environment. What we will do is to connect the inserted subnet by merging transitions, and merging transitions is described by the parallel composition operator $\|_A$. As we know from the previous chapter, parallel composition may duplicate transitions; thus, our general approach to refinement encompasses situations as in Figure 4.3, where 'send data' is duplicated.

Above, we have claimed that in a top-down design we expect that a refinement step preserves the relevant behaviour; this is only halfways true. If a place is refined as in the above example, then the refined net can perform actions like 'transmit data' that do not occur in the unrefined net; thus, if behaviour depends on the actions a system can perform, the two nets cannot have the same behaviour. In such a situation we would rather expect that we can determine the behaviour of the refined net from the refinement net and the behaviour of the unrefined net. In other words, if nets with the same behaviour are refined in the same way, then the resulting nets should have the same behaviour again; i.e. we would like to have a semantic equivalence that is a congruence with respect to refinement. This approach (for the refinement of transitions) is studied in the next chapter.

In this chapter, we insist that refinement should indeed be behaviour-preserving; consequently, the interior transitions of a refinement net for a place p must be λ-labelled. The exterior transitions, which are to be merged with the transitions adjacent to p, are labelled in order to direct the insertion.

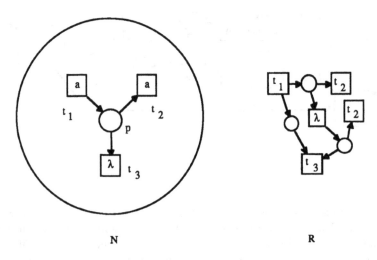

N   R

Figure 4.4

Figure 4.4 shows the outline of a net N and a refinement net R. The intention is to refine the place p by deleting the place, adding the net R, and merging the transition labelled t_1 with t_1, the two transitions labelled t_2 with two copies of t_2 etc. The labels of the merged transitions are to be taken from the net N, such that 'seen' from the environment N nothing changes. The result $N[p \to R]$ is shown in Figure 4.5.

In order to describe such a refinement using the operator $\|_A$, one might think that the labels of t_1, t_2, t_3 should appear in the net R and that they should form the synchronization set A. But this is obviously not possible. We cannot have $\lambda \in A$, and $a \in A$ would imply that all a-labelled transitions of R are merged with all a-labelled transitions of N in the same way, i.e. they are merged also with transitions not adjacent to p, and t_1 and t_2 cannot be distinguished.

The way around this problem is to view N as a relabelling of a net N^* where the transitions adjacent to p have different visible labels not appearing elsewhere in N^*. Now we can refine p in N^* using the parallel composition operator and relabel the result to get the refined net $N[p \to R]$.

Definition 4.1.1 Let N_p be a net with the single place p and the transition set $T_p = {}^\bullet p \cup p^\bullet$; let the labelling be the identity and $\alpha(N_p) = T_p \neq \Sigma$.

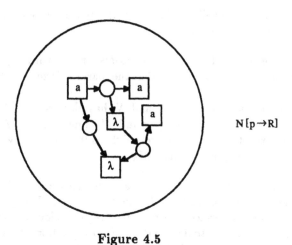

N[p→R]

Figure 4.5

An N_p-*refinement net* R is a net with $\alpha(R) = T_p$. An N_p-*host* is a net N such that $p \in S_N$, the subnet induced by $\{p\} \cup {}^\bullet p \cup p^\bullet$ is N_p except for the labelling, and $\alpha(N) \cap T_p = \emptyset$.

The *refined net* $N[p \to R]$ is obtained as follows. Let N^* be obtained from N by changing the labelling on ${}^\bullet p \cup p^\bullet$ to the identity and deleting p. Now $N = (N^* \|_{T_p} N_p)[f] \setminus A$ for an appropriate relabelling and hiding, and $N[p \to R] = (N^* \|_{T_p} R)[f] \setminus A$.

Requiring that the labelling of N_p is the identity implies that $T_p \subseteq \Sigma$. The condition $\alpha(N) \cap T_p = \emptyset$ ensures that Definition 4.1.1 meets the intentions described above.

Refinement nets that are suitable to refine the place p in any context net are studied in the next section.

We will describe next the refinement of transitions in a fashion similar to the refinement of places. For this, we need a composition operator that works by merging places. To introduce such an operator, one could label places and let the merging be guided by this labelling in analogy to the operator $\|_A$. Instead, to keep technicalities simple, we will assume that the nets we want to compose have just those places that should be merged in common.

In Section 4.3 we will study this merge-operator in a broader setting. But, since we are ultimately interested in the refinement of transitions, we will always assume that the nets to be merged are not symmetric, i.e. one of them is the context net, the other is a daughter net that we might want to exchange.

We will again consider the deadlocking behaviour of composed nets and how it is preserved by exchanging daughter nets. We require that daughter nets have λ-labelled transitions only; this is necessary in order to study deadlocks with respect to the transitions of the context net. Also, a daughter net has effects on the environment only via the common places; in themselves, the transitions of the daughter net are invisible from the environment.

This way we will only capture the refinement of λ-transitions in our approach. But in Section 4.4, where we will study the structure of those nets that are suitable to replace a transition in any context, we will see that there is a natural way to use those nets for the refinement of visible transitions, too.

Definition 4.1.2 Let P be a finite set. A *P-host* is a net N with $P \subseteq S_N$. A *P-daughter-net* N_1 is a net with place set $P \dot\cup S_1$ such that the marking is zero on P and all transitions are λ-labelled.

The *place composition* $N \parallel\!\!\!|_P N_1$ of N and N_1 has place set $S_N \dot\cup S_1$ and transition set $T_N \dot\cup T_1$, and the arc weight, initial marking and labelling is that of N or N_1 whichever is defined – with the exceptions that the arc weight is 0 if both arc weights are undefined and that the initial marking on P equals M_N.

Denoting the place set of N_1 by $P \dot\cup S_1$ is a slight misuse of notation, since it implies that S_1 is not the entire place set for a P-daughter-net N_1. It is justified by the use of such nets, where P is the interface to a host and 'really' belongs to this host.

We have used the symbol $\parallel\!\!\!|_P$ to denote the place composition. This operation is defined on nets and should not be confused with the composition $\parallel\!\!\!|_A$ with synchronization over $A \subseteq \Sigma$ defined on strings from Σ^∞. Now, the refinement of λ-transitions can be defined as follows.

Definition 4.1.3 Let N_h be a net consisting of a λ-labelled transition h and the place set $P = {}^\bullet h \cup h^\bullet$ such that ${}^\bullet h \neq \emptyset \neq h^\bullet$, ${}^\bullet h \cap h^\bullet = \emptyset$, and the initial marking is zero.

An N_h-λ-host N satisfies $h \in T_N$ and the subnet induced by $\{h\} \cup {}^\bullet h \cup h^\bullet$ equals N_h except for the marking. An N_h-*refinement net* R is a P-daughter-net where no arcs lead to ${}^\bullet h$ or from $h^\bullet$.

Let N be an N_h-λ-host, and let N^* be obtained from N by deleting h; then $N = N^* \parallel\!\!\!|_P N_h$ and the *refined net* $N[h \to R]$ is $N^* \parallel\!\!\!|_P R$.

In this definition, we only consider the refinement of a transition h that is not on a loop. Thus, P consists of the input places ${}^\bullet h$ of h and the output places $h^\bullet$. Similarly, we require that R treats ${}^\bullet h$ as input, i.e. that there are no arcs leading to ${}^\bullet h$, and analogously for $h^\bullet$. The refinement of arbitrarily labelled transitions will be defined in Section 4.4.

4.2   Deterministic nets and the refinement of places

The refinement technique for places we have described in the previous section replaces the injectively labelled λ-free net N_p by some other net and applies relabelling and hiding afterwards. This poses a problem for the immediate application of the results of Section 3.2, since failure semantics does not give a congruence with respect to hiding. Another point is that, with the results of Section 3.2, we might be able to ensure that a proper place refinement preserves the $\mathcal{F}$- or $\mathcal{FD}$-semantics; but some people might prefer the refined net to be bisimilar to the unrefined net. In this section, we will see that both problems can be solved, and to do this, we will first study nets that are deterministic in some sense.

Injectively labelled λ-free nets are of course deterministic in the sense that the marking reached after a firing sequence w is uniquely determined by the image of w, i.e. in the sense of automata theory. Deterministic nets in this sense are a very restricted class of nets; the corresponding nondeterminism cannot reliably be detected from any semantics we have considered here. Failure semantics, for example, detects nondeterminism if and only if we have $(wa, \emptyset), (w, \{a\}) \in \mathcal{F}(N)$ for some $w \in \Sigma^*$ and $a \in \Sigma$; in this case, w can either be performed such that a is possible or such that a is impossible.

In this section, we will study the class of $\mathcal{F}$- (and $\mathcal{FD}$-)deterministic nets, for which the failure semantics cannot detect any nondeterminism. This class is much broader than the class of deterministic nets. It allows favourable decidability results; this is important since, even if we only want to refine places of finite bounded nets, we are forced to work with the unbounded net N_p. Results about this class allow us to show that, if place refinement is required to preserve behaviour in quite a weak sense, then necessarily it will preserve behaviour in a very strong sense. Therefore, most sensible approaches to place refinement will be covered by our approach.

Definition 4.2.1 A net N is $\mathcal{F}$-*deterministic* if for all $(w, \{x\}) \in \mathcal{F}(N)$ we have $(wx, \emptyset) \notin \mathcal{F}(N)$. It is $\mathcal{FD}$-*deterministic* if for all $(w, \{x\}) \in F(N)$ we have $(wx, \emptyset) \notin F(N)$.

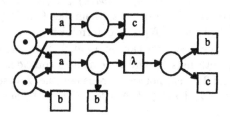

Figure 4.6

Figure 4.6 shows an example of an $\mathcal{F}$- and $\mathcal{FD}$-deterministic net. This example shows that our determinism notion ignores concurrency, i.e. it ignores that the net of Figure 4.6 can either perform a such that b can occur concurrently or such that b cannot occur concurrently.

Definition 4.2.1 translates the determinism notion for TCSP processes given in [Bro83] to nets. $\mathcal{F}$-determinism for transition systems is studied under the name initial determinacy in [Eng85]; this concept also appears e.g. in [Mil80] under the name determinacy and in [Old89] under the name external determinism. The results 4.2.3 ii) and 4.2.4 below have already been obtained in [Eng85], but for transition systems instead of Petri nets, for bisimulation instead of branching bisimulation, and not for the case of $\mathcal{FD}$-determinism. Since transition systems are special Petri nets, our results extend those of [Eng85], but the extension is only a small one since the failure semantics of a net can be determined from its reachability graph, which is a transition system.

If a net N has the divergence string w, then $(wa, \emptyset), (w, \{a\}) \in F(N)$ for all $a \in \Sigma$. Hence, $\mathcal{F}$- and $\mathcal{FD}$-determinism are related in a simple manner.

Proposition 4.2.2 *A net is $\mathcal{FD}$-deterministic if and only if it is divergence-free and $\mathcal{F}$-deterministic.*

If N_2 is a deadlock- (deadlock/divergence-) free implementation of an $\mathcal{F}$-deterministic net N_1, then it is intuitively clear that N_2 is $\mathcal{F}$-deterministic, too, since an implementation is intended to resolve some nondeterminism.

Theorem 4.2.3 *Let N_1, N_2 be nets.*

i) Let $\alpha(N_1) \cup \alpha(N_2) \notin \Sigma$. If N_1 is $\mathcal{F}$-deterministic and $N_1 \sqsubseteq_D N_2$, then N_2 is $\mathcal{F}$-deterministic and N_1 and N_2 are $\mathcal{F}$-equivalent (and thus language-equivalent). If N_1 is $\mathcal{F}D$-deterministic and $N_1 \sqsubseteq_{DD} N_2$, then N_2 is $\mathcal{F}D$-deterministic and N_1 and N_2 are $\mathcal{F}D$- and $\mathcal{F}$-equivalent.

ii) If N_1 and N_2 are $\mathcal{F}$-deterministic and language-equivalent, then they are $\mathcal{F}$-equivalent. If N_1 and N_2 are $\mathcal{F}D$-deterministic and language-equivalent, then they are $\mathcal{F}$- and $\mathcal{F}D$-equivalent.

Proof: i) The proof of the first part is by induction on the length of a history. If $(\lambda, X) \in \mathcal{F}(N_1)$ then $(a, \emptyset) \notin \mathcal{F}(N_1)$ for all $a \in X$, hence $(a, \emptyset) \notin \mathcal{F}(N_2)$ for all $a \in X$ and $(\lambda, X) \in \mathcal{F}(N_2)$ by Proposition 2.3.1. For the induction step, let $(wb, X) \in \mathcal{F}(N_1)$ with $w \in \Sigma^*$, $b \in \Sigma$; this implies $(w, \{b\}), (wba, \emptyset) \notin \mathcal{F}(N_1) \supseteq \mathcal{F}(N_2)$ for all $a \in X$; by induction $(w, \emptyset) \in \mathcal{F}(N_2)$, and Proposition 2.3.1 implies $(wb, \emptyset), (wb, X) \in \mathcal{F}(N_2)$. Thus, N_1 and N_2 are $\mathcal{F}$-equivalent, hence N_2 is $\mathcal{F}$-deterministic, too.

For the second part, the first part can be applied since N_1, hence also N_2, is divergence-free, and thus for both nets $\mathcal{F}$- and F-semantics coincide.

ii) By Proposition 4.2.2 it is enough to consider the $\mathcal{F}$-case. If $(w, X) \in \mathcal{F}(N_1)$, then $w \in L(N_1) = L(N_2)$ and $wa \notin L(N_1)$ for all $a \in X$, hence by Proposition 2.3.1 $(w, X) \in \mathcal{F}(N_2)$. $\square$

Theorem 4.2.4 *Let N_1, N_2 be $\mathcal{F}$-deterministic nets; N_1 and N_2 are branching bisimilar if and only if they are $\mathcal{F}$-equivalent. Let N_1, N_2 be $\mathcal{F}D$-deterministic nets; N_1 and N_2 are branching bisimilar with bisimilar divergence if and only if they are $\mathcal{F}$- or $\mathcal{F}D$-equivalent.*

Proof: By Proposition 4.2.2, we only have to consider the $\mathcal{F}$-case.
'$\Rightarrow$' is always true.
'$\Leftarrow$' Define $\mathcal{B} \subseteq [M_{N_1}\rangle \times [M_{N_2}\rangle$ by $(M_1, M_2) \in \mathcal{B}$ if $M_{N_1}[w\rangle\rangle M_1$ and $M_{N_2}[w\rangle\rangle M_2$ for some $w \in \Sigma^*$. Obviously $(M_{N_1}, M_{N_2}) \in \mathcal{B}$, and $M_1[\lambda\rangle\rangle M_1'$ implies $(M_1', M_2) \in \mathcal{B}$. Furthermore, if $(M_1, M_2) \in \mathcal{B}$ and $M_1[a\rangle\rangle M_1'$ for some $a \in \Sigma$, then for the corresponding w we have $(wa, \emptyset) \in \mathcal{F}(N_1) = \mathcal{F}(N_2)$, hence $(w, \{a\}) \notin \mathcal{F}(N_2)$ and $M_2[a\rangle\rangle M_2'$ for some M_2'; by definition of $\mathcal{B}$, this implies $(M_1', M_2') \in \mathcal{B}$. $\square$

Corollary 4.2.5 *Let N, N_1 and N_2 be nets with $\alpha(N_1) \cup \alpha(N_2) \neq \Sigma$, and let $A \subseteq \Sigma$.*

i) If N_1 is $\mathcal{F}$-deterministic and $N_1 \sqsubseteq_D N_2$, then $N \parallel_A N_1$ and $N \parallel_A N_2$ are branching bisimilar.

ii) If N_1 is $\mathcal{F}D$-deterministic and $N_1 \sqsubseteq_{DD} N_2$, then $N \parallel_A N_1$ and $N \parallel_A N_2$ are branching bisimilar with bisimilar divergence.

Proof: Apply Theorems 4.2.3, 4.2.4 and 3.4.3.   □

This result shows: if N_1 is $\mathcal{F}$- ($\mathcal{FD}$-) deterministic and we require that exchanging N_1 for N_2 in any deadlock- (deadlock/divergence-) free context preserves deadlock- (deadlock/divergence-) freeness, then it is ensured that in any context the exchange of N_1 and N_2 preserves behaviour in a very strong sense. For example, image-liveness is also preserved, which is not necessarily the case when exchanging arbitrary deadlock-equivalent nets, see Section 3.3.

An unfortunate fact is that the classes of $\mathcal{F}$- and $\mathcal{FD}$-deterministic nets are not closed under the operations we have considered so far.

Proposition 4.2.6 *The class of $\mathcal{F}$-deterministic nets and the class of $\mathcal{FD}$-deterministic nets are not closed under relabelling, hiding or $\|_A$ with $A \subseteq \Sigma$, $|\Sigma - A| \geq 2$. They are closed under $\|_\Sigma$.*

Proof: For $\|_A$ consider $a, b \notin A$ and nets with languages $\{\lambda, a\}$ and $\{\lambda, a, ab\}$.   □

Next we will obtain the announced decidability results.

Theorem 4.2.7 *It is decidable whether a given finite net is $\mathcal{F}$-deterministic or whether it is $\mathcal{FD}$-deterministic.*

Proof: By 4.2.2 and 3.2.9, we only have to show how to decide whether N is $\mathcal{F}$-deterministic. For given finite nets N_1, N_2 and finite $A \subseteq \Sigma$, let $combi(N_1, N_2, A)$ be obtained from $N_1 \|_{\Sigma - A} N_2$ by adding

  - a place s carrying one token, and for each $a \in A$ an empty place s_a,

  - a transition t on a loop with s

  - arcs of weight 1

      - from s to each transition of N_1 with label in A,

      - from each transition in N_2 with label in A to s,

      - for each $a \in A$ from each a-labelled transition of N_1 to s_a

      - for each $a \in A$ from s_a to each a-labelled transition in N_2. (See Figure 4.7)

With this construction, N is $\mathcal{F}$-deterministic if and only if t is live in $combi(N, N, \alpha(N))$: t cannot occur again if and only if the 'new' token, which is on s initially, is stuck in some s_x; this means we have fired a pre-image of some wx in the 'first' copy of N and a pre-image of w in the 'second' copy, such that x can be refused in the 'second' copy, i.e. $(wx, \emptyset) \in \mathcal{F}(N)$ and $(w, \{x\}) \in \mathcal{F}(N)$. Liveness is decidable [May81,Kos82].   □

Lemma 4.2.8 *Let N_1 and N_2 be finite nets with N_2 being $\mathcal{F}$-deterministic. Then it is decidable whether $L(N_1) \subseteq L(N_2)$.*

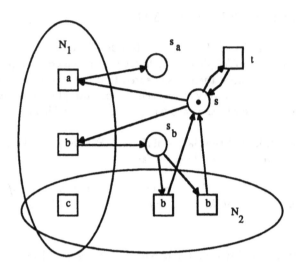

Figure 4.7 $combi(N_1, N_2, A)$ with $a, b \in A$, but $c \notin A$

Proof: The inclusion holds if and only if in $combi(N_1, N_2, \alpha(N_1) \cup \alpha(N_2))$ (see above) the transition t is live.   □

Theorem 4.2.9 *It is decidable whether given finite nets N_1, N_2 with N_1 being $\mathcal{F}$-deterministic are $\mathcal{F}$-equivalent. It is decidable whether given finite nets N_1, N_2 with N_1 being $\mathcal{F}D$-deterministic are $\mathcal{F}D$-equivalent.*

Proof: First, we decide whether N_2 is $\mathcal{F}$- or $\mathcal{F}D$-deterministic, too, see Theorem 4.2.7. Now we are done by Lemma 4.2.8 and Theorem 4.2.3 ii).   □

It should be remarked that this theorem just shows decidability in principle. Not only is the space and time complexity of deciding liveness prohibitively high, also the algorithm is so complex that it has never been implemented as far as I know.

Place refinement

In the following, we will apply our results to the case of place refinement. A net that is suitable in some sense to replace a place is called a module. The following definition of various types of modules is based either on the pre-orders we have studied in Section 3.2, i.e. in the D- and the DD-case; or it is based on an equivalence if we have not considered the corresponding pre-order, i.e. in the IR- and the IL-case. By Theorem 4.2.3, we could just as well use equivalences in all cases.

Definition 4.2.10 Let R be an N_p-refinement net for some net N_p as in Definition 4.1.1.

i) R is an N_p-*D-module* if for all deadlock-free N_p-hosts N the net $N[p \rightarrow R]$ is deadlock-free, too.

ii) R is an N_p-*DD-module* if for all deadlock/divergence-free N_p-hosts N the net $N[p \rightarrow R]$ is deadlock/divergence-free, too.

iii) R is an N_p-*IR-module* if for all N_p-hosts N and $s \in S_N - \{p\}$ s is invariantly reachable in N if and only if it is in $N[p \rightarrow R]$;

iv) R is an N_p-*IL-module* if for all N_p-hosts N the same actions are image-live in N and $N[p \rightarrow R]$.

We can characterize N_p-modules as follows.

Theorem 4.2.11 *An N_p-refinement net R is an N_p-D-module if and only if it is an N_p-IR-module if and only if it is an N_p-IL-module if and only if R and N_p are failure-equivalent.*

R is an N_p-DD-module if and only if it is a divergence-free N_p-D-module if and only if R and N_p are $\mathcal{F}D$-equivalent.

These results also hold if only the refinement of finite nets is considered in the definition of the modules.

Proof: If N is injectively labelled such that no transition of T_p is λ-labelled, then for suitable N^* we have $N = N^* \parallel_{T_p} N_p$ and $N[p \rightarrow R] = N^* \parallel_{T_p} R$. Thus, the proofs of Theorem 3.2.6 and Proposition 3.3.10 show: if R is an N_p-D- (N_p-DD-) module then $N_p \sqsubseteq_D R$ ($N_p \sqsubseteq_{DD} R$), and if R is an N_p-IR- or N_p-IL-module then N_p and R are failure-equivalent. Thus, by Theorem 4.2.3, they are in any case $\mathcal{F}$-equivalent, and in the DD-case also $\mathcal{F}D$-equivalent.

If, on the other hand, N_p and R are $\mathcal{F}$-equivalent (and $\mathcal{F}D$-equivalent in the DD-case), then by Theorem 4.2.4 N_p and R are branching bisimilar (with bisimilar divergence in the DD-case). Thus, N and $N[p \rightarrow R]$ are branching bisimilar (with bisimilar divergence for DD) by Theorem 3.4.3, since place refinement is based on $\parallel_{T_p}$, relabelling and hiding. Hence, these nets are deadlock-similar, and R is an N_p-D-module; in the DD-case, they are deadlock/divergence-similar and R is an N_p-DD-module; they are image-live-similar by Proposition 2.3.2, thus R is an N_p-IL-module. For the IR-case the labelling of N and $N[p \rightarrow R]$ is irrelevant, it is enough to know that N_p and R are IR-equivalent; this follows from Theorem 3.3.14 since bisimulation implies $\mathcal{F}^{++}$-equivalence by Proposition 3.3.13. $\qquad\square$

This result is remarkable, since the refinement technique involves hiding and $\mathcal{F}$-equivalence is not a congruence with respect to hiding. This problem has been evaded by using results on $\mathcal{F}$- and $\mathcal{F}D$-deterministic nets; again this is remarkable, since the refinement technique involves parallel composition, relabelling and hiding, and the classes of $\mathcal{F}$- and $\mathcal{F}D$-deterministic nets are not closed under any of these operations.

As a corollary of the last two theorems we get a decidability result.

Corollary 4.2.12 *It is decidable whether a given finite N_p-refinement net is an N_p-module of any given type.*

Finally, we list some properties that we can ensure simply by requiring that any deadlock-free net should be refined to a deadlock-free net or that refinement should preserve some other simple feature of behaviour.

Corollary 4.2.13 Let R be an N_p-module of any type and N be an N_p-host.

> i) N and $N[p \to R]$ are branching bisimilar (with bisimilar divergence if R is an N_p-DD-module).
>
> ii) For $t \in T_N - T_p$, t is live in N if and only if t is live in $N[p \to R]$.
>
> iii) The bisimulation from i) can be chosen such that related markings coincide on $S_N - \{p\}$; in particular, N and $N[p \to R]$ are marking-equivalent on $S_N - \{p\}$. The reachable markings of $N[p \to R]$ restricted to S_R are reachable markings of R.

Proof: i) see the proof of Theorem 4.2.11.

ii) Change the labelling of N such that in the resulting net N' all transitions except t are λ-labelled. Now t is live in N iff $lab_{N'}(t)$ is image-live in N' iff $lab_{N'}(t)$ is image-live in $N'[p \to R]$ by i) iff t is live in $N[p \to R]$.

iii) If in the proof of Theorem 3.4.3 $N_1 = N_1'$, we can choose $\mathcal{B}_1$ as the identity; since relabelling and hiding does not require a change of the bisimulation relation, the first part follows. The last part is not influenced by the relabelling and hiding occurring in the refinement, and follows from 3.1.3.   □

For an example of a module, we can consider the net of Figure 4.1 as N_p and the net of Figure 4.2 with the interior transitions labelled λ as R (both without the dangling arcs). It is not hard to see that R is an N_p-module; the easiest way is to check, that a bisimulation is given if we relate n tokens on s to each marking of R where the upper two and the right place carry n tokens in total and the upper two places together carry the same number of tokens as the lower two places. The net of Figure 4.3 does not show a module for the reason given at the beginning of the previous section.

Another example of a refinement with a possibly infinite module uses also the feature that refining p might split transitions adjacent to p. One can consider a place as a variable whose value is the number of tokens on the place. Another (safe) representation of a variable is by an (in general infinite) sequence of places $s_0, s_1, \ldots$, where a token on s_i represents the fact that the variable has the value i, see e.g. [Bes88b]. Figure 4.8 shows a net N with a place p and an N_p-module R such that in the refined net (Figure 4.9) the variable p is represented by safe places. For a (theoretical) application of this refinement see also [Vog91a].

Our approach to place refinement might look somewhat restricted for the following reason. Suppose we have shown with some effort that some R is an N_p-module of some type; afterwards we want to replace p' in some net N where the neighbourhood of p' is not N_p, but some similar $N_{p'}$. Now it seems that we have to repeat our efforts for the net R' which is equally similar to R.

First of all, assume that $N_{p'}$ is isomorphic to N_p except for the labelling, i.e. $N_{p'}$ is obtained from N_p by renaming p and the transitions, and by relabelling the transitions accordingly; naturally, the same relabelling of R yields on $N_{p'}$-module R' of the same

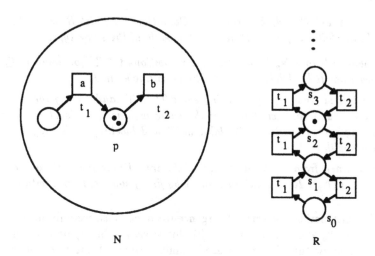

Figure 4.8

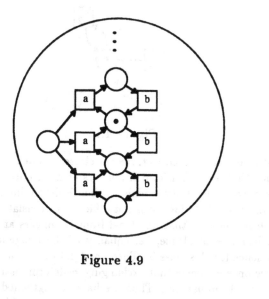

Figure 4.9

type. Some less trivial modifications of modules are listed in the next theorem; all can be shown using the characterization in Theorem 4.2.11.

Theorem 4.2.14 *Let N_p, N_p' be nets as in Definition 4.1.1, let R be an N_p-module of some type. In the following cases R' is an N_p'-module of the same type:*

i) *N_p' is obtained from N_p by deleting the transitions $t \in T'$ for some $T' \subseteq T_p$; R' is obtained from R by deleting all transitions with label in T'.*

ii) *N_p' is obtained from N_p by choosing some $T' \subseteq T_p$ and adding for each $t \in T'$ a transition t' and connect it to p in the same way as t; R' is obtained from R by duplicating each transition with label in T' and labelling each copy of a t-labelled transition with t'.*

iii) *N_p' is obtained from N_p by a change of the initial marking that can be effected by firing $w \in L(N_p)$; R' is obtained from R by firing some sequence with image w.*

Proof: All constructions preserve divergence-freeness; thus, we only have to show the $\mathcal{F}$-equivalence of N_p' and R'. For i) and ii) this is easy. For iii) observe that from $\mathcal{F}$-determinism we know that N_p and R are bisimilar. Hence, firing the sequence in R we can fire a sequence with the same image w in N_p to end up in bisimilar states. Since there is only one such sequence in N_p, we get that N_p' and R' are bisimilar. $\square$

This result shows that, for the refinement of places in nets of arc weight 1, it is in some sense sufficient to consider N_p-modules for N_p as shown in Figure 4.10; for every other N_p', some N_p'-modules can be constructed from these N_p-modules.

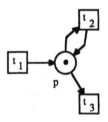

Figure 4.10

$\mathcal{F}$-deterministic nets are related to three net classes studied in the literature. [And83] discusses nets such that any label from Σ appears at most once. The B-condition given there is just the same as being $\mathcal{F}$-deterministic, thus $\mathcal{F}$-deterministic nets properly contain the nets studied in [And83]. In fact, it is not difficult to translate an $\mathcal{F}$-deterministic net into an $\mathcal{F}$-equivalent one such that any label from Σ appears at most once. In [And83] the author considers B-equivalence, i.e. equality of the image-languages, which is the same as $\mathcal{F}$-equivalence for this class by Theorem 4.2.3 ii). Then, some results are shown which easily follow once we know that exchanging $\mathcal{F}$-deterministic, $\mathcal{F}$-equivalent subnets transforms a net to a bisimilar net. Thus, we have strengthened the results of [And83]. It is not discussed in [And83] in which way B-equivalence would be necessary.

Another class of nets we want to mention is the class of deterministic nets of [VN82] and [Pel87]: such a net has a λ-free labelling such that under each reachable marking for every $a \in \Sigma$ at most one a-labelled transition is enabled. This class is strictly included in the class of $\mathcal{FD}$-deterministic nets (which in turn is strictly included in the class of $\mathcal{F}$-deterministic nets): a deterministic net is non-divergent since it is λ-free; a firing sequence w is determined uniquely by its image, thus, if $(lab^*(w), \{x\}) \in \mathcal{F}(N)$, then for the unique marking M with $M_N[lab^*(w)\rangle\rangle M$ we have $\neg M[x\rangle\rangle$, and hence $(lab^*(w)x, \emptyset) \notin \mathcal{F}(N)$.

[VN82] and [Pel87] study deterministic nets from a formal-language-theory point of view. In [Pel87], it is shown that inclusion for the languages of deterministic nets is decidable, but it is not clear whether this result could be used for an alternative proof of Lemma 4.2.8.

When treating nets from the point of view of formal language theory, one often defines the language of a net with regard to a set of final markings. In [Pel87, proof of 4.2] it is shown that $\{a^n b^n, a^n b^{2n} \mid n \geq 1\}$ is not a language of a deterministic net in this sense. Since we are not particularly concerned with markings, it may be no surprise that $\mathcal{F}$-deterministic nets are stronger than deterministic ones from this point of view:

Figure 4.11 shows an $\mathcal{FD}$-deterministic net N. Its $\mathcal{F}$-semantics is essentially $\{(\lambda, \{b\}), (a^n, \emptyset), (a^n b^m, \{a\}) \mid n, m \geq 1\}$. Let M be the marking with $M(p_1) = 1$ and $M(p) = 0$ for all other places p. The firing sequences leading to M have as images $\{a^n b^n, a^n b^{2n} \mid n \geq 1\}$.

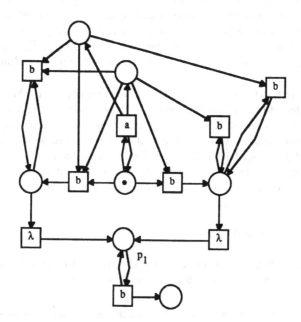

Figure 4.11

It is not clear whether the class of languages of type $L(N)$ definable by deterministic nets N (without final marking) is properly contained in the corresponding class definable by $\mathcal{F}$- or $\mathcal{FD}$-deterministic nets.

In [BR88], so-called intrinsically deterministic nets are studied. On the one hand, this approach is more general, since transitions do not have one label only, but may have several labels; on the other hand, the approach is more restrictive, since only finite bounded nets are treated. Both these differences are rather superficial. The essential point is the following: the definition of intrinsical determinism is more restrictive than our definition of $\mathcal{F}$-determinism; nevertheless, the result of [BR88], which states that for intrinsically deterministic nets the so-called external behaviour equivalence is the same as Milner's $\approx_2$, is weaker than Theorems 4.2.3 and 4.2.4 above.

Before we conclude this section by studying the place refinement of safe nets, let us have another look at the definition of $\mathcal{F}D$-determinism. According to our definition, an $\mathcal{F}D$-deterministic net must be divergence-free. This is a very reasonable approach for several reasons: using a divergence-free refinement net ensures that after a 'finite delay' we get a response from the inserted net – an important property; the nets N_p are divergence-free; requiring divergence-freeness, we obtained the result that $\mathcal{F}D$-deterministic and $\mathcal{F}D$-equivalent nets are in fact branching bisimilar with bisimilar divergence, and this gave a very close relation between N and $N[p \rightarrow R]$ (for an N_p-DD-module R) also in the case that N can diverge.

Alternatively, one could argue that we treat divergence as catastrophic and that consequently the refinement of diverging nets is of no interest. Instead, we should assume that we replace nets only in environments that ensure divergence-freeness; i.e. every environment of interest does not let a divergence string of the replaced net happen. This idea would lead to calling a net $\mathcal{F}D$-deterministic if $(wx, \emptyset) \in F(N)$ and $w \notin D(N)$ implies $(w, \{x\}) \notin F(N)$. If $w \in D(N)$, then of course $(wx, \emptyset), (w, \{x\}) \in F(N)$, but this is of no interest, since w cannot occur in a proper environment. We will not follow this alternative approach any further.

The situation with safe nets is as in this alternative approach. We may find that a net R is suitable to refine a place p whenever p is safe, but R might not be acceptable for place refinement in general. Thus, we will use R only for a restricted class of context nets, and it does not make sense to require that R is safe when we put it in an arbitrary context. In fact, the net N_p itself is not safe. Therefore, we could initiate a study of $\mathcal{F}U$- and $\mathcal{F}DU$-deterministic nets, defined similarly to the alternative definition discussed in the last paragraph. But it will turn out that this is not necessary.

Place refinement of safe nets

A simple net modification allows us to put the study of place refinement of safe nets into the framework of $\mathcal{F}$- and $\mathcal{F}D$-deterministic nets.

Definition 4.2.15 Let N_p be a net with the single place p and the transition set ${}^{\bullet}p \cup p^{\bullet}$; let the labelling be the identity and $\alpha(N_p) \neq \Sigma$; additionally, let all arc weights be 1 and the initial marking be 0 or 1 for p.

A *safe N_p-host* is an N_p-host that is a safe net. An N_p-refinement net R is a *safe N_p-α-module*, $\alpha \in \{D, DD, IR, IL\}$, if for all safe N_p-hosts N the net $N[p \rightarrow R]$ is safe and

– for $\alpha = D$: if N is deadlock-free, then $N[p \to R]$, is, too,

– for $\alpha = DD$: if N is deadlock/divergence-free, then $N[p \to R]$, is, too,

– for $\alpha = IR$: for all $s \in S_N - \{p\}$ s is invariantly reachable in N if and only if it is in $N[p \to R]$,

– for $\alpha = IL$: N and $N[p \to R]$ are image-live-similar.

From a safe N_p-host N, we obtain the net $\overline{N}$ by complementing the place p: we add a place $\overline{p}$ such that $M_{\overline{N}}(p) + M_{\overline{N}}(\overline{p}) = 1$, $W_{\overline{N}}(\overline{p}, t) = W_{\overline{N}}(t, \overline{p}) = 0$ if $t \notin T_p$ or t is on a loop with p, and $W_{\overline{N}}(\overline{p}, t) = W_N(t, p)$, $W_{\overline{N}}(t, \overline{p}) = W_N(p, t)$ otherwise, see e.g. [Dev90a]. Similarly, $\overline{N_p}$ is obtained from N_p by complementing p, and for an N_p-refinement net R we define $\overline{R} = \overline{N_p}[p \to R]$.

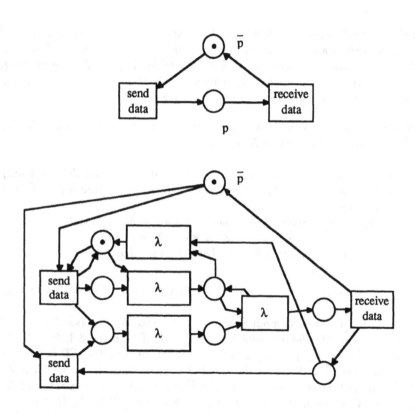

Figure 4.12

Figure 4.1 (without the dangling arcs) shows a net N_p as described in this definition; Figure 4.3 (without the dangling arcs and with the interior transitions turned to λ) is an N_p-refinement net R. Figure 4.12 shows the corresponding nets $\overline{N_p}$ and $\overline{R}$. We have two useful lemmas regarding $\overline{N}$ and $\overline{R}$.

Lemma 4.2.16 *Let N be a safe N_p-host. Then a branching bisimulation with bisimilar divergence $\mathcal{B}$ for N and $\overline{N}$ is given by $\{(M_1, M_2) \mid M_1, M_2$ are reachable markings that coincide on $S_N\}$. For all reachable markings M_2 of $\overline{N}$, we have $M_2(\overline{p}) + M_2(p) = 1$.*

Proof: The second part follows easily from the construction by an inductive proof. Thus $(M_1, M_2) \in \mathcal{B}$ implies for all $t \in T$ that $M_1[t\rangle$ iff $M_2[t\rangle$ by the safeness of p; also the markings reached after firing t obviously conincide again on S_N.   □

Lemma 4.2.17 *Let R be an N_p-refinement net such that $\overline{N_p}$ and $\overline{R}$ are $\mathcal{F}$- ($\mathcal{FD}$-) equivalent. Then a branching bisimulation (with bisimilar divergence) $\mathcal{B}$ for $\overline{N_p}$ and $\overline{R}$ is given by $\{(M_1, M_2) \mid M_1, M_2$ are reachable markings that coincide on $\overline{p}\}$.*

Proof: In the $\mathcal{FD}$-case, the divergence-freeness of $\overline{N_p}$ implies that of $\overline{R}$. Thus, $\overline{N_p}$ and $\overline{R}$ are $\mathcal{F}$-equivalent in any case, and we have to check whether $\mathcal{B}$ is a branching bisimulation. If $M_{\overline{R}}[w\rangle\rangle M_1$, then w is image-enabled in $\overline{N_p}$, and for $M_{\overline{N_p}}[w\rangle\rangle M_2$ the markings M_1 and M_2 coincide on $\overline{p}$; furthermore, M_2 is uniquely determined by $M_2(\overline{p})$. Thus, $\mathcal{B}$ is the branching bisimulation given in the proof of Theorem 4.2.4.   □

Now we are ready to obtain a characterization of safe N_p-modules, which is similar to the characterization for general N_p-modules.

Theorem 4.2.18 *An N_p-refinement net R is a safe N_p-D-module if and only if it is a safe N_p-IR-module if and only if it is a safe N_p-IL-module if and only if $\overline{R}$ is safe and $\overline{R}$ and $\overline{N_p}$ are $\mathcal{F}$-equivalent.*
   R is a safe N_p-DD-module if and only if it is a safe N_p-D-module and $\overline{R}$ is divergence-free if and only if $\overline{R}$ is safe and $\overline{R}$ and $\overline{N_p}$ are $\mathcal{FD}$-equivalent.
   These results also hold if only finite hosts are considered.

Proof: We show first that R is an N_p-module if and only if $\overline{R}$ is safe and $\overline{R}$ and $\overline{N_p}$ are $\mathcal{F}$- ($\mathcal{FD}$-) equivalent.
   '$\Rightarrow$' Let R be a safe N_p-module of any type. Since $\overline{N_p}$ is a safe N_p-host, $\overline{R}$ must be safe by definition. For the test nets N in the proofs of 3.2.6 and 3.3.10, $N \parallel_{T_p} \overline{N_p}$ is a safe N_p-host and $(N \parallel_{T_p} \overline{N_p})[p \rightarrow R] = N \parallel_{T_p} \overline{R}$. From this and the $\mathcal{FD}$-determinism of $\overline{N_p}$, we conclude that $\overline{N_p}$ and $\overline{R}$ must be $\mathcal{F}$- ($\mathcal{FD}$-) equivalent; see also 4.2.2 and 4.2.3.
   '$\Leftarrow$' Assume that $\overline{R}$ is safe and $\overline{N_p}$ and $\overline{R}$ are $\mathcal{F}$- ($\mathcal{FD}$-) equivalent. Let N be a safe N_p-host, let N^* be as in Definition 4.1.1 such that $N = (N^* \parallel_{T_p} N_p)[f] \setminus A$.
   By Lemma 4.2.16, $\mathcal{B}_1 = \{(M_1, M_2) \mid M_1, M_2$ are reachable markings that coincide on $S^* \cup \{p\}\}$ is a branching bisimulation with bisimilar divergence for $N^* \parallel_{T_p} N_p$ and $N^* \parallel_{T_p} \overline{N_p}$.
   By Lemma 4.2.17, we have a branching bisimulation (with bisimilar divergence) for $\overline{N_p}$ and $\overline{R}$, where reachable markings are related iff they coincide on $\overline{p}$. Hence, as in the proof of Theorem 3.4.3, we can find a branching bisimulation (with bisimilar divergence) $\mathcal{B}_2$ for $N^* \parallel_{T_p} \overline{N_p}$ and $N^* \parallel_{T_p} \overline{R}$ such that reachable markings are related iff they coincide

on $S^* \cup \{\bar{p}\}$. Recall that the reachable markings of $N^* \parallel_{T_p} \bar{R}$ restricted to $S_{\bar{R}}$ are reachable markings of $\bar{R}$, hence safe; thus, the safeness of $N^* \parallel_{T_p} N_p$ implies that of $N^* \parallel_{T_p} \bar{R}$.

Now we show that $\mathcal{B}_3 = \{(M_3, M_4) \mid M_3$ is a reachable marking of $N^* \parallel_{T_p} \bar{R}$ that coincides on $S^* \cup S_R$ with the reachable marking M_4 of $N^* \parallel_{T_p} R\}$ is a branching bisimulation with bisimilar divergence for $N^* \parallel_{T_p} \bar{R}$ and $N^* \parallel_{T_p} R$.

Whenever $(M_3, M_4) \in \mathcal{B}_3$, then $M_3[t\rangle$ implies $M_4[t\rangle$ and $M_4[t\rangle$ with $\bar{p} \notin {}^\bullet t$ implies $M_3[t\rangle$. Thus, we only have to consider the case that $(M_3, M_4) \in \mathcal{B}_3$, $M_4[t\rangle$ and $\bar{p} \in {}^\bullet t$. Let P be $S^* \cap {}^\bullet t$, and let t' be the label of t. Let M_1, M_2 be markings with $(M_1, M_2) \in \mathcal{B}_1$ and $(M_2, M_3) \in \mathcal{B}_2$; such markings do exist by the definition of a bisimulation, and they coincide on S^* and thus on P. Since t is enabled under M_4, P is marked by all of M_1, M_2, M_3 and M_4. Thus, t' is enabled under M_1; we have by construction of $\overline{N_p}$ that $p \in t'^\bullet - {}^\bullet t'$, and safeness of N implies $M_1(p) = 0$, hence $M_2(\bar{p}) = M_3(\bar{p}) = 1$. Therefore $M_3[t\rangle$, and $\mathcal{B}_3$ is a bisimulation as desired.

First, this implies safeness of $N^* \parallel_{T_p} R$, hence of $N[p \to R]$. Secondly, we can compose $\mathcal{B}_1$, $\mathcal{B}_2$ and $\mathcal{B}_3$ and apply Theorem 3.4.3 to get a branching bisimulation (with bisimilar divergence) for N and $N[p \to R]$. Thus, deadlock- (deadlock/divergence-) freeness carries over from N to $N[p \to R]$, the same actions are image-live in both nets, and the form of the bisimulation shows that the same places of $S^* = S_N - \{p\}$ are invariantly reachable in N and $N[p \to R]$.

Finally, let us consider the characterization involving that $\bar{R}$ is divergence-free. On the one hand, if R is a safe N_p-D-module and $\bar{R}$ is divergence-free, then $\bar{R}$ is safe and $\mathcal{F}$-equivalent to $\overline{N_p}$ (as we have already shown); since $\overline{N_p}$ is divergence-free, too, $\bar{R}$ and $\overline{N_p}$ are even $\mathcal{F}D$-equivalent.

On the other hand, if $\bar{R}$ is a safe and $\mathcal{F}D$-equivalent to $\overline{N_p}$, then the divergence-freeness of $\overline{N_p}$ implies that of $\bar{R}$ and thus, that $\bar{R}$ and $\overline{N_p}$ are $\mathcal{F}$-equivalent; hence, R is a safe N_p-D-module (as we have already shown). $\qquad\square$

Corollary 4.2.19 *It is decidable whether a given finite N_p-refinement net is a safe N_p-module of any given type.*

Proof: Apply the previous theorem. First decide whether $\bar{R}$ is safe. To do this, start constructing the reachability graph of $\bar{R}$; either you reach a marking that is not safe, or the reachability graph is finite and an inspection shows the safeness of $\bar{R}$. If $\bar{R}$ is safe, we can apply Theorems 3.2.6 and 3.2.8. $\qquad\square$

It should be remarked that this decision algorithm is practical in the sense that it is realistic to implement it; it does not rely on the decidability of liveness.

Refining a safe place with a safe module ensures the same properties as listed in Corollary 4.2.13. Safe modules can be modified in the same ways as listed in Theorem 4.2.14.

For N_p as shown in Figure 4.1, the net of Figure 4.3 is a safe N_p-module by Theorem 4.2.18 – provided we change the labelling of the interior transitions to λ. This shows that not every safe N_p-module is a general N_p-module, which was to be expected. Also, not every N_p-module is a safe N_p-module. Refining a place p in a safe net N by a module R ensures that all places in $S_N - \{p\}$ are safe in $N[p \to R]$ by Corollary 4.2.13 iii); but the places in S_R might not be safe.

4.3   Composition by merging places

While the parallel composition $\|_A$ describes the synchronous communication via actions from A, the place composition $\||_P$ introduced in Definition 4.1.2 describes the asynchronous communication via some channels P. The composition by merging places has found little attention in the literature, but see e.g. [Sou91,Che91]; these papers consider unlabelled nets, whereas we are interested in labelled nets.

In this section, we initiate research into place composition similarly to the approach in Chapter 3 for the synchronous case. What we present here is just a beginning since we restrict the application of $\||_P$ to the case where one component has internal transitions only. As we will see in the following section, this is enough to study the refinement of transitions as in [Vog87] in a broader framework.

To begin with, we will define several external equivalences for P-daughter-nets, which again require that exchanging equivalent nets preserves some basic feature of behaviour. Observe that, in this setting with asynchronous communication, it does not make sense to speak of a fully abstract congruence: on the one hand, we are interested in equivalences of P-daughter-nets; on the other hand, we always compose a P-daughter-net with a P-host, and the result is, in general, not a P-daughter-net.

Definition 4.3.1 Let N_1 and N_2 be P-daughter-nets. They are P-α-*equivalent*, $\alpha \in \{deadlock, deadlock/divergence, IL\}$, if $N \||_P N_1$ and $N \||_P N_2$ are α-similar for all P-hosts N. They are P-*IR-equivalent*, if for all P-hosts N the same places of $S_N - P$ are invariantly reachable in $N \||_P N_1$ and $N \||_P N_2$.

This definition is, of course, in perfect analogy to the definitions in Chapter 3. There we saw that, in the case of synchronous communication, the external equivalence based on deadlocks differs from the external equivalences based on image-liveness and invariant reachability. In the case of asynchronous communication, these three variants coincide; hence, we will only study P-deadlock- and P-deadlock/divergence-equivalence in this and the next section.

Theorem 4.3.2 *P-deadlock-, P-IL- and P-IR-equivalence coincide for P-daughter-nets.*

Proof: Let N_1 and N_2 be P-daughter-nets and N be a P-host.

First, assume that N_1 and N_2 are P-deadlock-equivalent. Let a be image-live in $N \||_P N_1$. Then, $N \setminus (\Sigma - \{a\}) \||_P N_1$ is deadlock-free, and thus also $N \setminus (\Sigma - \{a\}) \||_P N_2$. Hence, a is image-live in $N \||_P N_2$. We conclude that N_1 and N_2 are P-IL-equivalent.

Next, assume that N_1 and N_2 are P-IL-equivalent. Let $s \in S_N - P$ be invariantly reachable in $N \||_P N_1$. Construct $\tilde{N}$ from N by hiding all actions and adding a new, a-labelled transition which is on a loop with s. Then, a is image-live in $\tilde{N} \||_P N_1$, thus also in $\tilde{N} \||_P N_2$. Hence, s is invariantly reachable in $N \||_P N_2$. We conclude that N_1 and N_2 are P-IR-equivalent.

Finally, assume that N_1 and N_2 are P-IR-equivalent. Let $N \||_P N_1$ be deadlock-free. Construct $\tilde{N}$ from N by adding a new place s, a new λ-labelled transition t, and arcs of weight 1 from every visible transition to s and from s to t. Then, s is invariantly reachable in $\tilde{N} \, \|\|_P \, N_1$, thus also in $\tilde{N} \, \|\|_P \, N_2$. Hence, $N \, \|\|_P \, N_2$ is deadlock-free. We conclude that N_1 and N_2 are P-deadlock-equivalent.   □

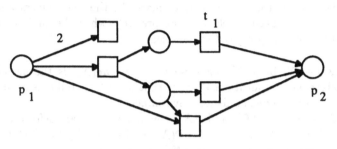

Figure 4.13

In order to study the external equivalences, it is important to describe how a P-daughter-net reacts to stimuli of the environment, i.e. to tokens put on the places of the interface. For this description, the interplay between P-daughter-net and environment is relevant. Consider the net of Figure 4.13, which has the interface $P = \{p_1, p_2\}$. (Recall that all transitions of P-daughter-nets are λ-labelled; these labels are omitted.) If we (taking the part of the environment) put a token onto p_1, then eventually two tokens will be produced on p_2. If after this production we put a second token onto p_1, we will get another two tokens on p_2. If, instead of waiting for two tokens on p_2, we already react to the first token by putting a second token onto p_1, then it might happen that all in all we only get two tokens on p_2 instead of four. (This case may happen if the first token on p_2 was produced by t_1.) As a third possibility, we could put two tokens onto p_1 in the beginning, and then we might get no tokens on p_2; this cannot happen in the other two cases.

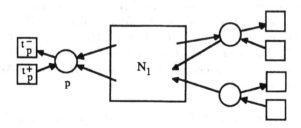

Figure 4.14

Thus, the relevant behaviour of the environment is to put tokens onto the interface places and to check that tokens have been produced by the P-daughter-net; this check can be done by removing the tokens. In order to make this potential behaviour of the environment visible in our action-oriented approach, we add for each interface place p two transitions t_p^+ and t_p^- that respectively add and remove one token from p, see Figure 4.14.

These transitions must be visible, and for simplicity we denote the actions they perform again by t_p^+ and t_p^-. In other words, the transitions t_p^+ and t_p^- represent a standard environment of N_1, and by adding this environment we can modify any semantics of nets to a semantics of P-daughter-nets as described in the next definition. This way, we can capture *when* tokens are put onto the interface places by the environment and *when* the P-daughter-net produces tokens itself, or more precisely when the environment notices that the P-daughter-net has produced some tokens.

Definition 4.3.3 Let N_1 be a P-daughter-net. Then $env(N_1)$ is the net obtained by adding for each place $p \in P$ transitions t_p^+, t_p^- with labels t_p^+, t_p^- and arcs of weight 1 from t_p^+ to p and from p to t_p^-. (*env* stands for 'environment'.)

The $PLan$-, $P\mathcal{F}$-, $P\mathcal{FD}$-, $P\mathcal{F}^+$- and $P\mathcal{F}^{++}$-*semantics* of N_1 are defined by $PLan(N_1)$ $= L(env(N_1))$, $P\mathcal{F}(N_1) = \mathcal{F}(env(N_1))$, $P\mathcal{FD}(N_1) = \mathcal{FD}(env(N_1))$ etc. P-daughter-nets N_1, N_2 are P-*bisimilar*, if $env(N_1)$ and $env(N_2)$ are bisimilar.

To get a first result, we look at what happens to the language and the markings of a net when we exchange two subnets. We need two lemmas first, where the proof of the first lemma is obvious.

Lemma 4.3.4 *Let N_1 be a P-daughter-net, M a marking of $env(N_1)$, $p \in P$, $t \in T_1$. Then $M[t_p^- t\rangle$ implies $M[t t_p^-\rangle$ and $M[t t_p^+\rangle$ implies $M[t_p^+ t\rangle$.*

To translate firing sequences of some net $N \amalg_P N_1$ to firing sequences of $env(N_1)$, we introduce a mapping ψ. Let N_1 be a P-daughter-net, N be a P-host, $P = \{p_1, \ldots, p_n\}$. Then ψ preserves the transitions of N_1 and translates each transition t of N to a sequence of transitions t_p^+ and t_p^-, $p \in P$, that has the same effect on the interface as t; formally, the homomorphism $\psi^* : (T \cup T_1)^* \to (\Sigma \cup T_1)^*$ is defined by $\psi(t) = t$ for $t \in T_1$ and

$$\psi(t) = (t_{p_1}^-)^{W(p_1,t)} \ldots (t_{p_n}^-)^{W(p_n,t)} (t_{p_1}^+)^{W(t,p_1)} \ldots (t_{p_n}^+)^{W(t,p_n)} \text{ for } t \in T.$$

Similarly, φ translates the firing sequences of $N \amalg_P N_1$ to action sequences of $env(N_1)$; in contrast to ψ, it deletes the transitions of N_1; formally, $\varphi^* : (T \cup T_1)^* \to \Sigma^*$ is the homomorphism defined by $\varphi(t) = \lambda$ for $t \in T_1$ and $\varphi(t) = \psi(t)$ for $t \in T$. Furthermore, $id_T^* : (T \cup T_1)^* \to T^*$ is defined by $id_T(t) = \lambda$ for $t \in T_1$ and $id_T(t) = t$ for $t \in T$.

With these mappings, we can describe how the dynamics of $N \amalg_P N_1$ is related to the dynamics of N and $env(N_1)$.

Lemma 4.3.5 *Let N be a P-host, N_1 be a P-daughter-net and M_0, M_0' be markings of $N \amalg_P N_1$. Let N' be obtained from N by deleting P. Let M, M' be the markings M_0, M_0' restricted to the places of N', and M_1, M_1' be the markings M_0, M_0' restricted to the places of $env(N_1)$. Let $w \in (T \cup T_1)^*$.*

   i) *$M_0[w\rangle M_0'$ if and only if $M[id_T^*(w)\rangle M'$ and $M_1[\psi^*(w)\rangle M_1'$.*

   ii) *$M[id_T^*(w)\rangle M'$ and $M_1[\varphi^*(w)\rangle\rangle M_1'$ implies $M_0[w'\rangle M_0'$ for some $w' \in (T \cup T_1)^*$ with $id_T^*(w') = id_T^*(w)$.*

Proof: i) The proof is by induction on the length of w, hence it is enough to consider the case $w = t \in T \cup T_1$. If $t \in T_1$, then t only depends on the marking of $S_1 \cup P$ and only changes the marking of $S_1 \cup P$, $id_T^*(t) = \lambda$ and $\psi^*(t) = t$. If $t \in T$, then $\psi(t)$ needs the same tokens on P as t; in its first part, $\psi(t)$ removes the same tokens from P as t, and, in its second part, it adds the same tokens, such that t and $\psi(t)$ have the same effect on P; furthermore $id_T(t) = t$. The claim follows.

ii) Let $id_T^*(w) = t_1 \ldots t_n$ with $t_i \in T$ for $i = 1, \ldots, n$. By Lemma 4.3.4, we can rearrange any firing sequence underlying $\varphi^*(w)$ to obtain some $w_0 \varphi(t_1) w_1 \ldots \varphi(t_n) w_n$ with $w_i \in T_1^*$ for $i = 0, \ldots, n$. Thus, we can apply i) for $w' = w_0 t_1 w_1 \ldots t_n w_n$.   □

Theorem 4.3.6 *Let N_1, N_2 be $PLan$-equivalent P-daughter-nets, N a P-host. Then, $N \lllr_P N_1$ and $N \lllr_P N_2$ are marking-equivalent on $S_N - P$ and language-equivalent.*

Proof: We can change N to N' by removing the tokens from P and adding a λ-transition that can fire once and puts the tokens onto P we have just removed. Obviously, $N \lllr_P N_i$ and $N' \lllr_P N_i$ are branching bisimilar with bisimilar divergence, marking-equivalent on $S_N - P$ and have the same invariantly reachable places in $S_N - P$. Thus, without loss of generality, we may assume that P is empty under M_N. This is useful for the application of Lemma 4.3.5, since under this assumption the initial marking of $N \lllr_P N_i$ restricted to the places of $env(N_i)$ is the initial marking of $env(N_i)$.

For any $w \in FS(N \lllr_P N_1)$ we have $M_N[id_T^*(w)\rangle$ and $M_{env(N_1)}[\psi^*(w)\rangle$, hence we have $M_{env(N_1)}[\varphi^*(w)\rangle$ by Lemma 4.3.5 i). By assumption and 4.3.5 ii), we find $w' \in FS(N \lllr_P N_2)$ with $id_T^*(w') = id_T^*(w)$. Since the transitions of N_1 and N_2 are internal and do not effect the marking of $S_N - P$, the claim follows.   □

In our setting, the result on the markings of nets $N \lllr_P N_i$ cannot be strengthened to all of S_N. The nets N_1 and N_2 shown in Figure 4.15 are even P-bisimilar, $P = \{p_1, p_2, p_3\}$; the P-bisimulation relates two markings if and only if

- they coincide on p_1,

- the places s_1 and p_2 of N_1 carry together as many tokens as the places s_3, s_4 and p_2 of N_2,

- the places s_1, s_2, p_3 of N_1 carry together as many tokens as the places s_3, p_3 of N_2.

But when we compose these P-daughter-nets with the P-host N of Figure 4.16, a marking with a token on p_2 and empty p_3 can be reached in $N \lllr_P N_1$ only.

Let us discuss a number of examples in order to see that all the equivalences we have defined in Definition 4.3.3 are different. First of all, if the P-daughter-net N_1 can do nothing, while N_2 can remove a token from some $p \in P$, then N_1 and N_2 are of course $PLan$-, but not $P\mathcal{F}$-equivalent.

Figure 4.17 shows two $P\mathcal{F}$-equivalent nets for $P = \{p_1, p_2\}$. To see that they are $P\mathcal{F}$-equivalent requires a lengthy case analysis. The main points are the following. In

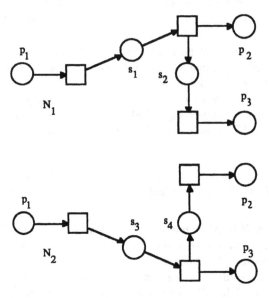

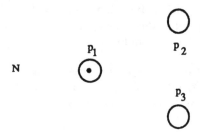

Figure 4.15

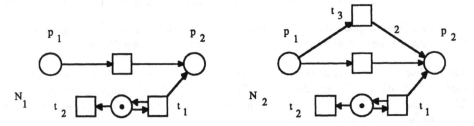

Figure 4.16

Figure 4.17

order to get maximal failure sets, only firing sequences of $env(N_i)$ that leave p_1 empty have to be considered. Any image firing sequence of $env(N_2)$ can also be obtained in $env(N_1)$ (using t_1 if $env(N_2)$ uses t_3) ending up with the same marking and refusal sets; the converse is clear anyway.

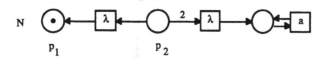

Figure 4.18

Since failure semantics is so useful for the synchronous case, one could hope that $P\mathcal{F}$-equivalence might give an internal characterization of P-deadlock-equivalence. But the nets of Figure 4.17 are not P-deadlock-equivalent. Figure 4.18 shows a P-host N such that $N \lVert_P N_1$ can deadlock, while $N \lVert_P N_2$ cannot. In $N \lVert_P N_1$, we can 'switch off' t_1 by firing t_2, and now the only possible behaviour is to let the remaining token cycle between the places p_1 and p_2, which does not involve any visible action. In $N \lVert_P N_2$, we can also switch off t_1; but if the a-labelled transition is not enabled already, then we can always move a token to p_1 in case p_1 is empty, put two tokens onto p_2 using t_3, and finally activate the action a. Thus, ordinary failure semantics is too weak to deal with deadlocks in the asynchronous case.

In terms of possible refusals, the difference between N_1 and N_2 is the following. While both, $env(N_1)$ and $env(N_2)$, can refuse the single actions $t_{p_1}^-$ and $t_{p_2}^-$ initially (switch off t_1), $env(N_2)$ cannot refuse the sequence $t_{p_1}^+ t_{p_2}^- t_{p_2}^-$, but $env(N_1)$ can. For this reason, we have introduced the $P\mathcal{F}^+$- and $P\mathcal{F}^{++}$-semantics. In fact, N_1 and N_2 are not $P\mathcal{F}^{++}$-equivalent; $(\lambda, \{(t_{p_1}^+ t_{p_2}^-)^n t_{p_2}^- \mid n \in \mathbb{N}_0\}) \in P\mathcal{F}^+(N_1) \subseteq P\mathcal{F}^{++}(N_1)$, but this pair is not in $P\mathcal{F}^{++}(N_2)$, since any $(t_{p_1}^+ t_{p_2}^-)^n$ or $(t_{p_1}^+ t_{p_2}^-)^n t_{p_1}^+$ leads to a marking of N_2 where $t_{p_1}^+ t_{p_2}^- t_{p_2}^-$ or $t_{p_2}^- t_{p_1}^+ t_{p_2}^- t_{p_2}^-$ cannot be refused.

The nets of Figure 4.19 are $P\mathcal{F}^{++}$-equivalent; again, we only give a sketch of the considerations that are necessary to see this. First of all, observe that these nets are $PLan$-equivalent. Now, consider some (w, X) in one of the $P\mathcal{F}^{++}(N_i)$ and a prefix v of $w' \in X$ according to Definition 3.3.12. If the firing sequence underlying wv in one $env(N_i)$ allows a transition that transports a token from p_1 to p_2, then we can assume that the transitions of N_i are switched off after this sequence in order to get a maximal refusal set, and exactly the same can be done in the other $env(N_j)$. Otherwise, the only difference can occur, when the middle transition of N_2 has fired in the firing sequence underlying wv in $env(N_2)$; furthermore, some w' with $vw' \in X$ must require two transports of a token from p_1 to p_2. Thus, w' has a prefix v' requiring one transport and we can image-fire wvv' in $env(N_1)$ in such a way that the situation corresponds to the one in $env(N_2)$. It is easier to see that N_1 and N_2 are not $P\mathcal{F}^+$-equivalent; $(\lambda, \{t_1^+ t_1^+ t_2^- t_2^-\}) \in P\mathcal{F}^+(N_2) - P\mathcal{F}^+(N_1)$.

Next, Figure 4.20 shows two nets which are $P\mathcal{F}^+$-equivalent, but not P-bisimilar. In order to get maximal refusal sets, we can assume that after all image firing sequences, which are the same for $env(N_1)$ and $env(N_2)$, both nets are switched off, hence $P\mathcal{F}^+$-equivalence follows. But $env(N_1)$ has no possiblity to get in a situation bisimilar to

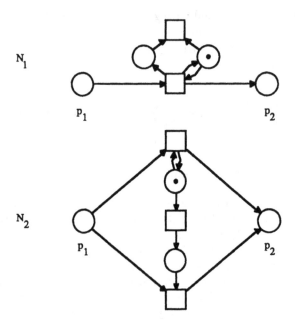

N_1

p_1 p_2

N_2

p_1 p_2

Figure 4.19

$env(N_2)$ with the lower place marked.

   Finally, let us take a P-daughter-net like N_1 shown in Figure 4.20; if we add an initially marked loop to N_1 such that the loop is disconnected from N_1, then only $P\mathcal{FD}$-equivalence distinguishes the resulting net from N_1. If we add such a loop to two totally different nets, then only $P\mathcal{FD}$-equivalence identifies the two resulting nets.

   This concludes our comparison of the equivalences defined in 4.3.3. The next example shows that $P\mathcal{FD}$-equivalence is too weak to imply P-deadlock/divergence-equivalence. Figure 4.21 shows two $P\mathcal{FD}$-equivalent P-daughter-nets N_1 and N_2, which can be seen as follows. First, both, $env(N_1)$ and $env(N_2)$, are divergence-free. Regarding the PF-semantics, the only difference could occur for some $wt'w' \in FS(env(N_2))$. For such a sequence, we have $ww' \in FS(env(N_1))$ reaching the same marking on $P = \{p_1, p_2\}$. Either, at least one of p_1 or p_2 is empty under this marking; then t is disabled in both nets and the same sets of actions can be refused. Or p_1 and p_2 are marked; then $t_{p_1}^-$ and $t_{p_2}^-$ cannot be refused in both nets, and again the same sets of actions can be refused – essentially, $\emptyset$ is the only refusal set.

   But N_1 and N_2 are not P-deadlock/divergence-equivalent; Figure 4.21 shows a P-host N such that $N \parallel_P N_1$ is deadlock- and divergence-free, whereas $N \parallel_P N_2$ can deadlock by firing t'.

   Again, the difference can be described by considering the refusal of sequences. We have that $(t_{p_1}^+ t_{p_2}^+, \{t_{p_2}^- t_{p_2}^-\})$ is in $P\mathcal{F}^+(N_2)$, but not in $P\mathcal{F}^+(N_1)$. In a sense, this sequence $t_{p_2}^- t_{p_2}^-$ does not describe a true interplay between the P-daughter-nets and their environment, it simply describes a potential marking of the place p_2 (or rather its removal). Before we

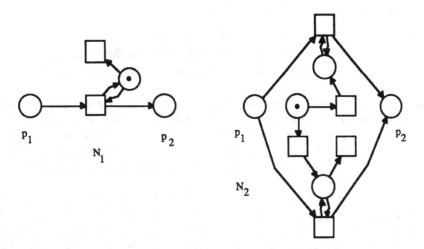

Figure 4.20

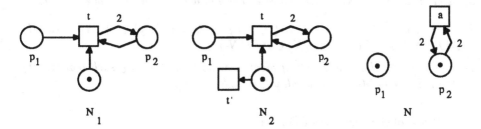

Figure 4.21

define a suitable variant of the $P\mathcal{F}D$-semantics, which considers markings of P, let us observe the difference to the case depicted in Figure 4.17. In that case, the net $env(N_2)$ could never refuse the sequence $t_{p_1}^+ t_{p_2}^- t_{p_2}^-$; this sequence does not describe a potential marking, but a potential reaction of N_2 (namely, producing two tokens on p_2) to an action of the environment (namely, putting a token onto p_1), i.e. it describes a true interplay.

In the next definition, we consider pairs consisting of an action sequence and a marking of the interface P; the marking gives an upper bound on some marking we might reach via the action sequence; thus, removing more tokens than indicated by the marking can be refused, and therefore the marking is similar to a refusal set. After the definition, we will characterize this new semantics in a way that is more similar to the variants of failure semantics we have considered so far. This characterization allows us to compare the new semantics with the semantics defined in 4.3.3, whereas the formulation of the definition is easier to grasp – I hope.

Definition 4.3.7 Let N_1 be a P-daughter-net. The $\mathcal{MF}_PD$-semantics of N_1 is defined as $\mathcal{MF}_P\mathcal{D}(N_1) = (\mathcal{MF}_P(N_1), PD(N_1))$, where

$$\mathcal{MF}_P(N_1) = \{(w,M) \mid w \in \Sigma^*,\ M : P \to \mathbb{N}_0,\ \text{and}$$
$$w \in PD(N_1) \text{ or there exists } M' \text{ such that } M_{env(N_1)}[w\rangle M',$$
$$\neg M'[t\rangle \text{ for all } t \in T_1 \text{ and } M'(p) \le M(p) \text{ for all } p \in P\}.$$

Observe that the marking M' in this definition is a stable state in the sense that no internal transitions are enabled under M'. This stable-state condition is in contrast to our previous variants of failure semantics, and it may appear as a very natural condition. But in case of divergence, it might be impossible to reach a stable state; for this reason, it is not sensible to consider refusals in stable states only, unless divergence is treated explicitly.

The P-daughter-nets of Figure 4.21 are distinguished by $\mathcal{MF}_P\mathcal{D}$-equivalence: for M defined by $M(p_1) = M(p_2) = 1$, we have $(t_{p_1}^+ t_{p_2}^+, M)$ in $\mathcal{MF}_P(N_2)$ but not in $\mathcal{MF}_P(N_1)$.

In the following result, we characterize $\mathcal{MF}_P\mathcal{D}$-semantics in terms of refusals of action sequences; as a consequence, we see that $\mathcal{MF}_P\mathcal{D}$-equivalence is stronger than $P\mathcal{FD}$-equivalence.

Proposition 4.3.8 *Let N_1 be a P-daughter-net and $w \in \Sigma^*$ such that $w \notin PD(N_1)$.*

i) *Let $M : P \to \mathbb{N}_0$, and let $X = \{(t_p^-)^n \mid p \in P,\ n = M(p) + 1\}$. Then $(w,M) \in \mathcal{MF}_P(N_1)$ if and only if $(w,X) \in P\mathcal{F}^+(N_1)$.*

ii) *$(w,Y) \in P\mathcal{F}(N_1)$ if and only if there is some $M : P \to \mathbb{N}_0$ such that $(w,M) \in \mathcal{MF}_P(N_1)$ and $M(p) = 0$ for all $p \in P$ with $t_p^- \in Y$.*

Proof: i) '$\Rightarrow$' obvious from the definitions.

'$\Leftarrow$' Extending a firing sequence by internal transitions can only enlarge the corresponding refusal sets; since $w \notin PD(N_1)$, we can extend each firing sequence underlying w until a stable state is reached. Thus, $(w,X) \in P\mathcal{F}^+(N_1)$ implies that there is some M' such that: $M_{env(N_1)}[w\rangle M'$, $\neg M'[t\rangle$ for all $t \in T_1$ and $\neg M'[v\rangle$ for all $v \in X$. The latter condition implies that $M'(p) \le M(p)$ for all $p \in P$.

ii) '$\Rightarrow$' As in the first part of the proof, we may assume that we have reached a stable state after w. Thus, for each $p \in P$ with $t_p^- \notin Y$, we can add some $(t_p^-)^n$ to Y in order to get a set X with $(w,X) \in P\mathcal{F}^+(N_1)$. Application of i) yields the result.

'$\Leftarrow$' obvious from the definitions. $\square$

Corollary 4.3.9 *For P-daughter-nets, $\mathcal{MF}_P\mathcal{D}$-equivalence implies $P\mathcal{FD}$-equivalence.*

Proof: Apply Proposition 4.3.8 ii). $\square$

Apart from this implication, $\mathcal{MF}_P\mathcal{D}$-equivalence is incomparable to the equivalences defined in 4.3.3; the reason is the same as the one given above for $P\mathcal{FD}$-equivalence. Collecting the results of our considerations, we get the following.

Theorem 4.3.10 *P-bisimulation implies $P\mathcal{F}^+$-equivalence. $P\mathcal{F}^+$-equivalence implies $P\mathcal{F}^{++}$-equivalence. $P\mathcal{F}^{++}$-equivalence implies $P\mathcal{F}$-equivalence. $P\mathcal{F}$-equivalence implies $P Lan$-equivalence. $M\mathcal{F}_P\mathcal{D}$-equivalence implies $P\mathcal{F}\mathcal{D}$-equivalence. No other implications hold in general.*

Proof: The implications follow from the results of Chapter 3 and Corollary 4.3.9; the other part we have just seen from the above examples.   □

Characterization of the external equivalences and their undecidability

We have just studied a number of semantic equivalences. Now we relate them to the external equivalences we are interested in. For the following theorem, note that the implication stated in Part i) follows immediately from Part ii) and Proposition 4.3.10. But my proof that P-deadlock-equivalence implies $P\mathcal{F}^{++}$-equivalence uses infinite context nets. Thus it fails, if we only work with finite context nets, in which case we would have a weaker notion of P-deadlock-equivalence as we have it here. Therefore, Part i) is interesting in its own right.

Theorem 4.3.11   *i) P-deadlock-equivalence implies $P\mathcal{F}$-equivalence. This also holds if only finite hosts are considered in the definition of P-deadlock-equivalence.*

ii) P-daughter-nets are P-deadlock-equivalent if and only if they are $P\mathcal{F}^{++}$-equivalent. $P\mathcal{F}^{++}$-equivalence also implies P-deadlock-equivalence, if only finite hosts are considered in the definition of P-deadlock-equivalence.

iii) Let N_1, N_2 be P-daughter-nets such that $env(N_1)$ and $env(N_2)$ are enabling-finite. N_1 and N_2 are P-deadlock/divergence-equivalent if and only if they are $M\mathcal{F}_P\mathcal{D}$-equivalent. This also holds if only finite hosts are considered in the definition of P-deadlock/divergence-equivalence.

Proof: i) This can be shown using P-hosts similar to the context nets shown in Figure 3.7. The set A is in our case the set $\{t_p^+, t_p^- \mid p \in P\}$ and we have to add P to the context net such that the transition labelled t_p^+ puts one token onto p, the transition labelled t_p^- removes one token from p, $p \in P$.

For clarity, we give an explicit construction of a suitable P-host. Suppose, N_1 and N_2 are P-deadlock-equivalent. Let $A = \alpha(env(N_1)) = \alpha(env(N_2)) = \{t_p^-, t_p^+ \mid p \in P\}$, $(w, X) \in P\mathcal{F}(N_1)$, $w = w_1 \ldots w_m$ with $w_i \in A$. Obviously, X does not contain any t_p^+, $p \in P$. Consider the net N (see Figure 4.22) defined by

$$S = P \cup \{s_j, s \mid j = 1, \ldots, m+2\}$$
$$T = \{t_j, t, t_p^- \mid j = 1, \ldots, m+1, \ t_p^- \in A \cap X\}$$

W takes values in $\{0, 1\}$ and is 1 for the pairs

- $(s_j, t_j), (t_j, s_{j+1})$ for $j = 1, \ldots, m+1$

- (t_j, p) for $j = 1, \ldots, m,$ $w_j = t_p^+$

- (p, t_j) for $j = 1, \ldots, m,$ $w_j = t_p^-$

- $(s, t), (t, s), (s, t_{m+1})$

- $(s_{m+2}, t_p^-), (p, t_p^-), (t_p^-, s)$ for $t_p^- \in A \cap X$

M is 1 on s_1 and s, and 0 everywhere else. All transitions are internal except for t.

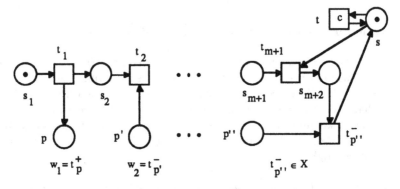

Figure 4.22

The only way that $N \ \|\!\|\!\|_P \ N_i$ can deadlock is to block t by firing $t_1 \ldots t_{m+1}$ interspersed with some transitions from N_i; this has the same effect on N_i as image-firing w in $env(N_i)$. More precisely, the interspersed sequence must correspond to a firing sequence of $env(N_i)$ with image w that yields a marking M_i with $\neg M_i[t_p^-\rangle)$ for all $t_p^- \in A \cap X$; otherwise we could mark s again. Therefore $N \ \|\!\|\!\|_P \ N_1$ can deadlock, thus $N \ \|\!\|\!\|_P \ N_2$ can, too. Hence $(w, X \cap A) \in P\mathcal{F}(N_2)$, and also $(w, X) \in P\mathcal{F}(N_2)$, since $env(N_2)$ does not have a transition with label in $X - A$. Therefore $P\mathcal{F}(N_1) = P\mathcal{F}(N_2)$.

ii) a) To show $P\mathcal{F}^{++}$-equivalence, we can use P-hosts similar to the context nets of Figure 3.19. We have to add the place set P to the context net; each of the transitions in the sequence $w_1 \ldots w_n$ and in the tree encoding the set X corresponds to some t_p^+ or t_p^- by construction, and is connected to p accordingly; but this time, these transitions are not labelled t_p^+ or t_p^-, instead t is the only visible transition.

b) Suppose, we are given some $P\mathcal{F}^{++}$-equivalent P-daughter-nets N_1 and N_2 and a P-host N. We want to show that $N \ \|\!\|\!\|_P \ N_1$ can deadlock if and only if $N \ \|\!\|\!\|_P \ N_2$ can. First, we change N to a net $\tilde{N}$ in the following way (see Figure 4.23):

- Rename every $p \in P$ into a new p'.

- For every $p \in P$, add

    - the empty place p

    - new λ-labelled transitions t_p^+ and t_p^-

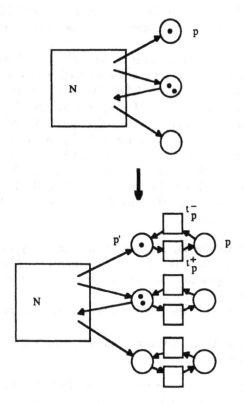

Figure 4.23

– arcs (t_p^+, p), (p, t_p^-), (t_p^-, p') and (p', t_p^+) of weight 1.

It is not hard to see that $N \parallel_P N_i$ can deadlock if and only if $\tilde{N} \parallel_P N_i$ can, $i = 1, 2$. Hence, we may assume that the P-host N satisfies the following properties:

– P is initially empty in N.

– Each transition of N that accesses a place in P either only removes one token from one place in P and does not add any token to any place in P, or it adds one token to one place in P and does not remove any token from any place in P.

Suppose that in $N \parallel_P N_1$ we can fire some $w \in FS(N \parallel_P N_1)$ such that, under the marking M_0 reached after w, we have a deadlock. Let Y be the set of all sequences $u \in T^*$ such that $M[u\rangle$ in N' (M, N' as in Lemma 4.3.5) and u contains a visible transition. Using 4.3.5 ii), we see that $\neg M_1[\varphi^*(u)\rangle\rangle$ for each $u \in Y$ (M_1 as in 4.3.5). Hence, we have $(\varphi^*(w), X) \in \mathcal{F}^+(env(N_1)) \subseteq \mathcal{F}^{++}(env(N_1)) = \mathcal{F}^{++}(env(N_2))$ with $X = \{\varphi^*(u) \mid u \in Y\}$.

First, assume $X = \emptyset$. By Lemma 4.3.5, we can fire some w' in $N \parallel_P N_2$ such that $id_T^*(w') = id_T^*(w)$. If we restrict the marking reached after w' to $S_N - P$, we get M; since $Y = \emptyset$, we see that $N \parallel_P N_2$ can deadlock, too.

Hence, assume that $X \neq \emptyset$. Applying the definition of $\mathcal{F}^{++}$ (Definition 3.3.12) to $(\varphi^*(w), X) \in \mathcal{F}^{++}(env(N_2))$, we find a prefix v of some element w' of X and a corresponding marking M_2' reached after $\varphi^*(w)v$.

Since there is some $u' \in T^*$ with $M[u')$ in N' and $\varphi^*(u') = w'$, we conclude from the second of the above properties of N that there is some prefix u of u' with $\varphi^*(u) = v$; let M' be defined by $M[u)M'$. Define the marking M_0' of $N \, \|_P \, N_2$ such that M_0' coincides with M' on $S_N - P$ and with M_2' on $S_2 \cup P$. We have $M_{N'}[id_T^*(w)u)M'$ and $M_{env(N_2)}[\varphi^*(w)v))M_2'$; by Lemma 4.3.5 ii), we find $w'' \in (T \cup T_2)^*$ such that $M_{N\|_P N_2}[w'')M_0'$ and $id_T^*(w'') = id_T^*(w)u$.

We want to show that M_0' is a deadlock for $N \, \|_P \, N_2$. Assume to the contrary, that $M_0'[u_1)$ and u_1 contains a visible transition. By Lemma 4.3.5 i), this implies $M'[id_T^*(u_1))$ in N', hence $M[u(id_T^*(u_1)))$; by definition of Y, we have $u(id_T^*(u_1)) \in Y$, thus $v(\varphi^*(u_1)) \in X$. On the other hand, $M_0'[u_1)$ implies $M_2'[\varphi^*(u_1)))$ in $env(N_2)$, also by Lemma 4.3.5 i) and by definition of φ. This is a contradiction to the choice of v and M_2' according to the definition of the $\mathcal{F}^{++}$-semantics.

iii) '$\Rightarrow$' The constructions for this implication are again very similar to those above and in the proof of Theorem 3.2.6. In order to test whether $(w, M) \in \mathcal{MF}_P(N_2)$, where $w \notin PD(N_2)$, we apply Proposition 4.3.8 and use the net shown in Figure 4.22, except that each arc pt_p^- has weight $M(p) + 1$.

'$\Leftarrow$' Let N be a P-host; as in the proof of Theorem 4.3.6 we may assume that P is empty under M_N.

If $N \, \|_P \, N_1$ can diverge, let us consider an infinite firing sequence w having a finite image. There are two cases to consider.

a) A prefix of $\varphi^*(w)$ is a divergence string of $env(N_1)$. Hence, a prefix v of $\varphi^*(w)$ is a divergence string of $env(N_2)$. If $v = v't_p^-$, then v' is a divergence string, too; therefore, we may assume that v ends with some t_p^+ (or is empty). If divergence can occur at some stage, it can also occur after some additional t_p^+; therefore; we may assume that the next transition after v in $\varphi^*(w)$ is a t_p^- (or that $v = \varphi^*(w)$). Thus, we may assume that there exists some prefix w' of $id_T^*(w)$ with $\varphi^*(w') = v$. With Lemma 4.3.5 ii), we conclude that $N \, \|_P \, N_2$ can diverge, too.

b) No prefix of $\varphi^*(w)$ is in $D(env(N_1)) = D(env(N_2))$. In this case, $id_T^*(w)$ ends with infinitely many internal transitions. There are two subcases. Either $\varphi^*(w)$ is finite and from the $\mathcal{MF}_P$-part of the $\mathcal{MF}_P\mathcal{D}$-semantics we conclude that $\varphi^*(w)$ is in $L(env(N_2))$. Or $\varphi^*(w)$ is infinite, and from the $\mathcal{MF}_P$-part of the $\mathcal{MF}_P\mathcal{D}$-semantics we conclude that every finite prefix of $\varphi^*(w)$ is in $L(env(N_2))$; by Corollary 3.3.18, we get $\varphi^*(w) \in L^\omega(env(N_2))$. Applying Lemma 4.3.5 ii) once or repeatedly, we can construct some $w' \in FS^\omega(N \, \|_P \, N_2)$ with $id_T^*(w') = id_T^*(w)$, thus $N \, \|_P \, N_2$ can diverge in this case, too.

Now suppose that $N \, \|_P \, N_1$ and $N \, \|_P \, N_2$ are divergence-free, but $N \, \|_P \, N_1$ can deadlock, i.e. after some $w \in FS(N \, \|_P \, N_1)$ we reach some marking M_1 such that no transition is enabled. Let $M : P \to \mathbb{N}_0$ coincide with M_1 restricted to P. Then $(\varphi^*(w), M) \in \mathcal{MF}_P(N_1) = \mathcal{MF}_P(N_2)$ and $\varphi^*(w) \notin PD(N_1) = PD(N_2)$. By Lemma 4.3.5 ii), we find $w' \in FS(N \, \|_P \, N_2)$ reaching a marking M_2 such that

- $id_T^*(w') = id_T^*(w)$, i.e. M_2 coincides on $S_N - P$ with M_1,

- M_2 restricted to P is less or equal to M, i.e. to M_1 restricted to P,

- $\neg M_2[t\rangle$ for all $t \in T_2$.

Since no transition of N is enabled under M_1 and M_1 bounds M_2 from above on $S_N \cup P$, we conclude that M_2 is a deadlock of $N \parallel\!\!\!\mid_P N_2$.   □

The above results, namely Theorems 4.3.11, 4.3.10 and 4.3.6, also show that the exchange of P-deadlock-equivalent nets does not only preserve deadlock-freeness, but also e.g. the language; we will not follow this line of thought any further. With $P\mathcal{F}^{++}$-equivalence, we have found for P-deadlock-equivalence an internal characterization, which does not require to consider all possible context nets. But the $P\mathcal{F}^{++}$-semantics is admittedly somewhat unwieldy; furthermore, we have the undecidability result below for P-deadlock- and P-deadlock/divergence-equivalence. For these reasons, we will restrict attention to some subclass of P-daughter-nets in the next section, where we will also get a stronger result for behaviour preservation.

Theorem 4.3.12 *P-deadlock- and P-deadlock/divergence-equivalence are undecidable for finite P-daughter-nets, if $|P| \geq 3$. This also holds if the equivalences are defined with respect to finite hosts only.*

Proof: It is undecidable whether for given finite λ-free nets N_1 and N_2 we have $L(N_1) = L(N_2)$ [Hac76b]. Let N be a finite λ-free net, $\{a_1, \ldots, a_k\} = \alpha(N)$. We may add a place s with one token and replace each a_i-labelled transition t by a sequence of i a-labelled transitions followed by a b-labelled transition such that the first transition removes the same tokens as t plus the token from s and the last transition adds the same tokens as t plus a token to s. Furthermore, we may add a run place s' (i.e. a place with one token that is on a loop with all transitions) and a λ-transition that removes this token. Call the modified net N' (see Figure 4.24). Finite λ-free nets N_1 and N_2 have the same language if and only if N_1' and N_2' have the same language.

We modify these nets further to get P-daughter-nets. Let $s(a), s(b), o \in P$ be different places. Add P to N' and arcs of weight 1 from $s(x)$ to each x-labelled transition and from each x-labelled transition to o, where $x \in \{a, b\}$. Change all labels to λ and call the resulting net N'' (see Figure 4.25). We show that N_1'' is P-deadlock-equivalent to N_2'' if and only if N_1'' is P-deadlock/divergence-equivalent to N_2'' if and only if N_1' and N_2' have the same language, which settles the claim.

Since each transition of N_1'' or N_2'' removes a token from P or from s', $env(N_1'')$ and $env(N_2'')$ are divergence-free. Therefore, we have by Theorem 4.3.11 iii) and Proposition 4.3.8: if N_1'' and N_2'' are $P\mathcal{F}^+$-equivalent, then they are P-deadlock/divergence-equivalent; if they are P-deadlock/divergence-equivalent, then they are $PLan$-equivalent. The same holds for P-deadlock-equivalence in place of P-deadlock/divergence-equivalence, using 4.3.11 ii).

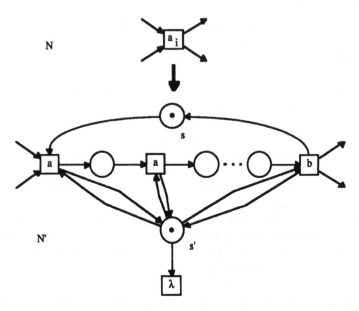

Figure 4.24

Define a homomorphism $\chi^* : \{a, b\}^* \to \{t^+_{s(a)}, t^+_{s(b)}, t^-_o\}^*$ by $\chi(a) = t^+_{s(a)} t^-_o$ and $\chi(b) = t^+_{s(b)} t^-_o$. We have $w \in L(N'_i)$ if and only if $\chi^*(w) \in PLan(N''_i)$. Therefore, P-deadlock/divergence-equivalence of N''_1 and N''_2 implies language equivalence of N'_1 and N'_2, and the same holds for P-deadlock-equivalence.

Now assume that N'_1 and N'_2 are language-equivalent. Then, for each firing sequence of $env(N''_1)$ there is one of $env(N''_2)$ with the same image and such that the markings reached coincide on P – and vice versa. To get maximal refusal sets for the $P\mathcal{F}^+$-semantics, N''_1 and N''_2 must be switched off in the end by removing the token from s'. Which sequences of actions are possible now in $env(N''_1)$ and $env(N''_2)$ depends on the marking of P only. Therefore, N''_1 and N''_2 are $P\mathcal{F}^+$-equivalent, hence P-deadlock/divergence- and

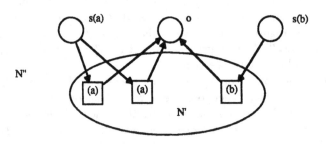

Figure 4.25

P-deadlock-equivalent. □

4.4 I,O-nets and the refinement of transitions

In this section, we want to develop a theory of P-daughter-nets that are deterministic in some sense; the aim is to obtain similar results for the refinement of transitions as those we have obtained in Section 4.2 for the refinement of places. Thus, we are looking for a notion of determinism that has the following properties.

– It should be decidable whether a P-daughter-net is deterministic.

– P-deadlock- and P-deadlock/divergence-equivalence should be decidable for deterministic nets.

– The notion of determinism should allow to treat the refinement of transitions, i.e. nets N_h in the sense of Definition 4.1.3 should be deterministic.

– If the exchange of some given deterministic nets always preserves behaviour in some weak sense (i.e. deadlock- or deadlock/divergence-freeness), then it should hold that their exchange also preserves behaviour in some strong sense (preferably stronger than each of our variants of failure semantics).

Unfortunately, the λ-labelled transition h of a net N_h 'decides' nondeterministically whether it fires or not (provided it is enabled). Suppose, by firing transitions t_p^+ with $p \in {}^\bullet h$, we have put just enough tokens on the places in the preset of h to make h enabled. Now, depending on the internal action of firing h, transitions t_p^- with $p \in {}^\bullet h$ may be refused or may be fired, i.e. the actions t_p^- with $p \in {}^\bullet h$ may occur or may be refused after the same image firing sequence of $env(N_h)$. This shows that $env(N_h)$ is not $\mathcal{F}$- or $\mathcal{FD}$-deterministic.

The situation is different for transitions t_p^- with $p \in h^\bullet$. If transition h is enabled, then – no matter whether h fires first or not – any of these transitions can be fired as the next visible transition, i.e. the actions t_p^- with $p \in h^\bullet$ cannot be refused. More generally, $env(N_h)$ is deterministic with respect to the actions t_p^- with $p \in h^\bullet$; in other words, observing an image firing sequence of $env(N_h)$, we can conclude whether t_p^- with $p \in h^\bullet$ is image-enabled after this sequence or not. Thus, ${}^\bullet h$ and $h^\bullet$ are treated differently with respect to determinism. Therefore, we will study in this section only those P-daughter-nets N_1 where P is split up into non-empty sets I of input places and O of output places; this means that no arcs of N_1 lead to I or from O. We have already required this for N_h-refinement nets in Definition 4.1.3.

There are three features that we observe for nets N_h and that play an important rôle in the proofs of the results below. As already remarked, N_h is deterministic with respect to $h^\bullet$, and this leads to the notion of output-determinism. Secondly, N_h produces only bounded output from a given finite input, and this leads to the notion of output-boundedness. Thirdly, N_h is non-absorbing, meaning that it always produces some additional output if it consumes some additional input; hence, it does not just absorb the input without any positive effect. (Given some additional input, it might happen to be unable to consume this input; naturally, it does not produce any additional output in this case.)

These notions are formalized in our first definition.

Definition 4.4.1 Let P be the disjoint union of nonempty sets I and O. A P-daughter-net N_1 is an I, O-net if it satisfies the following conditions:

- There are no arcs in N_1 leading to I or from O.
- N_1 is *output-deterministic*, i.e. $\forall p \in O : (w, \{t_p^-\}) \in P\mathcal{F}(N_1) \Rightarrow (wt_p^-, \emptyset) \notin P\mathcal{F}(N_1)$.
- N_1 is *output-bounded*, i.e. for all $p \in O$ and $(w, \emptyset) \in P\mathcal{F}(N_1)$ there is some $n \in \mathbb{N}_0$ with $(w(t_p^-)^n, \emptyset) \notin P\mathcal{F}(N_1)$.
- N_1 is *non-absorbing*, i.e. for $X_I = \{t_p^- \mid p \in I\}$, $w, w' \in \{t_p^+ \mid p \in I\}^*$ with $w' \neq \lambda$ we have: if $(w, X_I), (ww', X_I) \in P\mathcal{F}(N_1)$ then there are $p \in O$ and $n \in \mathbb{N}$ such that $(w(t_p^-)^n, \emptyset) \notin P\mathcal{F}(N_1)$ and $(ww'(t_p^-)^n, \emptyset) \in P\mathcal{F}(N_1)$.

N_1 is a P-*divergence-free* I, O-net if additionally $env(N_1)$ is divergence-free.

The formal definitions of output-determinism and output-boundedness are straightforward translations from the verbal explanations given above. In human words, the last part of Definition 4.4.1 says the following. The action sequence w describes that some tokens are put onto the input places, and $(w, X_I) \in P\mathcal{F}(N_1)$ says that now the transitions of N_1, which are internal, can fire in such a way that the input is empty afterwards. The action sequence ww' puts more tokens onto the input places, and these can also be consumed completely by N_1. Now, the requirements for p and n are that, given the input described by w, it cannot happen that we find n or more tokens on p, while this can happen if we have the larger input described by ww'.

It is easily checked that a net N_h with ${}^\bullet h = I$ and $h^\bullet = O$ is a P-divergence-free I, O-net. Also, given some disjoint I_i, O_i-nets N_i, $i = 1, \ldots, n$, the union of these nets is an I, O-net with $I = \bigcup_i I_i$ and $O = \bigcup_i O_i$; thus, in our framework, we can also deal with the simultaneous refinement of several transitions as studied in [Mül85].

Another example of an I, O-net N_1 is shown in Figure 4.26. This net is output-deterministic; given some image firing sequence w of $env(N_1)$, there are several cases. Either w visibly (i.e. using $t_{p_3}^+$) puts more tokens onto p_3 than it removes, in which case $t_{p_3}^-$ can fire immediately and the action $t_{p_3}^-$ cannot be refused. Or w removes more tokens from p_3 than it puts visibly, in which case the difference must be 1, one of the internal transitions must have fired, and the action $t_{p_3}^-$ cannot occur next. In the case that w visibly puts the same number of tokens onto p_3 as it removes, there are two subcases. Either w puts more tokens onto one of p_1 and p_2 than it removes visibly, in which case the corresponding internal transition has already fired or can fire next, thus action $t_{p_3}^-$ can occur next and cannot be refused. Or this is not the case, then the marking reached after w is the initial marking and $t_{p_3}^-$ cannot occur.

N_1 is output-bounded, since $env(N_1)$ can perform at most one more $t_{p_3}^-$ after some w than there are $t_{p_3}^+$ in w. Finally, N_1 is non-absorbing: if we have $w, ww' \in \{t_p^+ \mid p \in I\}^*$ as required in Definition 4.4.1, then $(ww', X_I) \in P\mathcal{F}(N_1)$ implies that ww' consists of

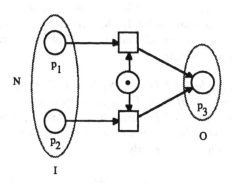

Figure 4.26

at most one action, i.e. $w = \lambda$, and w' is $t_{p_1}^+$ or $t_{p_2}^+$; we have $(t_{p_3}^-, \emptyset) \notin PF(N_1)$ but $(t_{p_1}^+ t_{p_3}^-, \emptyset), (t_{p_2}^+ t_{p_3}^-, \emptyset) \in PF(N_1)$.

This example also shows that there is some nondeterminism left in I,O-nets. Not only can N_1 decide whether it consumes a token from p_1, e.g. $(t_{p_1}^+ t_{p_2}^+ t_{p_1}^-, \emptyset) \in PF(N_1)$ and $(t_{p_1}^+ t_{p_2}^+, \{t_{p_1}^-\}) \in PF(N_1)$; it also chooses between tokens on p_1 and p_2, i.e. we have $(t_{p_1}^+ t_{p_2}^+, \{t_{p_1}^-\}), (t_{p_1}^+ t_{p_2}^+, \{t_{p_2}^-\}) \in PF(N_1)$, but $(t_{p_1}^+ t_{p_2}^+, \{t_{p_1}^-, t_{p_2}^-\}) \notin PF(N_1)$.

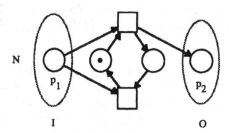

Figure 4.27

Figure 4.27 shows another net N_1 that illuminates our definition. Again, N_1 can choose whether to consume a token on p_1 or to do nothing, but otherwise it is very predictable. When consuming tokens from p_1, it strictly alternates between producing a token on p_2 or showing no reaction at all. This net is output-deterministic and output-bounded, but it is not non-absorbing. For $w = w' = t_{p_1}^+$ we have

- $(t_{p_1}^+, \{t_{p_1}^-\}), (t_{p_1}^+ t_{p_1}^+, \{t_{p_1}^-\}) \in PF(N_1)$

- $(t_{p_1}^+ (t_{p_2}^-)^n, \emptyset) \notin PF(N_1) \Rightarrow n \geq 2 \Rightarrow (t_{p_1}^+ t_{p_1}^+ (t_{p_2}^-)^n, \emptyset) \notin PF(N_1)$.

If in Definition 4.4.1 we replaced $PF(N_1)$ by $F(env(N_1))$ throughout, we would get the PFD-version of an I,O-net. From output-determinism we conclude that this PFD-version would coincide with the notion of a P-divergence-free I,O-net.

Observe that we have used $PF(N_1)$ in the definition of I, O-nets, although PF-equivalence is in general too weak to imply the external equivalences we are interested in. As a first justification, let us remark that the class of (P-divergence-free) I, O-nets is closed under the equivalences of interest.

Proposition 4.4.2 *Let N_1 be an I, O-net, N_2 be a P-daughter-net such that there are no arcs leading to I or from O. If N_1 is P-deadlock-equivalent to N_2, then N_2 is an I, O-net. If N_1 is P-divergence-free and P-deadlock/divergence-equivalent to N_2, then N_2 is a P-divergence-free I, O-net. These statements are true even if the external equivalences are defined with respect to finite hosts only.*

Proof: The first statement is a corollary to Theorem 4.3.11 i). The second statement follows from Theorem 4.3.11 iii) and Corollary 4.3.9; note that we only need one implication of 4.3.11 iii), and that in the proof of this implication the precondition regarding enabling-finiteness was not used. (We will come back to this issue below.) □

Since P-daughter-nets communicate with their environments via the places in P, it would be desirable to have a semantics that explicitly refers to the possible markings of P. We will now develop such a semantics for I, O-nets, and this semantics will turn out to be useful for our further results. First, we define $M_{\max}(w)$, which is the maximal output we can obtain by elongating w.

Definition 4.4.3 Let N_1 be an I, O-net and $w \in FS(env(N_1))$. The marking $M_{\max}(w) : O \to \mathbb{N}_0$ of O is defined by $M_{\max}(w)(p) = \max\{M(p) \mid M_{env(N_1)}[ww')M$ with $w' \in T_1^*\}$ for $p \in O$.

Since N_1 is output-bounded, the maximum in this definition is well-defined. Since N_1 is a subnet of $env(N_1)$, this definition also applies to firing sequences of N_1.

Before defining the announced semantics of I, O-nets, we prove a technical lemma. It states that the maximal output of an I, O-net only depends on the number of tokens we put onto I and not on internal choices of the net; also, in some sense, the input an I, O-net can consume does not depend on internal choices.

Lemma 4.4.4 *Let N_1 be an I, O-net and let $w_1, w_2 \in FS(env(N_1))$ with $lab^*_{env(N_1)}(w_1) = lab^*_{env(N_2)}(w_2) \in \{t_p^+ \mid p \in I\}^*$.*

i) $M_{\max}(w_1) = M_{\max}(w_2)$

ii) If I is empty after firing w_1 and we have some $w_2' \in T_1^*$ such that $M_{env(N_1)}[w_2w_2')M_2$ and M_2 coincides on O with $M_{\max}(w_2)$, then I is empty under M_2.

Proof: Let v be the image of w_1 and w_2.

i) Assume that $k = M_{\max}(w_1)(p) < M_{\max}(w_2)(p)$ for some $p \in O$. Let $M_{env(N_1)}[w_1w_1')M_1$, $M_{env(N_1)}[w_2w_2')M_2$ such that $M_{\max}(w_1')(p) = M_1(p)$, $M_{\max}(w_2)(p) = M_2(p)$ and $w_1', w_2' \in T_1^*$. These sequences show that $(v(t_p^-)^k, \{t_p^-\}), (v(t_p^-)^{k+1}, \emptyset) \in PF(N_1)$, which contradicts output determinism.

ii) Assume to the contrary; then we may assume that the last transitions of v have put the tokens onto I that are not removed by $w_2 w_2'$. Eliminating these from $w_2 w_2'$ we get a firing sequence w with image a proper prefix v' of v and we have: $(v', X_I), (v, X_I) \in PF(N_1)$, and by i) no firing sequence with image v can put more tokens onto any $p \in O$ than w does, hence $(v(t_p^-)^n, \emptyset) \in PF(N_1)$ implies $(v'(t_p^-)^n, \emptyset) \in PF(N_1)$ for all $p \in O$, $n \in \mathbb{N}$. This is a contradiction, since N_1 is non-absorbing. $\square$

Based on this lemma, we now define the IOF-semantics of an I,O-net; it describes the input-output behaviour by giving pairs consisting of a marking of I that can be consumed and a marking of O that is the corresponding maximal output. This semantics has the flavour of a failure semantics, since any such pair corresponds to an image firing sequence leading to a marking where all transitions t_p^-, $p \in P$, can be refused.

Definition 4.4.5 For an I,O-net N_1 we define its IOF-*semantics* $IOF(N_1)$ as the set of those pairs $(M_1 : I \to \mathbb{N}_0, M_2 : O \to \mathbb{N}_0)$ that satisfy the following property: if we modify N_1 by putting additional tokens onto I according to M_1, then some firing sequence w of the modified net empties I and puts tokens onto O as described by M_2 such that no elongation of w puts more tokens onto any $p \in O$.

This definition is restricted to I,O-nets, where due to output-boundedness we can find for every M_1 that 'can be consumed' some M_2 such that $(M_1, M_2) \in IOF(N_1)$. It would be interesting to develop a similar semantics for all P-daughter-nets without arcs to I or from O and to see what sort of behaviour can be dealt with with such a semantics.

The next proposition states in terms of the IOF-semantics that each consumable input of an I,O-net determines a unique output and that the output is strictly monotonic in the input.

Proposition 4.4.6 *Let N_1 be an I,O-net, $(M_1, M_2), (M_1', M_2') \in IOF(N_1)$.*

i) If $M_1 = M_1'$, then $M_2 = M_2'$.

ii) If $M_1 \leq M_1'$ (componentwise) and $M_1 \neq M_1'$, then $M_2 \leq M_2'$ and $M_2 \neq M_2'$.

Proof: i) Lemma 4.4.4 i)

ii) Let w correspond to (M_1, M_2), w' to (M_1', M_2') according to the definition. Let $w_0 \in \{t_p^+ \mid p \in I\}^*$ contain t_p^+ exactly $M_1'(p)$ times for all $p \in I$. Thus, we can fire $w_0 w'$ in $env(N_1)$ and, under the marking reached, I is empty and O is marked according to M_2'. We can also fire $w_0 w$ to reach M_2 on O. Now $M_2 = M_2'$ would contradict 4.4.4 ii), and 4.4.4 i) shows that we can elongate $w_0 w$ to reach M_2' on O, i.e. $M_2 \leq M_2'$. $\square$

The next theorem shows that for I,O-nets – just as for F-deterministic nets – a large range of equivalences collapses. We already know that P-bisimilarity implies $PLan$-equivalence; mainly in order to prove the reverse implication we have introduced the IOF-semantics.

Theorem 4.4.7 *Let N_1, N_2 be I,O-nets. The following statements are equivalent:*

i) N_1 *and* N_2 *are P-bisimilar.*

ii) N_1 *and* N_2 *are PLan-equivalent.*

iii) N_1 *and* N_2 *are IO$\mathcal{F}$-equivalent.*

Proof: i) $\Rightarrow$ ii) obvious

ii) $\Rightarrow$ iii) A pair $(M_1 : I \to \mathbb{N}_0, M_2 : O \to \mathbb{N}_0)$ and a sequence $w_1 w_2$, $w_1 \in \{t_p^+ \mid p \in I\}^*$, $w_2 \in \{t_p^- \mid p \in O\}^*$, correspond if w_1 contains t_p^+ $M_1(p)$ times for $p \in I$ and w_2 contains t_p^- $M_2(p)$ times for $p \in O$. We have $(M_1, M_2) \in IO\mathcal{F}(N_1)$ if and only if: for all (M_1', M_2') with $M_1' \leq M_1$, $M_2 \leq M_2'$, a sequence corresponding to (M_1', M_2') is in $L(env(N_1))$ if and only if $(M_1', M_2') = (M_1, M_2)$. This follows from the definition of $IO\mathcal{F}$ and Proposition 4.4.6. Thus, $IO\mathcal{F}(N_1)$ is determined by $L(env(N_1))$.

iii) $\Rightarrow$ i) We define a relation $\mathcal{B}$ on $[M_{env(N_1)}\rangle \times [M_{env(N_2)}\rangle$ by $(M_1, M_2) \in \mathcal{B}$ if and only if for some w we have $M_{env(N_1)}[w)\rangle M_1$ and $M_{env(N_2)}[w)\rangle M_2$ such that M_1 and M_2 coincide on I.

Obviously, the initial markings are related. Assume $(M_1, M_2) \in \mathcal{B}$ with the corresponding image firing sequence w and $M_1[t)M_1'$.

If $t = t_p^+, p \in P$, or $t = t_p^-, p \in I$, then $M_2[t)M_2$ and $(M_1', M_2') \in \mathcal{B}$. If $t = t_p^-, p \in O$, then consider the mapping $M : I \to \mathbb{N}_0$ giving for each $p \in I$ the number of tokens that were removed by transitions of T_1 while reaching M_1. By definition of $\mathcal{B}$, it gives also the number of tokens removed by transitions of T_2 while reaching M_2.

There is some M' with $(M, M') \in IO\mathcal{F}(N_1) = IO\mathcal{F}(N_2)$. One can determine an upper bound for M_1 on O from w and M'. By definition of $IO\mathcal{F}$ and the above results, we can fire some $w_0 \in T_2^*$ that takes no tokens from I and realizes the maximal possible marking on O. Hence $M_2[w_0 t)M_2'$ and $(M_1', M_2') \in \mathcal{B}$ in this case.

Finally, we have the case $t \in T_1$. Let $(M, M') \in IO\mathcal{F}(N_1)$ as above; furthermore $((M + W(., t))|_I, M'') \in IO\mathcal{F}(N_1) = IO\mathcal{F}(N_2)$ for some M''. By the above results, especially 4.4.4 ii), we have $M_2[w_0)M_2'$ for some $w_0 \in T_2^*$ that removes $W(p, t)$ tokens from each $p \in I$; thus again $(M_1', M_2') \in \mathcal{B}$. $\square$

As a corollary to the preceding proof, we can strengthen Corollary 3.3.18: for the case of I, O-nets, we can determine the infinite image firing sequences from the finite ones without any preconditions like enabling-finiteness or divergence-freeness.

Corollary 4.4.8 *Let N_1 be an I, O-net, $v \in \Sigma^\omega$. Then $v \in L^\omega(env(N_1))$ if and only if all finite prefixes of v are in $L(env(N_1))$.*

Proof: '$\Rightarrow$' obvious.

'$\Leftarrow$' For a marking M of I, we call a firing sequence w of $env(N_1)$ an M-sequence, if the marking reached after w coincides on I with M.

For every finite prefix u of v we have a firing sequence $w(u)$ of $env(N_1)$ with image u. For a prefix u of u' which in turn is a finite prefix of v, we define a marking $M(u, u')$ of I as follows. Let w' be the minimal prefix of $w(u')$ with image u; then we obtain $M(u, u')$ by restricting the marking reached after w' to I. For each $p \in I$, $M(u, u')(p)$ is bounded by the number of t_p^+ occurring in u; thus, for fixed u there are only finitely many $M(u, u')$.

Using these markings $M(u, u')$, we define an infinite graph. As vertices we take pairs $(u, M(u, u'))$, and we have a directed edge from $(u, M(u, u'))$ to $(ua, M(ua, u'))$, whenever $a \in \Sigma$ and the markings are defined. This graph is locally finite, since there are only finitely many $M(u, u')$ for fixed u. By König's Lemma, we can find an infinite path starting with $(\lambda, M(\lambda, u'))$, where the marking in this pair is the empty marking of I for all u'.

Consider a directed edge as defined above. The sequence $w(u')$ demonstrates that there exists an $M(u, u')$-sequence with image u that can be extended to an $M(ua, u')$-sequence with image ua. We apply the previous proof, Part iii) $\Rightarrow$ i), to obtain a bisimulation for $env(N_1)$ to itself; using this bisimulation, we conclude that in fact every $M(u, u')$-sequence with image u can be extended to an $M(ua, u')$-sequence with image ua. Thus, starting with the $M(\lambda, u')$-sequence λ, we can go along the above infinite path in order to construct inductively an infinite firing sequence with image v. $\square$

Using this corollary, we can prove a second corollary of Theorem 4.4.7.

Corollary 4.4.9 *I,O-nets are P-deadlock-equivalent if and only if they are PLan-equivalent. P-divergence-free I,O-nets are P-deadlock/divergence-equivalent if and only if they are PLan-equivalent. This also holds if the external equivalences are defined with respect to finite hosts only.*

Proof: By Theorem 4.3.11, we know that P-deadlock-equivalence lies between $P\mathcal{F}$- and $P\mathcal{F}^{++}$-equivalence; these lie between P-bisimilarity and PLan-equivalence. Hence, the first statement follows from Theorem 4.4.7.

Also by Theorem 4.3.11, we know that P-deadlock/divergence-equivalence coincides with $\mathcal{M}\mathcal{F}_P\mathcal{D}$-equivalence, but only under the assumption of enabling-finiteness. This assumption was used in the proof of Theorem 4.3.11 in order to apply Corollary 3.3.18; for I,O-nets we can apply Corollary 4.4.8 instead, hence do not have to assume enabling-finiteness. Therefore, Proposition 4.3.8 shows that P-deadlock/divergence-equivalence lies between $P\mathcal{F}$- and $P\mathcal{F}^+$-equivalence for P-divergence-free I,O-nets. Thus, the second statement follows as above. $\square$

Now we can show: if the exchange of some I,O-nets guarantees a weak behaviour preservation, then it automatically preserves behaviour in quite a strong sense.

Theorem 4.4.10 *Let N_1, N_2 be I,O-nets, N a P-host. If N_1 and N_2 are P-deadlock-equivalent, then $N \,\|\|_P\, N_1$ and $N \,\|\|_P\, N_2$ are bisimilar. If N_1 and N_2 are P-divergence-free and P-deadlock/divergence-equivalent, then $N \,\|\|_P\, N_1$ and $N \,\|\|_P\, N_2$ are bisimilar with bisimilar divergence.*

Proof: The bisimulation $\mathcal{B}$ is defined by: $(M_1, M_2) \in \mathcal{B}$ if and only if M_1 and M_2 coincide on I and, for some w_1, w_2 with $id_T^*(w_1) = id_T^*(w_2)$, we have $M_{N\|\|_P N_1}[w_1\rangle M_1$ and $M_{N\|\|_P N_2}[w_2\rangle M_2$.

The proof is similar to the one of Theorem 4.4.7, if we translate sequences from T^* to sequences from $\{t_p^+, t_p^- \mid p \in P\}^*$ via φ^* as in the last section. Observe that an infinite

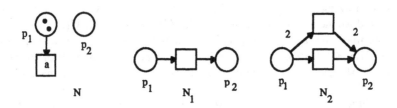

Figure 4.28

firing sequence of λ-transition of $N \parallel_P N_i$ contains infinitely many transitions of T, if N_i is P-divergence-free.   □

The following example shows that we cannot guarantee that $N \parallel_P N_1$ and $N \parallel_P N_2$ are branching bisimilar under the assumptions of the theorem. It is not hard to see that N_1 and N_2 as shown in Figure 4.28 are externally equivalent I, O-nets (see also Theorem 4.4.16 below). But, if we fire the transition 'with arc weight 2' in $N \parallel_P N_2$, then the only way to simulate this in $N \parallel_P N_1$ is to fire the only transition of N_1 twice; already the first firing changes the marking essentially, since now a can occur only once.

To complete our knowledge of I, O-nets we want to show next that I, O-nets are decidable and that the external equivalences are decidable for I, O-nets.

Theorem 4.4.11 *It is decidable whether a given finite P-daughter-net is an I, O-net.*

Proof: It is obvious, how to check the first condition. To check whether a finite P-daughter-net N_1 is output-deterministic, we simply have to decide whether t is live in $combi(env(N_1), env(N_1), \{t_p^- \mid p \in O\})$ analogously to the proof of 4.2.7.

To check whether N_1 is output-bounded, consider N_1' which is obtained from N_1 as follows (see Figure 4.29): add places s_1, s_2, where s_1 has one token and s_2 is empty; add a transition that can move the token from s_1 to s_2; add a transition on a loop with s_1 that puts one token onto each input place; merge the output places into one place o, i.e. each transition of N_1' puts as many tokens onto o as it puts onto the original output places all in all; add a transition t on a loop with s_2 that removes one token from o. Now, N_1 is not output-bounded if and only if there is some infinite firing sequence of N_1' containing t infinitely often. This can be decided by [VJ85], compare the proof of 3.2.9.

Finally, we have to check whether N_1 is non-absorbing. Assume that we already know that N_1 is output-deterministic and output-bounded. We construct an unlabelled net N from two disjoint copies of N_1, denoted by N_1 and N_1', as follows (see Figure 4.30):

$$S = \{s, s' \mid s \in S_1 \cup P\} \cup \{s(p) \mid p \in I\} \cup \{s_1(p), s_2(p) \mid p \in O\} \cup \{s_1, s_2, s_3\}$$

$$T = \{t, t' \mid t \in T_1\} \cup \{t_1(p), t_2(p), t_3(p) \mid p \in I\} \cup \{t_1(p), t_2(p) \mid p \in O\} \cup \{t_1, t_2, z\}$$

$$M_N(s) = M_N(s') = M_{N_1}(s), \quad s \in S_1,$$

$$M_N(s) = 1 \quad \text{for } s = s_1 \text{ or } s = s_3 \text{ or } s = s(p) \text{ with } p \in I,$$

$$M_N(s) = 0 \quad \text{for } s = p \text{ or } s = p' \text{ with } p \in P, \ s = s_1(p) \text{ or } s = s_2(p) \text{ for } p \in O \text{ or}$$

$$s = s_2.$$

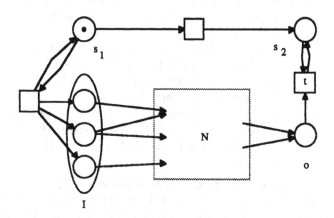

Figure 4.29

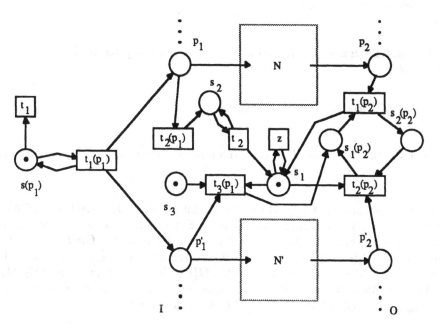

Figure 4.30

$$W(x,y) = \begin{cases} W_1(x,y) & \text{for } (x = s \text{ and } y = t) \text{ or } (x = s' \text{ and } y = t') \\ & \text{or } (x = t \text{ and } y = s) \text{ or } (x = t' \text{ and } y = s'), \\ & s \in S_1 \cup P,\ t \in T_1 \\ 1 & \text{for } (x,y) \in \{(s(p),t_1),\ (s(p),t_1(p)),\ (t_1(p),s(p)), \\ & (t_1(p),p),\ (t_1(p),p'),\ (p,t_2(p)),\ (t_2(p),s_2), \\ & (p',t_3(p)),\ (s_1,t_3(p)),\ (s_3,t_3(p)) \mid p \in I\} \\ & \cup \{(t_3(p_1),s_1(p_2)) \mid p_1 \in I, p_2 \in O\} \\ & \cup \{(p,t_1(p)),\ (p',t_2(p)),\ (t_1(p),s_1),\ (s_1,t_2(p)),\ (s_1(p),t_1(p)), \\ & (t_1(p),s_2(p)),\ (s_2(p),t_2(p)),\ (t_2(p),s_1(p)) \mid p \in O\} \\ & \cup \{(s_2,t_2),\ (t_2,s_2),\ (t_2,s_1),\ (s_1,z),\ (z,s_1)\} \\ 0 & \text{otherwise} \end{cases}$$

We claim that z is live in N if and only if N_1 is non-absorbing. To sketch the proof, observe that z is dead if and only if: t_1 has fired, hence the two copies of N_1 have been given some finite input. The first copy has taken all this input, while the second has left some tokens on I (or rather I') and some $t_3(p), p \in I$, has fired. Since N_1 is output-deterministic, the first copy is able to produce as many tokens on any output place as the second (Lemma 4.4.4 i)). Now, intuitively, the first copy has tried to 'bring z back to life', the second has tried to kill it again. z is dead, which means that the second copy has succeeded; hence, the first could not produce any additional token compared to the tokens produced by the second copy.   □

Corollary 4.4.12 *It is decidable whether a given finite P-daughter net is a P-divergence-free I, O-net.*

Proof: Apply Theorem 4.4.11 and Theorem 3.2.9.   □

Theorem 4.4.13 *It is decidable whether two given finite I, O-nets are P-deadlock-equivalent.*

Proof: We use Theorem 4.4.7 and Corollary 4.4.9 and check $IOF(N_1) = IOF(N_2)$ for given I, O-nets N_1, N_2. For this, we use the net of Figure 4.30 with the following changes: N_1 plays the rôle of the first, N_2 the rôle of the second copy of N_1. Omit $s_3, t_3(p), p \in I$. Reverse the arcs $(t_1(p), s_1), (s_1, t_2(p))$ and add a token on $s_1(p), p \in O$. (See Figure 4.31)

Now, z is live if and only if for any $(M_1, M_1') \in IOF(N_1)$ there is an $(M_2, M_2') \in IOF(N_2)$ with $M_2 \leq M_1$ and $M_2' \geq M_1'$. Using the same construction with N_1 and N_2 interchanged and applying Proposition 4.4.6, we get the result.   □

Corollary 4.4.14 *It is decidable whether two given finite P-divergence-free I, O-nets are P-deadlock/divergence-equivalent.*

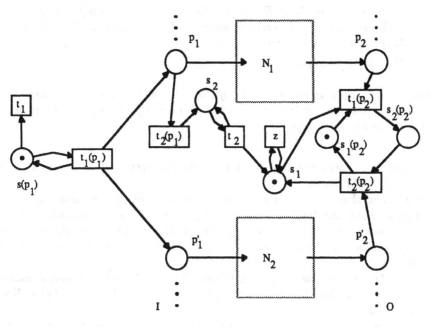

Figure 4.31

Proof: Apply Theorem 4.3.11 iii), Corollary 4.4.9 and Theorem 4.4.13.   □

Transition refinement

Now we will spell out what the above results mean for the refinement of transitions.

Definition 4.4.15 Let R be an N_h-refinement net for some net N_h as in Definition 4.1.3 with $^\bullet h = I$, $h^\bullet = O$, $I \cup O = P$. R is an N_h-D-, N_h-DD- or N_h-IL-module if for all N_h-λ-hosts N the nets N and $N[h \to R]$ are deadlock-, deadlock/divergence- or IL-similar. R is an N_h-IR-module if for all N_h-λ-hosts N the same places of $S_N - P$ are invariantly reachable in N and $N[h \to R]$.

Since N_h is an I, O-net we get the following characterization and decidability results.

Theorem 4.4.16 An N_h-refinement net R is an N_h-D-, N_h-IL- or N_h-IR-module (an N_h-DD-module)

  - if and only if R is an I,O-net (a P-divergence-free I,O-net) and R and N_h are $PLan$-equivalent
  - if and only if

      i) For every $t \in T_R$, the transition t is dead in $R-I$ or there exists some $n(t) \in \mathbb{N}_0$ such that $W_R(p,t) = n(t) \cdot W_{N_h}(p,h)$ for all $p \in I$.

ii) For $n \in \mathbb{N}_0$, let R_n be R with $n \cdot W_{N_h}(p, h)$ tokens on p for all $p \in I$. For each $n \in \mathbb{N}_0$ and each reachable marking M of R_n there is $M' \in [M)$ such that $M'(p) = n \cdot W_{N_h}(h, p)$ for all $p \in O$.

(iii) Each R_n is divergence-free for $n \in \mathbb{N}_0$.)

These characterizations are also valid if the external equivalences are defined with respect to finite nets only. It is decidable whether a given finite N_h-refinement net R is an N_h-module of any given type.

Proof: The first characterization and the decidability result follow immediately from 4.4.2, 4.4.9 and 4.4.11-4.4.13.

For the second characterization we consider the two implications separately.

'$\Rightarrow$' R is an I, O-net and $IO\,\mathcal{F}$-equivalent to N_h. If some $t \in T_R$ contradicts i) take a shortest firing sequence of $R - I$ containing such a t. If we put just enough tokens onto I to fire this sequence in R, we get some $(M_1, M_2) \in IO\,\mathcal{F}(R)$ that cannot be in $IO\,\mathcal{F}(N_h)$, a contradiction.

For ii) let M_1 describe the additional tokens of R_n on I, M_2 the required tokens on O, i.e. $M_2(p) = M'(p)$ for all $p \in O$. Then $(M_1, M_2) \in IO\,\mathcal{F}(N_h) = IO\,\mathcal{F}(R)$. Now ii) follows from Lemma 4.4.4.

Finally, if some R_n is not divergence-free, then in $env(R)$ we can put the appropriate number of tokens onto I by some firing sequence from $\{t_p^+ \mid p \in I\}^*$, thus R is not P-divergence-free in this case.

'$\Leftarrow$' By i) each firing sequence of R_n empties I or is a firing sequence of R_{n-1}, too. No firing sequence of R_n can put more tokens onto any $p \in O$ as described by M' in ii), since these tokens cannot be removed and thus ii) would be violated.

With this in mind it is easy to see that R is output-bounded and output-deterministic. Also R must be non-absorbing: if w and w' are as required in Definition 4.4.1, then w and ww' put tokens onto I to give the initial markings of some R_n and R_m with $n < m$, thus the condition required for R to be non-absorbing follows from ii) and the above remark.

Hence R is a (P-divergence-free) I, O-net. Property i) ensures that for every $(M_1, M_2) \in IO\,\mathcal{F}(R)$ M_1 is the initial marking on I for some R_n and ii) together with the above remark implies $IO\,\mathcal{F}(R) = IO\,\mathcal{F}(N_h)$. Thus R and N_h are $PLan$-equivalent by Theorem 4.4.7. $\square$

Especially the second characterization of this theorem is easy to work with. Consider the net N_h and the P-daughter-nets N_1 and N_2 of Figure 4.32. At first sight N_1 might seem suitable to replace the transition h. But a second look reveals that N_1 violates Condition i) of Theorem 4.4.16; in fact, it is not even an I, O-net since it can absorb tokens from p_2. And indeed, refining the deadlock-free N_h-λ-host shown in Figure 4.33 gives a net that can deadlock. The reason for this deadlock is that N_1 takes its input in a distributed fashion; it consumes the second token on p_1 without checking whether there is a second token on p_2. One could reformulate Condition i) as: an N_h-module does not allow distributed input.

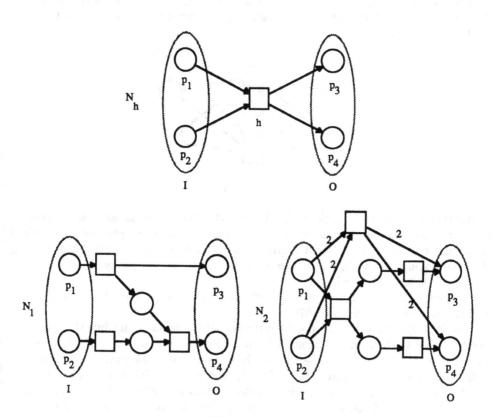

Figure 4.32

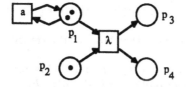

Figure 4.33

The second characterization of Theorem 4.4.16 also makes it easy to show that N_2 is an N_h-module of any type. This demonstrates that distributed output is allowed. N_2 also contains a transition that allows to 'speed up' h – firing this transition has the same effect as firing h twice.

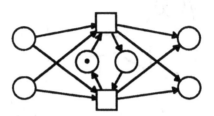

Figure 4.34

Another N_h-module N_1 is shown in Figure 4.34. Here the transition h is replaced by two transitions that fire alternatingly. In this case, N_1 is not empty under the initial marking; furthermore N_1 does not return to its initial marking after simulating a firing of h. One could say that N_1 has some memory, it remembers the parity of the number of simulations it has performed.

Theorem 4.4.16 also shows that the N_h-deadlock-modules we have defined here coincide with the modules studied in [Vog87].

Condition i) of this theorem points to the transitions of an N_h-module that are most closely related to the original transition h: the transitions that access I. If h is visible in a host N, these transitions are the natural choice for being visible in the refined net; i.e. each $t \in T_R$ should be labelled $lab_N(h)^{n(t)} \in \Sigma$ since t starts $n(t)$ simulations of h. This would require to extend our net model in order to allow a labelling with arbitrary action sequences. Instead of doing this, we will restrict the refinement of visible transitions to the case that, for all transitions t of the refinement net R, we have $n(t) \in \{0,1\}$, i.e. to the case where R does not allow a 'speed up' as N_2 of Figure 4.32.

Definition 4.4.17 An N_h-refinement net R is *simple* if for all $t \in T_R$ we have $\bullet t \cap I = \emptyset$ or $W_R(p,t) = W_{N_h}(p,h)$ for all $p \in I$.

A net N is an N_h-*host*, if $h \in T_N$ and the subnet induced by $\{h\} \cup \bullet h \cup h\bullet$ equals N_h except for the marking and the labelling.

The refinement $N[h \to R]$ of an N_h-host N with a simple refinement net R is defined as for N_h-λ-hosts except that every $t \in T_R$ with $\bullet t \cap I \neq \emptyset$ is labelled $lab_N(h)$ in $N[h \to R]$.

We list a number of properties that hold for the transition refinement with modules.

Theorem 4.4.18   *i) If N is an N_h-λ-host and R is an N_h-D- (N_h-DD-) module, then N and $N[h \to R]$ are bisimilar (with bisimilar divergence).*

   ii) If N is an N_h-host and R is an N_h-D- (N_h-DD-) module and simple, then N and $N[h \to R]$ are branching bisimilar (with bisimilar divergence).

In all these cases:

iii) Let $t \in T_N - \{h\}$. Then t is live in N if and only if t is live in $N[h \rightarrow R]$.

iv) N and $N[h \rightarrow R]$ are marking-equivalent on $S_N - O$. More precisely: For any reachable marking M of N there is a reachable marking M' of $N[h \rightarrow R]$ such that M and M' coincide on all of S_N. For any reachable marking M' of $N[h \rightarrow R]$ there is a reachable marking M of N such that M and M' coincide on $S_N - O$ and $M' \leq M$ on O.
Every reachable marking of $N[h \rightarrow R]$ coincides on S_R with a reachable marking of $R - (I \cup O)$.

Proof: i) follows from Theorem 4.4.10 immediately. We can also define a bisimulation $\mathcal{B} \subseteq [M_N\rangle \times [M_{N[h \rightarrow R]}\rangle$ as follows. Let $\xi^* : (T - \{h\} \cup T_R)^* \rightarrow T^*$ be defined by $\xi(t) = t$ for $t \in T$ and $\xi(t) = h^{n(t)}$ for $t \in T_R$ and $n(t)$ as in 4.4.16 i) (if t is dead in $R - I$ let $n(t) = 0$). Now $(M_1, M_2) \in \mathcal{B}$ if $M_{N[h \rightarrow R]}[w\rangle M_2$ and $M_N[\xi^*(w)\rangle M_1$ for some w.

If $(M_1, M_2) \in \mathcal{B}$, then M_1 and M_2 coincide on $S_N - O$ and $M_2 \leq M_1$ on O. By 4.4.16, we can in every case fire some $w \in T_R^*$ with $\xi^*(w) = \lambda$ such that for $M_2[w\rangle M_2'$ we have that M_2' and M_1 coincide on all of S_N. With this, it is not hard to see that $\mathcal{B}$ is in fact a bisimulation (with bisimilar divergence) and that it is a branching bisimulation under the conditions of ii). Also iv) follows now. Furthermore, $(M_1, M_2) \in \mathcal{B}$ and $M_2[t\rangle$ implies $M_1[t\rangle$ for $t \in T_N - \{h\}$, while $M_1[t\rangle$ implies that $M_2[wt\rangle$ for some $w \in T_R^*$ with $\xi^*(w) = \lambda$; hence iii) follows.   □

Remark: In this proof, ξ relates the transitions of T_R with sequences from $\{h\}^*$. One can see that labelling these transitions with the corresponding sequence from $\{lab_N(h)\}^*$ would make $\mathcal{B}$ a bisimulation as we have defined it in Section 2.3 also in the case that R is not simple.   □

As in the case of N_p-modules, we can modify N_h and an N_h-module R of any type in a corresponding way to get N_h' and an N_h'-module R'. We can delete the same place in both nets or duplicate the same place in both nets. Even if deletion of places makes I or O empty, the resulting net R' guarantees behaviour preservation in the sense of Theorem 4.4.18. Also, we can merge two input places or two output places and add up their arc weights. Furthermore, if we want to refine a transition h that is on a loop with some place s in N, we can split up s into an input place s_I and an output place s_O, and take an N_h-module R for the resulting net N_h. When we merge s_I and s_O and use the resulting net R' to refine h in N in the same way as above, then again behaviour preservation in the sense of Theorem 4.4.18 is guaranteed.

Thus, if N_h is of arc weight 1 and has a single input and a single output place, then we can construct from each N_h-module a module for the refinement of any given transition by applying the above transformations.

Transition refinement of safe nets

The transition refinement of safe nets deserves special attention. Unfortunately it is not clear, how our general approach can be adapted to this case. The reason is that the test nets we have used to prove Theorem 4.3.11 i) are safe in the interior, but are not necessarily safe on P.

Instead, we will directly show a characterization of modules for the transition refinement in safe nets, and we will need essentially only one test net for this proof. The characterization corresponds to the second characterization of Theorem 4.4.16.

It should be remarked that it would have been possible to prove this part of 4.4.16 and also Theorem 4.4.18 directly. The reason for studying I, O-nets first was not to make these proofs easier; the aim was to exhibit those features that are important for the case of transition refinement, those features that allow nice decidability results and behaviour preservation in a strong sense, namely: output-determinism, output-boundedness and non-absorption. Corresponding results for safe nets would be very desirable.

In the following definition, we will restrict ourselves to studying simple N_h-refinement nets R. One can show that if we omit the word 'simple' in the following definition, then, for every safe N_h-module R and every safe N_h-λ-host N, each transition t that violates the simplicity of R is dead in $N[h \to R]$. This is to be expected considering the above results, but the proof is somewhat involved; therefore, we prefer to keep things simple.

Definition 4.4.19 Let N_h (as defined in Definition 4.1.3) have arc weight 1. An N_h-$(N_h$-λ-) host is a *safe N_h- (N_h-λ-) host*, if it is safe as a net.

A simple N_h-refinement net R is a *safe N_h-D-, N_h-DD-, N_h-IL- or N_h-IR-module* if it has arc weight 1 and for all safe N_h-λ-hosts N

- the net $N[h \to R]$ is safe and

- N and $N[h \to R]$ are deadlock-, deadlock/divergence- or IL-similar or the same places of $S_N - P$ are invariantly reachable in N and $N[h \to R]$.

Let R be a simple N_h-refinement net of arc weight 1. A transition $t \in T_R$ with $\bullet t \cap I \neq \emptyset$ is called a *start transition*. We obtain $\overline{R}$ from R by adding arcs of weight 1 from every output place to every start transition, putting one token onto each output place and deleting I.

Let us fix some $a \in \Sigma$; we obtain $\tilde{N}_h$ from N_h by adding an a-labelled transition t_0, arcs of weight 1 from each output place to t_0 and from t_0 to each input place, and one token on every output place. We put $\tilde{R} = \tilde{N}_h[h \to R]$.

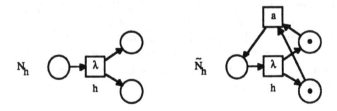

Figure 4.35

Figure 4.35 shows an example of nets N_h and $\tilde{N}_h$; Figure 4.36 shows a corresponding example of nets R, $\tilde{R}$ and $\overline{R}$.

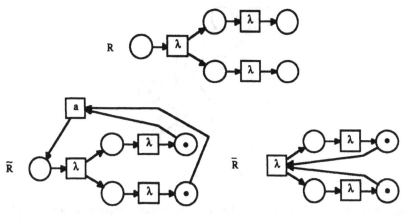

Figure 4.36

Theorem 4.4.20 *Let R be a simple N_h-refinement net of arc weight 1. R is a safe N_h-D-, N_h-IL- or N_h-IR-module if and only if*

i) *$\tilde{R}$ is safe and a is image-live*

   if and only if

ii) *$\overline{R}$ is safe and under no reachable marking all start transitions are dead.*

R is a safe N_h-DD-module if and only if

iii) *R is a safe N_h-D-module and $\tilde{R}$ is divergence-free*

   if and only if

iv) *R is a safe N_h-D-module and every infinite firing sequence of $\overline{R}$ contains infinitely many start transitions.*

Proof: Obviously $\overline{R}$ can be obtained from $\tilde{R}$ by short-circuiting the a-labelled transition. Hence, ii) follows easily once we have shown i), and iv) follows easily once we have shown iii).

Here we will only show that the conditions on $\tilde{R}$ are necessary. The rest of the proof will be postponed until later (see the proof of Theorem 4.4.22).

$\tilde{N}_h$ is a safe and deadlock/divergence-free N_h-λ-host and a is image-live. Thus $\tilde{R} = \tilde{N}_h[h \to R]$ must be safe – and in the DD-case divergence-free –, and a must be image-live, also in order to make $\tilde{R}$ deadlock-free.

For the IR-case we modify $\tilde{N}_h$ slightly, see Figure 4.37: we add places s, s' and an internal transition t_0', a token on s and arcs from s to t_0', from t_0' to s', from s' to the a-labelled transition t_0 and from t_0 to s. This modified net is a safe N_h-λ-host and s is invariantly reachable, which does only depend on the image-liveness of a. If we refine h by R in this net, we must get a safe net where a is image-live. From this we conclude, that $\tilde{R} = \tilde{N}_h[h \to R]$ must be safe and a must be image-live also in the IR-case.   □

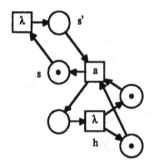

Figure 4.37

Corollary 4.4.21 *It is decidable whether any given finite N_h-refinement net R is a safe N_h-module of any given type.*

Proof: By Theorem 4.4.20 we can simply inspect the reachability graph of $\tilde{R}$ or $\overline{R}$. □

Now we will show how the behaviour is preserved if we refine a transition of a safe net by a safe module.

Theorem 4.4.22 *Let N be a safe N_h-host and R a safe N_h-D-module (a safe N_h-DD-module).*

  i) N and $N[h \to R]$ are branching bisimilar (with bisimilar divergence).

  ii) If $t \in T - \{h\}$, then t is live in N if and only if it is live in $N[h \to R]$. If h is live in N, then $t \in T_R$ is live in $N[h \to R]$ if and only if it is live in $\tilde{R}$ or $\overline{R}$.

  iii) N and $N[h \to R]$ are marking-equivalent on $S_N - O$. More precisely:
  For every reachable marking M of N there is a reachable marking M' of $N[h \to R]$ such that M and M' coincide on S_N. For every reachable marking M' of $N[h \to R]$ there is a reachable marking M of N such that M and M' coincide on $S_N - O$ and on O we have $M \leq M'$.
  Every reachable marking of $N[h \to R]$ coincides on S_R with a reachable marking of $\tilde{R}$ (or $\overline{R}$).
  If h is live in N, then $N[h \to R]$ and $\tilde{R}$ (or $\overline{R}$) are marking-equivalent on S_R.

Proof: We will only use the fact that R satisfies the characterization given in Theorem 4.4.20. Due to this self-restriction, we will also be able to give the missing part of the proof of that theorem along the way.

We define again $\xi^* : (T_N - \{h\} \cup T_R)^* \to T_N^*$ by $\xi(t) = t$ for $t \in T_N - \{h\}$, $\xi(t) = h$ if $t \in T_R$ is a start transition, and $\xi(t) = \lambda$ otherwise. Furthermore, $\nu^* : (T_N - \{h\} \cup T_R)^* \to T_R^*$ is defined by $\nu(t) = \lambda$ for $t \in T_N$, $\nu(t) = t$ for $t \in T_R$.

Let $\mathcal{B} \subseteq [M_N\rangle \times [M_{N[h \to R]}\rangle$ be defined by $(M_1, M_2) \in \mathcal{B}$ if $M_{N[h \to R]}[w\rangle M_2$ and $M_N[\xi^*(w)\rangle M_1$ for some w. Furthermore, $\mathcal{B}' \subseteq [M_N\rangle \times [M_{N[h \to R]}\rangle \times [M_{\overline{R}}\rangle$ is defined by $(M_1, M_2, M_3) \in \mathcal{B}'$ if $M_{N[h \to R]}[w\rangle M_2$, $M_N[\xi^*(w)\rangle M_1$, and $M_{\overline{R}}[\nu^*(w)\rangle M_3$ for some w.

We will show by induction on the length of w:

(∗) Every firing sequence of w of $N[h \to R]$ gives rise to some $(M_1, M_2) \in \mathcal{B}$ and $(M_1, M_2, M_3) \in \mathcal{B}'$, M_1 and M_2 coincide on $S_N - O$, M_2 and M_3 coincide on S_R and for each $p \in O$ either $M_3(p) = M_2(p) = 0 \wedge M_1(p) = 1$ or $M_3(p) = 1 \wedge M_1(p) = M_2(p)$.

This is clear for $w = \lambda$. Thus let w, M_1, M_2, M_3 be given as required and let $M_2[t\rangle M_2'$. If $t \in T_N - \{h\}$, then t is enabled under M_1, and for the resulting marking M_1' we see immediately that $(M_1', M_2') \in \mathcal{B}$, $(M_1', M_2', M_3) \in \mathcal{B}'$, M_1' and M_2' coincide on $S_N - O$, and M_2' and M_3 on S_R. For $p \in O$ either t leaves the marking unchanged and we are done, or t adds a token implying $M_1(p) = 0$ and thus $M_1'(p) = M_2'(p) = M_3(p) = 1$, or t removes a token implying $M_2(p) = 1$ thus $M_3(p) = 1$ and $M_1'(p) = M_2'(p) = 0$.

If $t \in T_R$ and $^\bullet t \cap I = \emptyset$, then the enabledness of t depends only on the marking of S_R, which is changed in $N[h \to R]$ and $\overline{R}$ in the same way, while the marking of $S_N - O$ remains unchanged in $N[h \to R]$ and, since $\xi(t) = \lambda$, in N. If t puts a token onto some $p \in O$, then we conclude from $M_3[t\rangle$ and the safeness of $\overline{R}$ that $M_3(p) = 0$, hence $M_2(p) = 0$, $M_1(p) = 1$ and after firing t p is marked in all nets; if t does not access p, then the marking of p remains unchanged in all three nets.

Finally, let $t \in T_R$ be a start transition. Since $M_2[t\rangle M_2'$ we have $M_1[h\rangle M_1'$ for some M_1'. This implies $M_1(p) = 0$ for all $p \in O$, hence $M_3(p) = 1$. Thus $M_3[t\rangle M_3'$ for some M_3'. Obviously M_1' and M_2' coincide on $S_N - O$ and M_2' and M_3' on S_R. Furthermore, for $p \in O$, $M_1(p) = 0$ implies $M_2(p) = 0$, thus $M_1'(p) = 1$ and $M_2'(p) = M_3'(p)$. This finishes the inductive proof.

(∗) shows that for every reachable marking M' of $N[h \to R]$ there is a reachable marking M of N such that M and M' coincide on $S_N - O$ and $M' \leq M$ on O; also that M' coincides on S_R with a reachable marking of $\overline{R}$ (or $\tilde{R}$). Especially, this shows that $N[h \to R]$ is safe, a fact that we need for the still missing half of the proof of Theorem 4.4.20. Furthermore, (∗) gives half of the proof that $\mathcal{B}$ is a branching bisimulation, namely that every move of $N[h \to R]$ can be simulated by N.

For the other half assume that $(M_1, M_2) \in \mathcal{B}$, $(M_1, M_2, M_3) \in \mathcal{B}'$ and $M_1[t\rangle M_1'$. By the assumptions on $\overline{R}$ in Theorem 4.4.20 ii), we can find some firing sequence v of $\overline{R}$ enabled under M_3 that contains no start transition, but enables some start transition. We have $M_2[v\rangle M_2''$, $M_3[v\rangle M_3''$, $(M_1, M_2'') \in \mathcal{B}$ (since $\xi^*(v) = \lambda$), $(M_1, M_2'', M_3'') \in \mathcal{B}'$ and $M_3''(p) = 1$ for all $p \in O$. This shows by (∗) that for the reachable marking M_1 of N we can find a reachable marking M_2'' such that M_1 and M_2'' coincide on all of S_N. If $t \in T_N - \{h\}$, then $M_2''[t\rangle M_2'$ and $(M_1', M_2') \in \mathcal{B}$; if $t = h$, then some start transition t' of R is enabled under M_2'' (since M_2'' coincides with M_3'' on S_R and $M_2''(p) = 1$ for $p \in I$), thus $M_2''[t'\rangle M_2'$ and $(M_1', M_2') \in \mathcal{B}$, since $\xi(t') = h$.

Thus, $\mathcal{B}$ is a branching bisimulation (all $t \in T_R$ that are not start transitions are λ-labelled in $N[h \to R]$). Every infinite firing sequence of N can be translated to an infinite firing sequence having the same image and passing through bisimilar markings. If $\tilde{R}$ is divergence-free and w is an infinite firing sequence of $N[h \to R]$ in which infinitely often a transition from T_R occurs, then infinitely often a start transition occurs in w; thus each infinite firing sequence of $N[h \to R]$ can be translated with the above results on $\mathcal{B}$ and $\mathcal{B}'$ to an infinite firing sequence of N having the same image and passing through bisimilar markings. Therefore, $\mathcal{B}$ is a branching bisimulation with bisimilar divergence, if

$\tilde{R}$ is divergence-free.

Since we have found a branching bisimulation (with bisimilar divergence), we conclude also that an N_h-refinement net R satisfying the conditions of Theorem 4.4.20 is indeed a safe N_h-D-, N_h-IL- or N_h-DD-module; together with (∗) we also have that such an R is a safe N_h-IR-module.

Furthermore, the above results show that transitions from $T_N - \{h\}$ are live in N if and only if they are live in $N[h \rightarrow R]$.

It remains to consider the case that h is live in N. Let $(M_1, M_2, M_3) \in \mathcal{B}'$ and let $M_3[v_1 t_1 v_2 \ldots t_n v_{n+1})M_3'$ in $\tilde{R}$, $n \in \mathbb{N}_0$, such that all t_i are start transitions and no v_i contains a start transition. Let $M_1[w_1 h w_2 h \ldots w_n h)M_1'$ such that h occurs in no w_i. By the above considerations we get $M_2[v_1 w_1 t_1 v_2 w_2 t_2 \ldots w_n t_n v_{n+1})M_2'$. This shows that every reachable marking of $\overline{R}$ coincides on S_R with some reachable marking of $N[h \rightarrow R]$ and that every live transition of $\overline{R}$ is live in $N[h \rightarrow R]$. Since in this case we can find for every reachable marking M_3 of $\overline{R}$ some $(M_1, M_2, M_3) \in \mathcal{B}'$, we also conclude from (∗) that every $t \in T_R$ that is live in $N[h \rightarrow R]$ is also live in $\overline{R}$.   □

Again, we can modify N_h by duplicating or removing places from $I \cup O$, and a corresponding modification of a safe N_h-module gives a safe N_h'-module for the modified net N_h'. Also we can refine transitions that are on a loop by merging an input and an output place of a safe module before replacing the transitions.

Thus, if distributed output is of no importance, then it is enough to consider safe N_h-modules R where N_h has just one input and one output place. The corresponding nets $\overline{R}$ have just one marked place representing the environment of R.

Carefully considering the above proof of Theorem 4.4.22 gives the following two corollaries.

Corollary 4.4.23 *Let N be a safe N_h-host, R a simple N_h-refinement net of arc weight 1 such that $\overline{R}$ is safe. Then $N[h \rightarrow R]$ is safe and $L(N[h \rightarrow R]) \subseteq L(N)$.*

Proof: This follows from (∗) and the succeeding considerations in the proof of 4.4.22. For these, only the safety of $\overline{R}$ was needed.   □

Corollary 4.4.24 *We call a simple N_h-refinement net R of arc weight 1 a safe and live N_h-module if for every safe and live N_h-host N the refined net $N[h \rightarrow R]$ is safe and live. R is a safe and live N_h-module if and only if $\overline{R}$ is safe and live.*

Proof: Since $\tilde{N}_h$ is a safe and live N_h-host, $\tilde{R}$ and thus $\overline{R}$ must be safe and live. The other implication follows from Theorems 4.4.20 and 4.4.22.   □

Transition refinement of nets have been studied in [Val79,SM83,Mül85,Vog87], and we have already mentioned the close relation of the approach presented here to that of [Vog87]. The main differences to the other three papers are: they define classes of nets that are suitable for transition refinement in some context, but they do not try to characterize such refinement nets; the considered refinement nets have just one start transition and just one end transition producing all the output, i.e. no initial conflict and no distributed output is allowed; some results on behaviour preservation are shown, but they are weaker than ours; finally, all four papers just discuss transition refinement and do not put this problem in a broader framework.

[Val79] gives results on the preservation of liveness, safeness and boundedness for the refinement of transitions that are not self-concurrent, i.e. especially for the transition refinement of safe nets. The refinement nets are given in the form $\overline{R}$ (see Definition 4.4.19), and they are required to be live, to enable only the start transition initially, and to have "no memory", i.e. to return to the initial marking after each simulation, see the above discussion on the net of Figure 4.34.

In [SM83] these requirements are dropped and transitions that are at most k times self-concurrently enabled for some fixed $k \in \mathbb{N}$ are refined. Also decidability results are shown.

Finally, [Mül85] defines a class of refinement nets that are similar to our N_h-D-modules, but with the above mentioned restrictions. Results on the preservation of liveness and boundedness are shown for the refinement of arbitrary transitions. These results are not completely covered by ours since the broader class of nets with capacities is considered. Furthermore, [Mül85] studies the simultaneous refinement of several transitions, a case that fits into our approach of exchanging I, O-nets as mentioned above.

Chapter 5

Action Refinement and Interval Words

In the previous chapter, we have studied the behaviour-preserving refinement of transitions. If a refinement introduces subactions that were not present originally, we cannot expect that the unrefined and the refined net are equivalent – such a refinement is called action refinement. Instead, we expect that equivalent nets are refined to nets that are equivalent again, i.e. we want an equivalence that is a congruence for action refinement. Such congruences are studied in this and the following chapter; this chapter studies linear-time and failure-type congruences, the next chapter is devoted to bisimulation-type congruences.

This chapter starts with a section that introduces the area of action refinement. Section 5.2 defines the action refinement for Petri nets that we will use, and it discusses why we have chosen this definition and why we restrict ourselves in this chapter to nets that are free of self-concurrency.

In Section 5.3 we show first that action refinement preserves self-concurrency-freeness, and this proof is based on what we call interval words. Secondly, we show that processes and semiwords give congruences for action refinement. Thirdly, we present a fully abstract partial order semantics based on what we call interval semiwords. This semantics demonstrates to what degree partial order semantics is necessary in order to get a congruence for action refinement.

In Section 5.4, we define a linear-time-, a failure- and a failure/divergence-semantics based on interval words and show full abstractness for all three. Finally, we show in Section 5.5 that interval words are a sort of sequential representation of interval semiwords. We discuss whether interval words or interval semiwords are to be preferred, and in the course of this discussion we prove that all the fully abstract congruences of this chapter are decidable for finite bounded nets.

5.1 Introduction to action refinement

Figure 5.1 shows the life of a philosopher as computer scientists see it, see [Dij71]. The philosopher sits down at a table and does something there repeatedly.

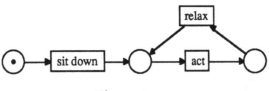

Figure 5.1

If we have a closer look at the philosopher's activity, i.e. refine 'act', we see that the activity consists in either thinking or eating, see Figure 5.2.

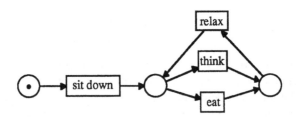

Figure 5.2

Eating – being the more complicated of the two activities – deserves further attention, and a second refinement reveals some of the involved subactivities, see Figure 5.3.

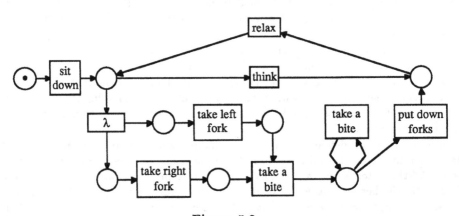

Figure 5.3

Apart from the labelling, both these transition refinements are refinements with modules as discussed in the previous chapter. But due to the labelling, these refinements are not behaviour-preserving according to most of our definitions; for example, action 'take left fork' can occur in the refined net of Figure 5.3, but not in the unrefined net of Figure 5.1.

One way out of this problem would be to change the labels of the refinement net to λ; but certainly, this is often inappropriate. In our example the system 'philosopher' will have to communicate with the system 'left fork', and this will be done by synchronization of the actions 'take left fork' and 'put down forks'. Therefore, these actions should not be internal to the philosopher.

In this situation, we cannot expect that refinement is behaviour-preserving, but we do expect that we can determine the behaviour of the refined net from that of the unrefined net and the refinement net. In other words, if we refine semantically equivalent nets – i.e. nets with the same behaviour – in the same way, then the refined nets should be equivalent again. Here, 'refinement in the same way' means replacing all transitions with the same label by (copies of) the same net; this is called *action refinement*. Thus, action refinement should preserve equivalence, and not behaviour as the refinement in the previous chapter; viewed the other way round, the semantic equivalence should be a congruence with respect to action refinement.

When we speak of a congruence for action refinement, we view action refinement as a family of unary operations. Each operation *ref* in this family is given by defining a refinement net $ref(a)$ for all actions a; application of *ref* to a net N replaces each transition t by a copy of $ref(lab(t))$. A congruence ensures that for equivalent nets N and N' the refined nets $ref(N)$ and $ref(N')$ are equivalent again. Another view would be to regard action refinement as one operation with infinitely many arguments; one argument is the net N we want to refine, the other arguments are, for all actions a, the refinement nets we want to use to refine a-labelled transitions. With this more general view, we would require more of a congruence: if the refinement nets $ref(a)$ and $ref'(a)$ are equivalent for all a, then $ref(N)$ and $ref'(N)$ should be equivalent for all nets N. A problem with this view is that refinement nets are only a special class of nets; thus, we would really need two equivalences. We will indicate below what can be done in this direction. Recently, the more general view has also been considered in [Dev91] and [JM91]; additionally, the latter paper gives a different definition of refinement, such that one does not have the problem that refinement nets are only a special class of nets.

It is no surprise that there are many congruences for action refinement, and we will come across quite a number of them in this and the next chapter. To make some sensible choice, we have to ensure that we do not distinguish nets unnecessarily, but also that we do not identify too many nets. For example, isomorphism is a congruence with too many distinctions; considering all nets to be equivalent gives a congruence with too few distinctions. Thus, in line with the general approach of this book, we are mostly interested in congruences that are fully abstract with respect to action refinement and some basic equivalence. The basic equivalences in this chapter are language-, $\mathcal{F}$- and $\mathcal{FD}$-equivalence – language equivalence, because it is for many people of some interest in itself and easier to handle, $\mathcal{F}$- and $\mathcal{FD}$-equivalence, because they are the right equivalences under the circumstances we have studied in Chapter 3.

Action refinement allows us to change the level of abstraction: what has been considered as one action in a first design step, is described in greater detail by a system of subactions after the refinement. Congruences with respect to action refinement support this top-down design of concurrent systems; therefore, they promise to be of practical importance. But they are also of additional theoretical interest, since they give an important argument for semantics with 'true concurrency', which are based on partial orders (see Section 2.2); this argument has already been mentioned in [Lam86] and [Pra86]. That none of the interleaving semantics we have defined so far can induce such a congruence can be seen from the classical and simple example shown in Figure 2.2, which we repeat for convenience as Figure 5.4 (see e.g. [CDMP87]). These two nets are equivalent whatever

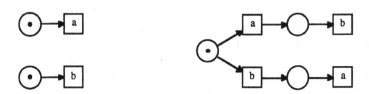

Figure 5.4

interleaving semantics we consider. But if we replace a be the sequence $a_1 a_2$, then only the first net allows the image firing sequence $a_1 b a_2$. In order to save interleaving semantics, one could extend the model by introducing priorities and require that a subaction like a_2 should have priority once a_1 has occurred. Then, $a_1 b a_2$ is not an admissible sequence any more. Such a solution is suggested in [GMM88] in a setting of process algebra.

As already mentioned in Section 2.2, the nets of Figure 5.4 are distinguished by their step languages, and the step language is often not regarded as modelling 'true concurrency'. The easiest example of nets with the same step language that are not partial-word- or process-equivalent is shown in Figure 2.3, repeated as Figure 5.5. This example shows

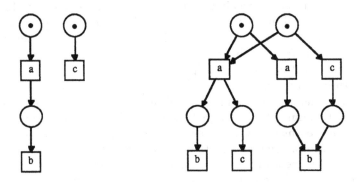

Figure 5.5

that step-language equivalence cannot serve as a congruence with respect to action refinement (see also [GG89a]). Again, this can be demonstrated with a simple splitting of actions: if we refine a by $a_1 a_2$, b by $b_1 b_2$, and c by $c_1 c_2$, then only the first net allows the image firing sequence $a_1 c_1 a_2 b_1 c_2 b_2$.

This example is also interesting, since the distinguishing action sequence is an interleaving observation of the refined net that associates an interval with each action of the original net. The intervals of a and b both overlap with the interval of c; therefore, this interleaving observation tells us that in the original net it is possible to perform a followed by b and – concurrently to both a and b – the action c. In this way, intervals induce partial orders, and these interval orders play an important rôle in this chapter. That they are useful when dealing with non-atomic actions of concurrent systems has already been suggested in [Lam86]. Recently, David Murphy has developed a semantic model for real-timed concurrent systems in a setting of event structures; he suggests to enrich the information on causality inherent in event structures by giving a time interval for each event occurrence [Mur90]; additionally, a corresponding interval process algebra is introduced. Also recently, the use of interval orders for the description of system runs has been advocated in [JK90].

The above examples may lead to the idea that a simple splitting of actions is enough to exhibit all necessary distinctions; such an approach is studied e.g. in [AH89]). But consider the nets of Figure 5.6. If we split the actions in both nets, the resulting nets are

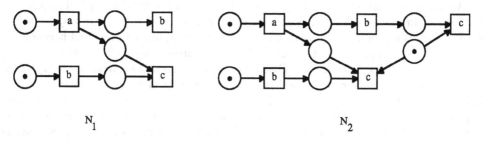

N_1 $\qquad\qquad\qquad\qquad\qquad\qquad\qquad\qquad$ N_2

Figure 5.6

language-equivalent: the additional c of N_2 can only occur in the splitted N_2 if a finishes, then some b starts, and then some b finishes before c starts; all such action sequences are possible in the splitted N_1, too. But if we refine b by a choice between $b_1 b_2$ and $b'_1 b'_2$, then only the second net can perform $b_1 a b'_1 b'_2 c b_2$. This shows that simple splitting is not enough in the presence of autoconcurrency; if two b-actions start concurrently and later some b ends, then with simple splitting we cannot tell which b ended; but this information is essential for the treatment of action refinement. More involved examples considering failure equivalence or bisimulation for the splitted nets can be found below or in [GV91].

In the next section, we will define action refinement of nets and discuss this definition. Section 5.3 gives results on action refinement and linear-time semantics, especially presenting a semantic equivalence that is fully abstract with respect to action refinement and language equivalence. This equivalence is based on so-called interval semiwords, a special class of partial words; thus, the full-abstractness result shows to which degree partial order semantics is necessary to deal with action refinement. The fully abstract equivalence and its extensions to failure-type equivalences are studied in detail in Section 5.4 and Section 5.5.

The results we represent in this chapter are very similar to those of [Vog91b]; here, we use an extended class of refinement nets and consider action refinement of nets without self-concurrency, not only of safe nets. [Vog91b] works directly with interval semiwords;

in this book, the approach is based on the one hand on processes, on the other hand on a representation of interval semiwords by special sequences; this representation prepares the approach of Chapter 6 to bisimulations that are suitable for action refinement.

Other congruence results for the action refinement of nets are given in [BDKP91] and [Dev90b]; these congruences are of bisimulation type, and we will return to them in the next chapter.

Process algebras usually do not offer a refinement operator. The main problem is that simple syntactic substitution does not work in the presence of synchronization. First studies of action refinement for process algebras without synchronization can be found in [AH89,NEL89,Eng90,Ace90], where [NEL89,Eng90] are the closest to our approach; in these two papers, also a linear-time semantics based on interval orders is given and full abstractness is shown. Recently, definitions for action refinement in full process algebras have been given in [AH91] and [GL91]; in both these papers, ST-bisimulation is considered as a suitable semantics, and we will study this semantics in the next chapter.

Action refinement for event structures in the sense of [NPW81] is studied in [GG89b, Gla90c] and in the next chapter. As it is natural for this model, actions are only refined by conflict-free, non-repetitive structures, and therefore the approach presented here is more general. The above papers contain results for equivalences of bisimulation type which we will study in the next chapter. In [GG89b], it is also shown that process equivalence is a congruence, a result we will establish in our more general framework in Section 5.3. [Gla90c] also gives a congruence result for ST-traces; since ST-traces are a (more or less) sequential presentation of interval semiwords, this result had already been established in the more general framework of safe nets in [Vog91b]; the special sequences we will use in Section 5.4 are very close to ST-traces.

The representation of interval semiwords by sequences raises the question whether partial order semantics is really necessary to deal with action refinement. In fact, it is claimed in [GG89a] that this is not the case. It is not easy to settle this issue since it is not so clear what a partial order semantics is. An obvious answer to the latter question is: a partial order semantics is a semantics that makes use of partial orders – say, in order to describe system runs. But partial orders can be encoded by sequences if these contain pointers of some kind, compare especially the causal trees of [DD89]; in fact, it is evident that everything we deal with in computer science or mathematics can be encoded sequentially. Hence, if we say that partial order semantics is necessary in order to deal with action refinement, we obviously do not simply refer to the explicit use of partial orders. Thus it may be better to say, as suggested by David Murphy, that the *power* of partial order semantics is necessary in order to deal with action refinement. Another weak point in the above 'obvious answer' is that sequences and step sequences can be seen as special partial orders; this view makes the language and the step language to two partial order semantics, which is not intented. To this, we can only say that a 'real' partial order semantics should also use partial orders that do not correspond to sequences or step sequences, i.e. partial orders where the co-relation is not transitive. This is very ad hoc, but at least it explains in what sense interval semiwords give a 'real' partial order semantics. And in this sense, our results prove that the power of partial order semantics is indeed necessary in order to deal with action refinement.

5.2   A technique of action refinement

In Chapter 4, we have already studied the refinement of transitions, and we have characterized modules for the behaviour-preserving refinement of a transition in an arbitrary context. One might expect that we could make use of the results of Chapter 4 for the study of action refinement, but this is not the case. After giving the definition of action refinement, we will discuss the differences between the refinement nets we use in this chapter and the modules of the previous chapter. At the end of this section, we will also discuss why we restrict ourselves to nets without self-concurrency.

General assumption All nets in this chapter are free of self-concurrency.

First, we define the refinement nets used for action refinement. They are given in the form denoted by $\overline{R}$ in Section 4.4, i.e. the environment is represented by one marked place called *idle*. In this chapter we simply speak of refinement nets as opposed to N_h- or N_p-refinement nets as defined in Chapter 4.

Definition 5.2.1 A *refinement net* is a pair $(R, idle)$, where $idle \in S_R$ carries one token initially, all arcs incident to *idle* have weight 1, and we have:

i) Only start transitions are enabled under M_R, where a *start transition* is a transition $t \in T_R$ with $t \in idle^\bullet$.

ii) For all reachable markings M of R we have that $M(idle) \geq 1$ implies $M = M_R$.

A transition t of R is called an *end transition* if $t \in {}^\bullet idle$. Let $init(R, idle) = \{a \in \Sigma \mid M_R[a\rangle)\}$ denote the set of *initially enabled* actions.

A *refinement function ref* associates with each $a \in \Sigma$ a refinement net $ref(a)$ such that no end transition is λ-labelled. Additionally, we define $ref(\lambda)$ as the refinement net consisting of a λ-labelled transition on a loop with the place *idle*. If *ref* is understood we will denote $ref(a)$ simply as $(N_a, idle_a)$ with $N_a = (S_a, T_a, W_a, M_a, lab_a)$.

The second definition explains how a refinement function is applied to a net. We replace each a-labelled transition t by a copy of $ref(a)$. The idea is that this copy may simulate t; firing a start transition starts such a simulation by removing the appropriate number of tokens from the preset of t; analogously, firing an end transition finishes a simulation.

Definition 5.2.2 Let *ref* be a refinement function and N be a net. The *action refinement* $ref(N)$ of N is defined by:

$$
\begin{aligned}
S_{ref(N)} &= S \,\dot{\cup}\, \{(t,s) \mid t \in T,\, s \in S_{lab(t)} - \{idle_{lab(t)}\}\} \\
T_{ref(N)} &= \{(t,t') \mid t \in T,\, t' \in T_{lab(t)}\} \\
W_{ref(N)}(s, (t,t')) &= W(s,t) \text{ if } t' \text{ is a start transition of } ref(lab(t)) \\
W_{ref(N)}((t,t'), s) &= W(t,s) \text{ if } t' \text{ is an end transition of } ref(lab(t)) \\
W_{ref(N)}((t,s), (t,t')) &= W_{lab(t)}(s,t') \\
W_{ref(N)}((t,t'), (t,s)) &= W_{lab(t)}(t',s) \\
W_{ref(N)} &\text{ is } 0 \text{ in all other cases}
\end{aligned}
$$

$$
\begin{aligned}
M_{ref(N)}(s) &= M_N(s) \\
M_{ref(N)}((t,s)) &= M_{lab(t)}(s) \\
lab_{ref(N)}((t,t')) &= lab_{lab(t)}(t')
\end{aligned}
$$

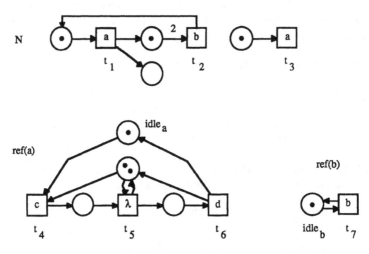

Figure 5.7

Figure 5.7 shows a net N and two refinement nets $ref(a)$ and $ref(b)$. Refining N we replace every a-transition by a copy of $ref(a)$; for example, in order to refine t_1, we split up $idle_a$ into an input and two output places, delete t_1, insert $ref(a)$ and merge the input place with the only place in ${}^\bullet t_1$, the output places with the two places in $t_1^\bullet$. The marking of these three places is that of N; the token on $idle_a$ means nothing for the action refinement, but it allows us to speak about the behaviour of $ref(a)$.

The result of the action refinement is shown in Figure 5.8. Observe that b-transitions are not really refined, due to the special form of $ref(b)$. Since $ref(\lambda)$ has to have this special form, we see that internal transitions cannot be refined; they are unobservable, while the purpose of action refinement is to deal with a visible change in the level of abstraction.

Whenever we want to refine only some actions, we can choose the refinement nets for the remaining actions to be analogous to $ref(b)$ in Figure 5.7. This way, the transitions with these labels remain unchanged (up to isomorphism). Therefore, it is no restriction that we refine all transitions in an action refinement; this uniform treatment of all transitions of N makes things technically a bit easier.

Also observe that, in the case that $ref(a)$ equals $ref(b)$ in Figure 5.7, applying ref relabels all a-transitions by b. Thus, relabelling is a special form of refinement; this is not true for hiding since we have required that every end transition of $ref(a)$, $a \in \Sigma$, must be visible. This requirement is related to the fact that $\mathcal{F}$-semantics does not induce a congruence with respect to hiding, see Section 3.4. Having in mind this fact, it is no

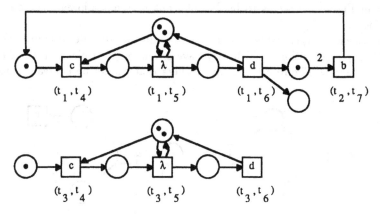

Figure 5.8

surprise that without this requirement our result below on a refined version of $\mathcal{F}$-semantics would be wrong. Many other semantic equivalences are congruences with respect to hiding. In these cases, one can simulate the use of refinement nets where some of the end transitions are internal: use a special action for the respective end transitions; refine with the modified nets; finally, hide the special action; this gives the same result as using the original refinement nets with some internal end transitions.

We will discuss now some of the restrictions we have made. To argue for the necessity of these restrictions, we will mostly proceed as follows. We have seen in Chapter 2 that nets can be distinguished by the choices they offer or by the concurrency they allow. But if two language-equivalent nets are λ-free, sequential and conflict-free, then – in our action oriented approach – we can see no reason whatsoever to distinguish these nets. If refinement of two such nets in the same way does not lead to language-equivalent nets, we conclude that the respective form of refinement is not appropriate.

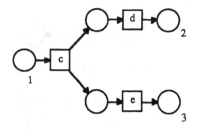

Figure 5.9

The first example of this type concerns distributed output. In Chapter 4, our studies of behaviour-preserving transition refinement revealed that distributed input may introduce new deadlocks, while distributed output is perfectly sensible. Thus, a refinement net as indicated in Figure 5.9 seems to be feasible, but it is not allowed according to our definitions above.

The first reason for this restriction is that it would be difficult to apply this refinement to arbitrary a-transitions, which may have any number of places in their pre- and postsets. Additionally, Figure 5.10 shows two nets that should be equivalent according

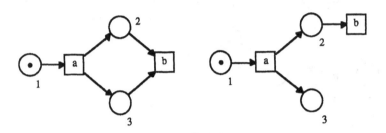

Figure 5.10

to our considerations above. But applying the refinement of Figure 5.9 yields the nets of Figure 5.11; only the second of these nets can perform the action sequence cdb.

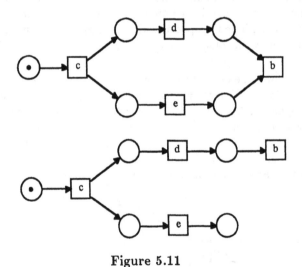

Figure 5.11

Regarding refinement nets with distributed input, it is in fact no valid argument in the context of this chapter, that they may introduce new deadlocks – we do not expect that action refinement necessarily preserves deadlock-freeness. But we can give an argument against distributed input that is similar to the argument above. Figure 5.12 shows two nets that again should be equivalent; but refining b by the net shown in Figure 5.13 yields the nets of Figure 5.14, where only the second net allows the action sequence cad.

It must be stressed that introducing new deadlocks is indeed feasible for action refinements. Our definition of a refinement net does allow the case that a refinement net

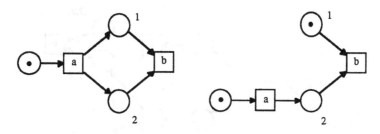

Figure 5.12

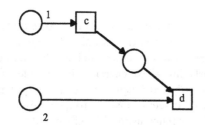

Figure 5.13

starts a simulation of a transition by firing a start transition, but is afterwards unable to finish this simulation. This is just a generalization; if such an effect is not wanted, one has to choose the refinement nets accordingly; our results would not change if we disallowed the introduction of deadlocks. By allowing this case, Definition 5.2.1 above is more general than the corresponding definition in [Vog91b]; this case is also allowed in the action refinement considered in [BDKP91].

In [GG90], a different technique for action refinement is presented. In this technique, refinement nets may have several input and output places; when a transition is refined, its preset is replaced by the cartesian product of this preset and the set of input places, and analogously for the postset. Refining a in the second net of Figure 5.10 with the

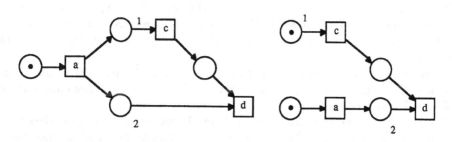

Figure 5.14

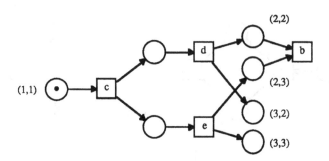

Figure 5.15

refinement net of Figure 5.9 gives the net shown in Figure 5.15.

We see that this net cannot perform the undesirable sequence *cdb*. Thus, this technique allows final concurrency, since the refinement net finishes its activities by performing *d* and *e* concurrently; similarly initial concurrency is possible. I would not say that the net of Figure 5.15 shows distributed output, since the output produced by *d* cannot be accessed before *e* has finished. It seems possible to generalize all our results to refinement nets with initial and final concurrency using this refinement technique. But some technical effort is required, and so far nearly no results are known for this technique.

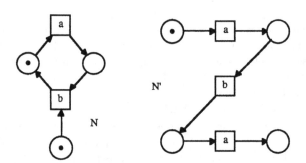

Figure 5.16

The next example concerns Condition i) of Definition 5.2.1. Figure 5.16 shows two nets that again should be equivalent. If we refine *a* in these nets with the net shown in Figure 5.17, then the first net can perform *d* only once, while the second can perform it twice.

Condition ii) of Definition 5.2.1 has two consequences. The first is that *idle* is a safe place in any refinement net. This ensures that a refinement net produces output at most as often as it has consumed input.

The second consequence is that refinement nets do not have a 'memory' in the sense we have discussed in Section 4.4. Figure 5.18 shows a net that may be used to replace action *a* alternatingly by a_1 or a_2. This net is more or less the same as that of Figure 4.34; it

Figure 5.17

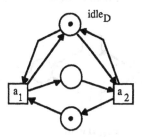

Figure 5.18

'remembers' the parity of the number of simulations it has performed. If we refine the nets in Figure 5.16 with this net, the resulting nets (see Figure 5.19) have different languages; the first can perform a_1ba_2, while the second can perform a_1ba_1.

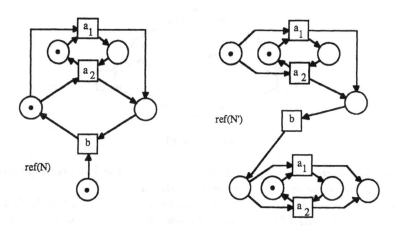

Figure 5.19

The last restriction we have to discuss is that we only consider nets without self-concurrency. Consider the nets of Figure 5.20. These nets are conflict-free; they are not sequential, but they do not only have the same language, but they are also step-language-, partial-word- and process-equivalent. Therefore, they are usually considered to be indistinguishable. Refining them with the net of Figure 5.21 yields nets where only

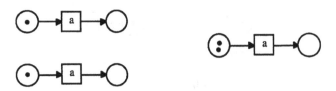

Figure 5.20

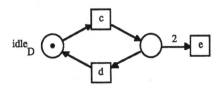

Figure 5.21

the second can perform *cce*. (A similar example is discussed in [GG90].)

Another difficulty with self-concurrency arises, if we want to work with processes or partial words. If we refine the nets of Figure 5.20 with the net shown in Figure 5.22, only

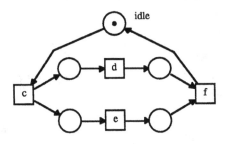

Figure 5.22

the second net allows the process sketched in Figure 5.23 and the corresponding image partial word. (A similar example is given in [BDKP91].)

Several consequences are possible. One can restrict refinement nets further to avoid these problems. In [BDKP91] and [Dev90b], congruence results based on processes are shown for nets with self-concurrency, where the refinement nets are state machines, i.e. all transitions have just one ingoing and one outgoing arc of weight 1 and initially only *idle* is marked. Such refinements allow one to replace an action by a choice between several actions or a sequence of several actions, and they include relabelling.

The second possiblity is to consider only nets without self-concurrency as we do it here, or to consider only safe nets (without isolated transitions); these nets form an important subclass of nets without self-concurrency. The latter choice is taken in [Vog91b] and also advocated in [GG90]. More liberally, one can consider all nets, but refine only those

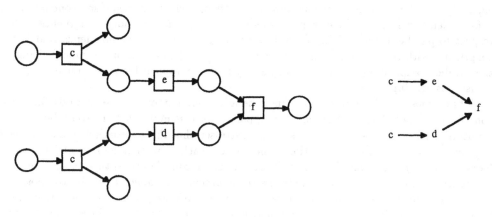

Figure 5.23

transitions that cannot occur self-concurrently. This is done in [BDKP91] and [Dev90b], and I do not see any problems to generalize the results of this chapter to such an approach; but the treatment would be less uniform and less clear.

Remark: There is a third possibility: we can also deny the existence of self-concurrency. Self-concurrency is quite a hybrid feature. In general, Petri nets describe concurrent systems with a static structure, and only self-concurrency allows us to describe some very limited form of dynamic structure: if we add more and more tokens to all the places in $^\bullet t$, we implicitly create more and more copies of the transition t, which can occur independently of each other.

We can eliminate self-concurrency from steps by requiring that every step can contain a transition at most once. Using this restricted form of steps in the definition of partial words gives exactly semiwords. To adapt processes, one can proceed as follows: freeness of self-concurrency can be implemented by adding for each transition a loop marked with one token. The use of semiwords is based on the view that every transition implicitly has such a loop. For the use of processes we have to make these loops explicit; for a net N, let $loop(N)$ be obtained by adding for each transiton a marked loop; now we can regard the processes of $loop(N)$, defined as usual, as processes of N without self-concurrency.

Also when refining transitions we have to make these loops explicit, at least in those cases where the refinement net contains start transitions that are not at the same time end transitions. Without any change in the definition of refinement, we could have the case that a start transition that is not an end transition fires twice in a row; this would correspond to a self-concurrent firing of the refined transition, which we do not want; also, it would lead to markings in the refinement net that we have not taken care of in Definition 5.2.1 ii). As a suitable modification of refinement, we can simply keep the place *idle* instead of deleting it as in Definition 5.2.2. It seems to be feasible to generalize the results of this chapter to arbitrary nets using these ideas.

A similar problem arises for parallel composition, as we have already seen at the end of Section 3.1. If parallel composition duplicates a transition that is self-concurrently enabled, then the two copies may fire independently of each other even if self-concurrency is forbidden. As a consequence, semiword equivalence is in general not a congruence with respect to parallel composition. Thus, in order to give a semantic equivalence that is a congruence with respect to action refinement and parallel composition we should change the parallel composition operator in such a way that the implicit loops become explicit if transitions are duplicated.

I have chosen not to work out this generalization, partly because it adds further complications and obscures the main ideas, partly because quite likely it would be counterintuitive for many readers. But it should be stressed that this approach would really be a generalization, and not a restriction to nets where each transition is on some loop; the loops in this approach would be implicit, and for some considerations this makes quite a difference. In Chapter 7, we will study the construction of unlabelled nets by composing atomic nets, nets with a single place only; in such a context, it is important that the implicit loops do not define an atomic net, but are implicitly present in all the atomic nets. □

As we have declared in the general assumption above, we restrict our considerations in this chapter to nets without self-concurrency. For these nets, partial words and semiwords coincide; we will use the word 'semiword' instead of 'partial word' in order to remind the reader of our general assumption and the alternative approach we have just sketched.

5.3   Action refinement and linear-time semantics

This section is devoted to linear-time congruences for action refinement based on partial orders. But our first aim is to show the well-definedness of action refinement in the setting of this chapter, i.e. to show that every refined net is free of self-concurrency; for this purpose we use an approach based on sequences, an approach we will study further in the next section.

ST-markings and interval sequences

When considering action refinement, it is not adequate any more to regard a transition firing as instantaneous. Instead, a transition t consists of a beginning t^+ and an end t^-; t^+ consumes the input of t, and t^- produces the output. We will call an enabled sequence over $\{t^+, t^- \mid t \in T_N\}$ an interval word, since a sequence like $t_1^- t_2^+ t_1^- t_2^-$ shows that the firing of t_1 and t_2 took place during two overlapping intervals. Interval words correspond more or less to what is called pre-ST-traces in [Gla90c] in a setting of event structures, and we will see later on that they are a sequential representation of the interval semiwords of [Vog91b].

If transitions have a beginning and an end, a system state cannot adequately be described by a marking alone; instead, it consists of a marking together with some transitions that have started, but have not finished yet. We call such a system state an ST-marking (S = Stellen, T = Transition (German)); ST-markings were introduced in [GV87] in a technically slightly different version.

Definition 5.3.1 An *ST-marking* of a net N is a pair $Q = (M, C)$, where M is a marking of N and C a finite set of transitions from T_N, those transitions that are *currently firing*, or just *current* for short. The *initial ST-marking* is $Q_N = (M_N, \emptyset)$.

$T_N^{\pm} = \{t^+, t^- \mid t \in T_N\}$ is the set of *transition parts*; t^+ is called the *start* of t, t^- the *end* of t. A transition part q is *enabled* under an *ST*-marking $Q = (M, C)$, denoted by $Q[q\rangle$, if

 - $q = t^+$ and $M[t\rangle$

   or

 - $q = t^-$ and $t \in C$.

If q is enabled, then the *follower ST-marking* $Q' = (M', C')$ obtained by *firing* q under Q is defined by

 - $M'(s) = M(s) - W_N(s, t)$, $s \in S_N$, and $C' = C \cup \{t\}$, if $q = t^+$

   and

 - $M'(s) = M(s) + W_N(t, s)$, $s \in S_N$, and $C' = C - \{t\}$, if $q = t^-$.

We write $Q[q\rangle Q'$.

An *interval sequence* of N is a sequence w from $(T_N^{\pm})^*$ such that, for every $t \in T$, the start and the end of t occur alternatingly in w beginning with t^+. If, for some t, every t^+ is followed by a t^-, then t is called *terminated*; if all transitions are terminated, then w is called *terminated*, too.

An interval sequence w is *enabled* under Q_N yielding Q, denoted by $Q_N[w\rangle Q$, if $w = \lambda$ and $Q = Q_N$ or $w = vq$ with $q \in T_N^{\pm}$ and $Q_N[v\rangle Q'[q\rangle Q$ for some Q'. We write $Q_N[w\rangle$ if we are not interested in Q, and we say that w is an *interval word* of N.

An *ST*-marking Q is *reachable* in N, if $Q_N[w\rangle Q$ for some interval word w; $[Q_N\rangle$ denotes the set of all reachable *ST*-markings.

When defining an interval sequence, we have required that for every $t \in T$ the occurrences of t^+ and t^- describe non-overlapping intervals starting at an occurrence of t^+, ending at the next t^- (except for the last occurrence of t^+, which might not be matched by a succeeding t^-). To make the above definition sensible, we should expect the following: for all reachable *ST*-markings Q, interval sequences v and transition parts q we have that vq is an interval sequence whenever $Q_N[v\rangle Q[q\rangle$. The problem we could have is that q may be some t^+ where t is not terminated in v; if this happened, it would also not be adequate to work with a *set* of current transitions. Luckily, we can show that this problem does not arise due to the absence of self-concurrency. We also show that we can obtain a reachable marking from a reachable *ST*-marking by 'undoing' the current transitions.

Proposition 5.3.2 *Let w be a sequence of transition parts of a net N.*

 i) *w is an interval word of N if and only if $w = \lambda$ or $w = vq$ for some transition part q such that $Q_N[v\rangle Q'[q\rangle$ for some ST-marking Q'.*

 ii) *If w is an interval word reaching $Q = (M, C)$, then*

- $M + \sum_{t \in C} W_N(.,t)$ *is a reachable marking of* N,
- *for all* $t \in T$, t *is terminated in* w *if and only if* $t \notin C$,
- *for all* $t \in T$, $Q[t^+)$ *implies* $t \notin C$.

Proof: It is clear from the definition that an interval word w is empty or can be written as vq as required. We show the remaining claims together by induction on the length of w, where the case $w = \lambda$ is clear. Thus, let $w = vq$ with q being a transition part and suppose that $Q_N[v)Q'[q)$ with $Q' = (M',C')$.

a) $q = t^-$

$Q'[t^-)$ implies that $t \in C'$, hence by induction we have that t is not terminated in v, and thus vt^- is an interval sequence and an interval word. Furthermore, $M + \sum_{t' \in C} W(.,t') = M' + W(t,.) - W(.,t) + \sum_{t' \in C'} W(.,t')$ is reachable from the marking $M' + \sum_{t' \in C'} W(.,t')$ by t, and this marking is reachable by induction. For each transition t' we have: t' is terminated in w if and only if (t' is terminated in v or $t' = t$) if and only if $t' \notin C = C'-\{t\}$.

b) $q = t^+$

$Q'[t^+)$ implies $M'[t)$. If t were in C', then t would be self-concurrently enabled under the reachable marking $M' + \sum_{t' \in C'} W(.,t')$, a contradiction. Thus, t is terminated in v and vt^+ is an interval sequence and an interval word. The marking $M + \sum_{t' \in C} W(.,t') = M' - W(.,t) + W(.,t) + \sum_{t' \in C'} W(.,t')$ is reachable by induction. Hence, $Q[t'^+)$ implies $t' \notin C$ – just as we have above concluded $t \notin C'$ from $Q'[t^+)$. Finally, $t' \in T$ is terminated in w if and only if ($t' \neq t$ and t' is terminated in v) if and only if $t' \notin C = C' \cup \{t\}$. $\square$

Next, we show that the firing step sequences of a net can be seen as special interval words and that reachable markings can be seen as special reachable ST-markings.

Proposition 5.3.3 *Let* N *be a net and* $w = \mu_1 \ldots \mu_n \in \mathcal{M}(T)^*$ *such that each* μ_i *corresponds to a subset of* T. *Put* $\mu_i^{\pm} = t_1^+ \ldots t_k^+ t_1^- \ldots t_k^-$, *where* $t_1 \ldots t_k$ *are the transitions in* μ_i *in some ordering. Then* $v = \mu_1^{\pm} \ldots \mu_n^{\pm}$ *is an interval word of* N *if and only if* $w \in FSS(N)$. *A marking* M *of* N *is reachable if and only if* $(M, \emptyset)$ *is a reachable* ST-*marking.*

Proof: This follows immediately from the definitions and Proposition 5.3.2 by induction on n. $\square$

Before going on, we relate reachable ST-markings to processes. The following theorem shows that, especially in view of processes, ST-markings are a very approporiate generalization of markings, since they correspond to arbitrary cuts in processes (for T-restricted nets); compare Theorem 2.2.1 and observe that in this chapter all transitions and hence all events of processes have a non-empty preset, since all nets are free of self-concurrency.

Theorem 5.3.4 *Let* $Q = (M,C)$ *be an* ST-*marking of a* T-*restricted net* N. *Then* Q *is reachable if and only if for some process* $\pi = (B,E,F,l)$ *of* N *there is a cut* $B' \cup E'$ *of* $(B \cup E, F^+, l)$ *with* $B' \subseteq B$, $E' \subseteq E$, $M = l^*(B')$ *and* $C = l(E')$. *The same equivalence holds for self-concurrence-free nets in general, if we replace the condition* $C = l(E')$ *by:* $C \subseteq l(E')$ *and* $e^{\bullet} = \emptyset$ *for all* $e \in E'$ *with* $l(e) \notin C$.

Proof: The case of T-restricted N follows immediately from the general case since we have $e^\bullet = \emptyset$ for all $e \in E$ if N is T-restricted.

'$\Rightarrow$' Apply Theorem 2.2.1 ii) to the reachable marking $M' = M + \sum_{t \in C} W(.,t)$ in order to obtain a process with a cut $B'' \cup E''$ such that $l^*(B'') = M'$ and $e^\bullet = \emptyset$ for all $e \in E''$. We may assume that B'' is maximal in the process; thus, to obtain π, we can add to the process a set E''' of events with $l(E''') = C$, and we can put $B' = B'' - {}^\bullet E'''$ and $E' = E'' \cup E'''$.

'$\Leftarrow$' Let $E''' = \{e \in E' \mid l(e) \in C\}$. Then, with $B'' = B' \cup {}^\bullet E'''$ and $E'' = E' - E'''$, we obtain a cut of π that corresponds to the marking $M + \sum_{t \in C} W(.,t)$, which is reachable by Theorem 2.2.1 ii). From the corresponding ST-marking, which has an empty set of current transitions, we can reach Q by firing all t^+ with $t \in C$. □

After the next lemma, we will be able to get the first insight into the behaviour of action refined nets; namely, we can determine their reachable ST-markings.

Lemma 5.3.5 *Let $(R, idle)$ be a refinement net.*

i) *If (M, C) is a reachable ST-marking of R and $M(idle) \geq 1$, then $M = M_R$ and $C = \emptyset$.*

ii) *If M is a reachable marking of R and $M[t + t'\rangle$, then neither t nor t' is a start or an end transition.*

Proof: i) $M + \sum_{t \in C} W_R(.,t)$ is a reachable marking by Proposition 5.3.2, thus it must be the initial marking by Definition 5.2.1 ii). If some $t \in C$ were a start transition, then $M(idle) \geq 1$ would imply $M_R(idle) > 1$, a contradiction; hence, no $t \in C$ can be a start transition. On the other hand, since $M_R[t\rangle$ for all $t \in C$, every $t \in C$ must be a start transition by Definition 5.2.1 i). Hence $C = \emptyset$ and $M = M_R$.

ii) If t' is a start transition, firing t^+ under M leads to an ST-marking that contradicts i). If t' is an end transition, firing $(t')^+ t^+ (t')^-$ leads to an ST-marking that contradicts i). □

Theorem 5.3.6 *Let N be a net and ref an action refinement.*

i) *If an ST-marking (M, C) of $ref(N)$ is reachable, then there are a reachable ST-marking (M_1, C_1) of N and reachable ST-markings (M_t, C_t) of $ref(lab(t))$ for $t \in T_N$ such that*

$$(M_t, C_t) = (M_{lab(t)}, \emptyset) \quad \text{iff} \quad t \notin C_1 \quad \text{for } t \in T_N$$
$$M(s) = M_1(s) \quad \text{for } s \in S_N$$
$$M(t, s) = M_t(s) \quad \text{for } s \in S_{lab(t)} - \{idle_{lab(t)}\}$$
$$C = \{(t, t') \mid t' \in C_t\}.$$

ii) *Suppose, (M, C), (M_1, C_1) and (M_t, C_t) are given as in i), and q is a transition part of $ref(N)$. Then, $(M, C)[q\rangle(M', C')$ if and only if (M', C') can be determined analogously to i) from (M_1', C_1') and (M_t', C_t'), $t \in T$, and we have one of the following cases:*

- $q = (t,t')^-$, t' is not an end transition, $t \in C_1$, $t' \in C_t$ and we have $(M'_t, C'_t) = (M_t + W_{lab(t)}(t',.)$, $C_t - \{t'\})$;
- $q = (t,t')^-$, t' is an end transition, $t \in C_1$, $C_t = \{t'\}$, we have $(M'_t, C'_t) = (M_t + W_{lab(t)}(t',.)$, $\emptyset)$ and $(M'_1, C'_1) = (M_1 + W_N(t,.)$, $C_1 - \{t\})$;
- $q = (t,t')^+$, t' is not a start transition, $t \in C_1$, $M_t[t')$, and we have $(M'_t, C'_t) = (M_t - W_{lab(t)}(.,t')$, $C_t \cup \{t'\})$;
- $q = (t,t')^+$, t' is a start transition, $t \notin C_1$, $M_1[t)$, $M_t[t')$, and we have $(M'_t, C'_t) = (M_t - W_{lab(t)}(.,t')$, $\{t'\})$ and $(M'_1, C'_1) = (M_1 - W_N(.,t)$, $C_1 \cup \{t\})$.

Unless explicitly required otherwise, (M'_1, C'_1) and all (M'_t, C'_t) coincide with (M_1, C_1) and (M_t, C_t), $t \in T_N$.

Proof: By induction on the length of an interval word w reaching (M, C), where in the case $w = \lambda$ we can clearly choose all ST-markings to be the initial ST-markings in order to satisfy i). For the induction step, it is clearly enough to show ii). Hence, assume we can determine some reachable ST-marking (M, C) of $ref(N)$ from (M_1, C_1) and (M_t, C_t), $t \in T_N$, and we have $(M, C)[q\rangle(M', C')$.

If $q = (t,t')^-$, then $t \in C_1$ and $t' \in C_t$. If t' is not an end transition, firing q only changes the marking on places (t, s). Thus, we only change (M_t, C_t) to $(M_t + W_{lab(t)}(t',.), C_t - \{t'\})$, and now we can determine (M', C') as desired. If t' is an end transition, then firing q additionally changes the marking on places in $t^\bullet$. $M'_t = M_t + W_{lab(t)}(t',.)$ marks $idle_{lab(t)}$ in this case. Thus $M'_t = M_{lab(t)}$ and $C'_t = C_t - \{t'\} = \emptyset$ by Lemma 5.3.5. We may put $M'_1 = M_1 + W_N(t,.)$ and $C'_1 = C_1 - \{t\}$. Now the required equations can be verified.

If $q = (t,t')^+$ and t' is not a start transition, then $M_t[t')$ and thus $t \in C_1$. Again we simply have to change (M_t, C_t), this time to $(M'_t, C'_t) = (M_t - W_{lab(t)}(.,t'), C_t \cup \{t'\})$. If t' is a start transition, then $M_1[t)$, hence $t \notin C_1$ by Proposition 5.3.2; this shows $M_t = M_{lab(t)}$ and this marking enables t' – especially $idle_{lab(t)}$ is marked, which we cannot immediately conclude from the marking of $ref(N)$. Now it suffices to make the following changes:

$$M'_1 = M_1 - W_N(.,t), \qquad C'_1 = C_1 \cup \{t\},$$
$$M'_t = M_{lab(t)} - W_{lab(t)}(.,t'), \qquad C'_t = \{t'\}.$$

We have just checked that one of the four cases for q given in the theorem applies; that each of them implies $(M, C)[q\rangle$ is easily checked. □

Observe that the inverse implication for the first part of this theorem does not hold in general, since some refinement nets might not be able to perform a simulation of a transition. As an extreme case, consider refinement nets without transitions; for a net with many reachable ST-markings $(M_1, \emptyset)$, we could choose $(M_t, C_t) = (M_{lab(t)}, \emptyset)$ and construct many ST-markings of the refined net as described in the first part; but in fact, only the initial ST-marking is reachable in the refined net.

As a corollary to the above theorem, we obtain the important result that action refinement of a net without self-concurrency with refinement nets without self-concurrency yields a refined net without self-concurrency. The analoguous result holds for the subclass of safe nets without isolated transitions.

Theorem 5.3.7 *Let N be a net, ref an action refinement; then $ref(N)$ is free of self-concurrency. If N and all $ref(a)$, $a \in \Sigma$, are safe and without isolated transitions, then $ref(N)$ is, too.*

Proof: Every reachable marking of $ref(N)$ combined with the empty set gives a reachable ST-marking (this holds even if $ref(N)$ were not self-concurrency-free). Therefore, if some (t, t') is enabled concurrently to itself under some reachable marking, we can find corresponding reachable ST-markings (M_1, C_1) and $(M_t, \emptyset)$ according to Theorem 5.3.6. If t' is a start transition, we get $M_1[t + t\rangle$; otherwise we get $M_t[t' + t'\rangle$. In view of Proposition 5.3.2, this is a contradiction.

The second part follows from Proposition 5.3.3 and Theorem 5.3.6. □

Partial order semantics and action refinement

Our next aim is to describe how the processes and semiwords of a refined net can be constructed from the processes and semiwords of the unrefined net. First, we give the notions 'process' and 'semiword' a specialized meaning if applied to refinement nets.

Definition 5.3.8 A *process* of a refinement net $(R, idle)$ is a process of R that contains at most one event that is labelled by a start transition, and analoguously a *semiword* of $(R, idle)$ is defined. A process (semiword) of $(R, idle)$ is *complete*, if it contains an event that is labelled by an end transition.

Event and action structures of processes of $(R, idle)$ and image semiwords of $(R, idle)$ are obtained from processes and semiwords of $(R, idle)$ as usual. An action structure or an image semiword of $(R, idle)$ is called *complete*, if one of its underlying processes or semiwords is complete.

To construct a process of a refined net, we replace each event in a process π of the unrefined net by a process of the corresponding refinement net. Special care has to be taken for the conditions that correspond to the initial marking of a refinement net. Assume we have two equally labelled events e_1 and e_2 of π; these events represent occurrences of the same transition t, and they are ordered in π since nets in this chapter are free of self-concurrency. Assume further, that no event between e_1 and e_2 corresponds to t as well. We will replace e_1 by a complete process $procref(e_1)$ of $ref(lab(t))$. This complete process starts from the initial marking on places (t, s) of $ref(N)$ and reproduces it in the end; the next access to these places is by the process $procref(e_2)$ that replaces e_2. Thus, we have to identify the maximal conditions of $procref(e_1)$ with the minimal conditions of $procref(e_2)$ in a suitable way. If $procref(e_2)$ does not need some of the initial tokens, then some condition of $procref(e_2)$ is minimal and maximal at the same time and it may get identified with a minimal condition of some $procref(e_3)$. Thus, the identification gives rise to equivalence classes of conditions of the inserted processes.

Definition 5.3.9 Let N be a net, $\pi = (B, E, F, l)$ a process of N and *ref* a refinement function. A *process refinement function procref* for π and *ref* associates to each $e \in E$ a process of $ref(lab_N(l(e)))$ such that $procref(e)$ is complete whenever e is not maximal in E with respect to F^+.

A labelled causal net $\pi' = (B', E', F', l')$ is called *derived from* π and *procref* if it is constructed as follows:

Put $B'' = B \dot\cup \bigcup_{e \in E} \{e\} \times B_{procref(e)}$. For every pair $(e_1, e_2) \in E \times E$ with $l(e_1) = l(e_2)$, $e_1 F^+ e_2$ and $(e_1 F^+ e_3 F^+ e_2 \Rightarrow l(e_3) \neq l(e_1))$, choose a label-preserving bijection from $\min(B_{procref(e_2)})$ to $\max(B_{procref(e_1)})$; identify (e_2, b_2) and (e_1, b_1) whenever b_2 is mapped to b_1 by such a bijection. Delete all conditions that are mapped to some $(e, idle_{lab(l(e))})$ or belong to the postset of an event $e \in E$ for which $procref(e)$ is not complete; the result is B'.

$$
\begin{aligned}
E' \;&=\; \bigcup_{e \in E} \{e\} \times E_{procref(e)} \\
F' \;&=\; \{(b, (e, e')) \mid (b, e) \in F \text{ and } l_{procref(e)}(e') \text{ is a start transition}\} \\
&\quad \cup \{((e, e'), b) \mid (e, b) \in F \text{ and } l_{procref(e)}(e') \text{ is an end transition}\} \\
&\quad \cup \{((e, b), (e, e')) \mid (b, e') \in F_{procref(e)}\} \\
&\quad \cup \{((e, e'), (e, b)) \mid (e', b) \in F_{procref(e)}\} \\
&\qquad\qquad (\text{where } (e, b) \text{ is rather a representative of an equivalence class} \subseteq B'') \\
l'(b) \;&=\; l(b) \quad \text{for } b \in B \\
l'(e, x) \;&=\; l_{procref(e)}(x) \quad \text{for } x \in B_{procref(e)} \cup E_{procref(e)}.
\end{aligned}
$$

Observe that for some processes π of N it may be impossible to find a process refinement function; this case can occur if some $ref(a)$ does not have complete processes.

An example of a process π of the net of Figure 5.7 and a derived labelled causal net are shown in Figure 5.24. Observe that the arc starting at the one condition indicated with an s could just as well start at the other one. These two conditions correspond to two tokens on the same place; these tokens together with a token on $idle_a$ form the initial marking of $ref(a)$ in Figure 5.7. Thus, the two conditions correspond to maximal conditions of $procref(e_1)$ and to minimal conditions of $procref(e_2)$; the two possible label-preserving bijections lead either to the process shown in Figure 5.24 or to the process that results from switching the arc to the other condition.

Furthermore observe, that non-maximal events must be replaced by complete processes, while a maximal event *may* be replaced by a complete process; in our example, we have chosen this option for the event labelled t_3.

Theorem 5.3.10 *Let N be a net, ref a refinement function. Then the processes of $ref(N)$ are those labelled causal nets that can be derived from processes π of N and process refinement functions procref for π and ref.*

Proof: It is not hard to see that a derived labelled causal net is a process of $ref(N)$. Observe in particular, that all events of a process π with label t are totally ordered, since N is free of self-concurrency. Thus for one of them, say e, $\{e\} \times \min(B_{procref(e)})$ represents the initial marking of $ref(N)$ on the places $(l(e), s)$. For every other e_2 of them, there is a unique e_1 such that the conditions in the definition of B' in 5.3.9 are fulfilled; $\min(B_{procref(e_2)})$ and $\max(B_{procref(e_1)})$ both represent the initial marking of $ref(lab(t))$, thus a label-preserving bijection exists. Consequently, conditions in the derived causal

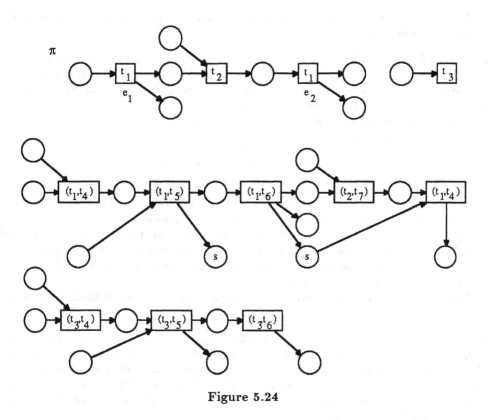

Figure 5.24

net are unbranched, and conditions with empty preset represent the initial marking of $ref(N)$.

Vice versa, assume we have a process π of $ref(N)$. Fix some $t \in T$; by Lemma 5.3.5 ii) and Theorem 5.3.6, the events of π that are mapped to some (t, t') with t' being a start or an end transition are totally ordered by F^+; start and end transitions alternate beginning with a start transition; all other events with label (t, t'') are between a start and the following end transition in the partial order F^+ (if there is a following end transition). With some care one can see that the events and conditions between a start and the following end transition (or after a start transition, if no end transition follows) together with these two events and some more conditions for the initial and the final marking form a process of $ref(lab(t))$. This process is complete if there is some end transition; otherwise, no events with label (t_0, t_0') with $t_0 \neq t$ follow. Furthermore, we can contract all events and conditions between a start and the following end transition (or after a start transition, if no end transition follows) including these two events to a new event labelled t; if we perform all these contractions, delete conditions that correspond to places of some refinement net, and possibly add some conditions in the postset of events that correspond to incomplete processes, then we obtain a process π' of N. The process π can be derived from π' and the processes of the refinement nets we have identified above.   □

As a next step, we transfer this result to event and action structures of processes and to semiwords and their images.

Definition 5.3.11 Let N be a net, *ref* a refinement function and $p = (E, <, l)$ the event structure of a process or a semiword of N. An *event structure refinement function evref* for p and *ref* associates to each $e \in E$ the event structure of a process of $ref(lab(l(e)))$ that is complete whenever e is not maximal in p.

We define the labelled partial order $evref(p) = (E', <', l')$ by

$$E' = \{(e, e_0) \mid e \in E, e_0 \in E_{evref(e)}\},$$
$$(e, e_0) <' (e', e'_0) \text{ if } e < e' \text{ or } e = e' \text{ and } e_0 <_{evref(e)} e'_0,$$
$$l'(e, e_0) = l_{evref(e)}(e_0).$$

A *semiword refinement function swref* for p and *ref* associates to each $e \in E$ a semiword of $ref(lab(l(e)))$ that is complete whenever e is not maximal.

If p is an action structure or an image semiword of N, then an *action structure refinement function acref*, an *image semiword refinement function imswref* resp., for p and *ref* associates to each $e \in E$ an action structure, an image semiword resp., of $ref(l(e))$ that is complete whenever e is not maximal.

The labelled partial orders $swref(p)$, $acref(p)$ and $imswref(p)$ are defined in the same way as $evref(p)$.

Theorem 5.3.12 *Let N be a net, ref a refinement function. The event structures of processes of $ref(N)$ are those labelled partial orders that can be determined as $evref(p)$, where p is the event structure of a process of N and evref is an event structure refinement function for p and ref. The semiwords of $ref(N)$ are those labelled partial orders that are more sequential than some $swref(p)$, where p is a semiword of N and swref a semiword refinement function for p and ref.*

$$Proc(ref(N)) = \{acref(p) \mid p \in Proc(N),$$
$$acref \text{ an action structure refinement function for } p \text{ and } ref\}$$
$$SW(ref(N)) = \{q \mid q \succeq imswref(p) \text{ for some } p \in SW(N),$$
$$imswref \text{ an image semiword refinement function for } p \text{ and } ref\}$$

Proof: If we derive a process π' of $ref(N)$ from a process π of N and some function *procref* according to Definition 5.3.9, then we can obtain $ev(\pi')$ also as $evref(ev(\pi))$, where *evref* is the event structure refinement function for $ev(\pi)$ and *ref* defined by $evref(e) = ev(procref(e))$. To see this observe:

- the maximal conditions of π and π' are irrelevant for $ev(\pi)$ and $ev(\pi')$;

- also irrelevant for $ev(\pi')$ is each condition of π' obtained by identification as described in Definition 5.3.9, i.e. obtained from a minimal condition of some *procref*(e) (representing some initial token of $ref(lab(l(e)))$) and from a maximal condition of some other *procref*(e'); for events $e <_{ev(\pi')} e'$ that are both adjacent to such a condition, we can find e_0 and e'_0 such that $e \leq_{ev(\pi')} e_0 <_{ev(\pi')} e'_0 \leq_{ev(\pi')} e'$, e_0 corresponds to an end transition and e'_0 corresponds to a start transition – as observed in the proof of Theorem 5.3.10;

- furthermore, the conditions of some $procref(e)$ labelled with $idle_{lab(l(e))}$ are irrelevant for $ev(procref(e))$, since $procref(e)$ contains at most one event labelled by a start transition;

- if $(e, e_0) <_{ev(\pi')} (e', e'_0)$, then a path in π' from (e, e_0) to (e', e'_0) either stays within $procref(e)$, in which case $e = e'$ and $e_0 <_{evref(e)} e'_0$, or it leaves $procref(e)$. If it leaves $procref(e)$, then by the second of these observations we may assume that it leaves $procref(e)$ at (an event labelled by) an end transition, enters some $procref(e'')$ at a start transition, thus $e <_{ev(\pi)} e''$, and so on, therefore $e <_{ev(\pi)} e'$.

Event structures of processes are special semiwords, event structure refinement functions are special semiword refinement functions, and all the least sequential semiwords are event structures of processes (Theorem 2.2.9). Thus, we conclude from the first part of the theorem that every semiword of $ref(N)$ can be constructed as described in the theorem. Now we have to prove that the construction always results in a semiword of $ref(N)$. For this, assume that p is a semiword of N, $swref$ is a suitable semiword refinement function and $q \succeq swref(p)$. Let p' be a least sequential semiword of N with $p \succeq p'$ and, for all $e \in E = E'$, let $swref'(e)$ be a least sequential semiword of $ref(lab(l(e)))$ with $swref(e) \succeq swref'(e)$. All the events that are maximal in p' are also maximal in p, and $swref'(e)$ is complete if and only if $swref(e)$ is; this implies that $swref'$ is in fact an event structure refinement function for p'. We conclude that $swref'(p')$ is the event structure of a process of $ref(N)$, that $q \succeq swref(p) \succeq swref'(p')$, and finally that q is a semiword of $ref(N)$. Therefore, the second statement of the theorem follows.

With these results the two equations follow; it does not matter whether we refine the event structure of a process or a semiword first and take the image afterwards, or whether we take the images first and refine them in a second step.   □

Corollary 5.3.13 *Process and semiword equivalence are congruences with respect to action refinement.*

Thus, we have found two congruences for action refinement, where process equivalence implies semiword equivalence but not vice versa (Corollary 2.2.11). But, both these congruences are not fully abstract with respect to action refinement and language equivalence, i.e. they make more distinctions than are necessary for a linear-time congruence. Therefore, Corollary 5.3.13 is only a limited argument in favour of partial order semantics.

It should be remarked that process equivalence and semiword equivalence are also congruences with respect to parallel composition, the other most important operator we study in this book; in order to apply Corollary 3.1.8, recall that all nets in this chapter are free of self-concurrency, hence semiwords and partial words coincide.

A fully abstract partial order semantics

We conclude this section by presenting a semantic equivalence that is fully abstract with respect to action refinement and language equivalence. We base it on a special class of semiwords called interval semiwords; with the announced full abstractness result, we will therefore show to which degree partial order semantics is necessary in order to deal with action refinement. (Regarding the meaning of this, recall the corresponding discussion in the introduction to this chapter.)

Definition 5.3.14 A partial order $(E, <)$ is an *interval order*, if for every $e \in E$ there is a closed real interval $I(e)$ such that $e < e'$ if and only if $x < y$ for all $x \in I(e)$, $y \in I(e')$.

A semiword $(E, <, l)$ of a net N is an *interval semiword* of N if $(E, <)$ is an interval order. Let $ISW(N)$ be the set of images of interval semiwords of N. Two nets are *interval-semiword-equivalent* if they have the same images of interval semiwords.

Thus an interval semiword can intuitively be seen as the observation of a system run where each firing takes some time; in such an observation, we can see that two transitions fire concurrently, namely when we observe that the corresponding intervals overlap. The image of an interval semiword is correspondingly an observation of action occurrences that take some time; here it is important that we do not just see actions starting and ending – when two a-actions overlap and one of them ends, we can see which one it is.

Interval orders can also be characterized combinatorially by forbidden substructures as shown in the following result of [Fis70].

Theorem 5.3.15 *A partial order $(E, <)$ (where E may also be countably infinite) is an interval order if and only if for all $w, x, y, z \in E$ we have: $w < x$ and $y < z$ implies $w \leq z$ or $y \leq x$.*

Another nice property is that we can recognize whether the image of a semiword is the image of an interval semiword. This is not so obvious, since the internal transitions are not visible on the image level, but they influence whether a semiword is an interval semiword or not.

Proposition 5.3.16 *Let $p = (E, <, l)$ be an image of a semiword of a net N. Then p is an image interval semiword if and only if $(E, <)$ is an interval order.*

Proof: '$\Rightarrow$' If q is an interval semiword with image p, then the real intervals that can be associated to the events of q show that $(E, <)$ is an interval order.

'$\Leftarrow$' Let $q = (E_q, <_q, l_q)$ be a semiword with image p. We want to show that we can enlarge $<_q$ to an interval order $<'$ such that

$$(*) \quad x <' y \text{ if and only if } x < y \text{ for all } x, y \in E.$$

Thus, let $<'$ be a partial order containing $<_q$ which is maximal with property $(*)$ (note, that $<_q$ has property $(*)$). We have to check that $<'$ is an interval order. Assume that for some $w, x, y, z \in E_q$ we have $w <' x$ and $y <' z$. If $z <' w$, then we have $y <' x$ and are satisfied. If $\neg z <' w$ and $\neg w \leq' z$, then by maximality of $<'$ there must be $w', z' \in E$ with $w' \leq' w$, $z \leq' z'$, but not $w' \leq z'$, see Figure 5.25. Analoguously, $\neg x <' y$ and $\neg y \leq' x$ implies that there are some $x', y' \in E$ such that $y' \leq' y$, $x \leq' x'$, but not $y' \leq x'$. Thus, we would have $w' <' x'$ and $y' <' z'$, hence by $(*)$ $w' < x'$ and $y' < z'$, but neither $w' \leq z'$ nor $y' \leq x'$, a contradiction to our assumption. Therefore, $w \leq' z$ or $y \leq' x$ and $\leq'$ is an interval order. $\square$

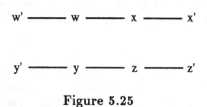

Figure 5.25

The next results state how the image interval semiwords of a refined net can be determined, namely in much the same way as semiwords, and that interval-semiword equivalence is fully abstract.

Theorem 5.3.17 *Let N be a net and ref a refinement function. The image interval semiwords of $ref(N)$ are those labelled interval orders that are more sequential than some $imswref(p)$, where p is an image interval semiword of N and $imswref$ is an image semiword refinement function for p and ref that assigns only image interval semiwords of the appropriate refinement nets.*

Proof: From the equation for $SW(ref(N))$ in Theorem 5.3.12 and Proposition 5.3.16, it is clear that the above construction yields image interval semiwords of $ref(N)$. The other half of the proof will be shown in the next sections.   □

Theorem 5.3.18 *Interval-semiword equivalence is fully abstract with respect to action refinement and language equivalence.*

Proof: The congruence result follows from Theorem 5.3.17, the other part will be shown in the next sections.   □

It is important that interval-semiword equivalence is also a congruence with respect to parallel composition.

Theorem 5.3.19 *Let $A \subseteq \Sigma$ and let N_1, N_2 be nets. Then*

$$ISW(N_1 \parallel_A N_2) \;=\; \{p \mid \text{there are } p_1 \in ISW(N_1), p_2 \in ISW(N_2), \text{ such that for the}$$
$$\text{labelled interval order } p \text{ we have } p \succeq q \in p_1 \parallel_A p_2\}$$

Interval-semiword equivalence is a congruence with respect to $\parallel_A$ for all $A \subseteq \Sigma$.

Proof: Use Theorem 3.1.7 ii) and Proposition 5.3.16. If $p \in ISW(N_1 \parallel_A N_2) \subseteq PW(N_1 \parallel_A N_2)$, then there exist $p_1 \in PW(N_1)$, $p_2 \in PW(N_2)$ and $q \in p_1 \parallel_A p_2$ such that $p \succeq q$. By definition of semiwords and $\parallel_A$, the labelled partial orders p_1 and p_2 may be assumed to be induced suborders of q and also of p; hence $p_1 \in ISW(N_1)$ and $p_2 \in ISW(N_2)$.

Vice versa, if p is in the right-hand side, then $p \in PW(N_1 \parallel_A N_2)$, thus p, being a labelled interval order, is in $ISW(N_1 \parallel_A N_2)$.   □

The last result of this section shows how interval-semiword equivalence is related to semiword- and step-language-equivalence.

Proposition 5.3.20 *Semiword equivalence implies interval-semiword equivalence, which implies step-language equivalence. Both implications cannot be reversed in general, not even for safe nets.*

Proof: Proposition 5.3.16 shows that the image interval semiwords can be determined from the image semiwords of a net, hence the first implication follows. Image firing step sequences are image interval semiwords – for all events in the same step we can choose the same interval; thus, they are those image interval semiwords with a transitive *co*-relation, hence the second implication follows.

Figure 5.26 shows four nets. $N_1 + N_2$ and $N_1 + N_3$ are interval-semiword-, but not semiword-equivalent; $N_1 + N_3$ and $N_1 + N_4$ are step-language-, but not interval-semiword-equivalent. $\square$

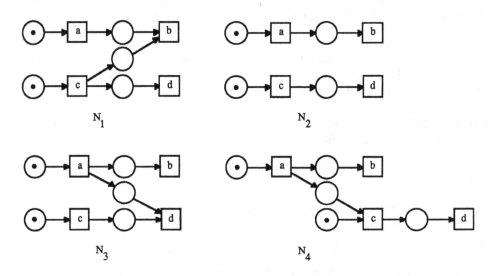

Figure 5.26

5.4   A closer look at interval words

In this section, we will base several congruences with respect to action refinement on interval words and show that they are fully abstract for language-, $\mathcal{F}$- and $\mathcal{F}\mathcal{D}$-equivalence.

Technical preliminaries

First, using Theorem 5.3.6 we will show how the interval words of a refined net can be determined from the interval words of the unrefined net. This result is not so easy to describe in a constructive style, which we employed in the corresponding results on processes and semiwords; instead we will use some sort of projections similarly as in Section 3.1.

Definition 5.4.1 Let N be a net, *ref* a refinement function. We define mappings $proj^*$ and $proj_t^*$, $t \in T_N$, from interval sequences of $ref(N)$ to sequences over $T_N^{\pm}$ and $T_{ref(lab(t))}^{\pm}$ by

$$proj((t,t')^+) = \begin{cases} t^+ & \text{if } t' \text{ is a start transition} \\ \lambda & \text{otherwise} \end{cases}$$

$$proj((t,t')^-) = \begin{cases} t^- & \text{if } t' \text{ is an end transition} \\ \lambda & \text{otherwise} \end{cases}$$

$$proj_t((t_0,t')^+) = \begin{cases} (t')^+ & \text{if } t = t_0 \\ \lambda & \text{otherwise} \end{cases}$$

$$proj_t((t_0,t')^-) = \begin{cases} (t')^- & \text{if } t = t_0 \\ \lambda & \text{otherwise} \end{cases}$$

An *interval word* of a refinement net $(R, idle)$ is an internal word of R where for at most one occurrence of some t^+ t is a start transition; it is called *complete* if some t^- occurs where t is an end transition.

We will first relate interval words of a refinement net $(R, idle)$ to the interval words of R.

Lemma 5.4.2 *Let $(R, idle)$ be a refinement net, w an interval sequence of R. Then w is an interval word of R if and only if it is the concatenation of interval words of $(R, idle)$ where all of these, except possibly the last one, are complete. A complete interval word of $(R, idle)$ finishes with the end of an end transition.*

Proof: Let w be an interval word of R. If w has no occurrence of a t^- with t an end transition, then it has at most one occurrence of a t^+ with t a start transition, since the start of a start transition consumes the token on *idle* and only the end of an end transition can put a token onto *idle* again. Thus we are done in this case.

Otherwise let w_1 be the prefix of w that ends with the first occurrence of a t^- where t is an end transition; $w = w_1 w_2$. By Lemma 5.3.5 i), the ST-marking reached after w_1 equals the initial ST-marking. Hence, w_2 is an interval word of R and we are done by induction.

For the reverse implication and the second claim, observe again that the ST-marking reached after firing the end of an end transition is the initial ST-marking. Since only start transitions are enabled now by definition, a complete interval word of $(R, idle)$ finishes with the end of an end transition. Hence, the concatenation described in the lemma is an enabled sequence of transition parts in R, thus an interval word of R by Proposition 5.3.2.
□

Now we show how the interval words of a refined net can be determined. (Observe the difference between an interval word of N_a and $(N_a, idle_a)$.)

Proposition 5.4.3 *Let N be a net, ref a refinement function. An interval sequence w of $ref(N)$ is an interval word of $ref(N)$ if and only if $proj^*(w)$ is an interval word of N and $proj_t^*(w)$ is an interval word of $N_{lab(t)}$ for all $t \in T_N$. In this case, the ST-marking (M, C) reached after w can be determined from the ST-markings (M_1, C_1) reached after $proj^*(w)$ and (M_t, C_t) reached after $proj_t^*(w)$ as described in Theorem 5.3.6.*

Proof: Follows by induction from Theorem 5.3.6.   □

Next, we have to lift interval words to the action level. At a first glance, it seems natural to replace each t^+ by $lab(t)^+$ and each t^- by $lab(t)^-$ in order to accomplish this lifting. But this would loose relevant information, since in a sequence $a^+a^+a^-$ we would not know which a^+ the a^- belongs to. On the level of transitions, a sequence $t^+t^+t^-$ cannot fire and, thus, each t^- in an interval word is uniquely matched to a preceding t^+. To keep this matching information on the action level, we add a second component; the resulting action interval sequences correspond to the interval sequences in [Gla90c]. This second component is a pointer connecting the matching starts and ends of actions; in the next section, we will see that these pointers encode a partial order.

Additionally, we define the ready set of an image interval word; this set consists of the visible actions that may happen next. The ready set will be useful for the failure semantics we will study below.

Definition 5.4.4 *The set $\Sigma^\pm = \{a^+, a^- \mid a \in \Sigma\}$ is the set of action parts, a^+ is the start, a^- the end of a. An action interval sequence w over some set E of events is an element of $(\Sigma^\pm \times E)^*$ such that there is a function $act : E \to \Sigma$ and for each $e \in E$:*

- *in w, there is a unique occurrence of some (a^+, e) with $a \in \Sigma$, and for this occurrence we have $a = act(e)$*
- *in w, there is at most one occurrence of some (a^-, e) with $a \in \Sigma$; if (a^-, e) occurs in w then it is preceded by (a^+, e), and e is called terminated in this case.*

If all events are terminated, then w is called terminated, too.

Two action interval sequences w_1 over E_1 and w_2 over E_2 are isomorphic if there is a bijection $f : E_1 \to E_2$ such that w_2 can be obtained from w_1 by replacing each (a^+, e) by $(a^+, f(e))$ and each (a^-, e) by $(a^-, f(e))$. As always, we do not distinguish between isomorphic action interval sequences.

Definition 5.4.5 *Let $w = q_1 \ldots q_n$ be an interval sequence of a net N, $q_i \in T^\pm$. Let E be a set and $start : E \to \{1, \ldots, n\}$ be an injection such that q_i is the start of a visible transition if and only if $i \in start(E)$. Let $act : E \to \Sigma$ satisfy $q_{start(e)} = t^+ \Rightarrow lab(t) = act(e)$. Define a partial function $end : E \to \{1, \ldots, n\}$ as follows:*

- *if $q_{start(e)} = t^+$, then the next occurrence of t^- in w is $q_{end(e)}$ – provided such an occurrence exists;*
- *if it does not exist, $end(e)$ is undefined.*

The *image* v of w is obtained from w by replacing $q_{start(e)}$ by $(act(e)^+, e)$ and $q_{end(e)}$ by $(act(e)^-, e)$ and deleting all q_i with $i \notin start(E) \cup end(E)$.

If w is an interval word of N reaching an ST-marking (M, C) and *Ready* $\subseteq \Sigma$ is the set of all actions that are image-enabled under $M + \Sigma_{t \in C, lab(t) = \lambda} W(t, .)$, then v is called an *image interval word* of N with *ready set Ready*, and w is an interval word *underlying* v *belonging to Ready*; (M, C) is an ST-marking *reached after* v.

The same way, we obtain from an interval word w of a refinement net $(R, idle)$ the *image interval word* v of $(R, idle)$ with *ready set Ready*. We call v *complete* if w is, *incomplete* otherwise; we call v *non-empty* if $w \neq \lambda$.

Let $IW(N)$ be the set of image interval words of N. We call nets N_1 and N_2 *interval-word-equivalent* if $IW(N_1) = IW(N_2)$.

Observe that an image interval word v of a refinement net may be complete and incomplete at the same time, since some interval word underlying v may be complete while another is not. Similarly, λ may be a non-empty image interval word. If we speak of an incomplete non-empty image interval word v (with ready set *Ready*) in the following, this means that some interval word $w \neq \lambda$ underlying v (and belonging to *Ready*) is not complete.

In order to define the ready set in the above definition, we cannot simply take all a with $M[a\rangle$. We also have to account for the internal transitions that are currently firing; their ending is invisible, but would be ignored by the usual image-firing rule, which we have not adapted to the present situation with action parts.

The following lemma is immediately implied by the definitions.

Lemma 5.4.6 *Let v be the image of an interval sequence of a net N constructed from w and E as in Definition 5.4.5. Then v is an action interval sequence over E.*

Next, we lift Proposition 5.3.3 to the action level and show that interval-word equivalence refines step-language equivalence.

Proposition 5.4.7 *Let N be a net and $w = \mu_1 \ldots \mu_n \in \mathcal{M}(T)^*$, such that each μ_i corresponds to a subset of T. Put $\mu_i^{\pm} = t_1^+ \ldots t_k^+ t_1^- \ldots t_k^-$, where $t_1 \ldots t_k$ are the transitions in μ_i in some ordering. Then the image of $v = \mu_1^{\pm} \ldots \mu_n^{\pm}$ is in $IW(N)$ if and only if $lab^*(w) \in SL(N)$.*

Proof: The reverse implication follows immediately from Proposition 5.3.3. For the remaining implication, we do not know that the sequence v of transition parts is an interval word, but only that some interval word v' has the same image as v. We would like to apply Proposition 5.3.3 to v', but v' possibly does not have the required form; namely, for an internal transition t it may be that occurrences of transition parts t^+ or t^- are not in the special ordering as required; also an occurrence of t^+ might not be followed by t^-, as required. We have to show that v' can be transformed to an interval word of the required form which has the same image. First note: if, for some internal t, an occurrence of t^+ in v' is not followed by t^-, we might transform v' to the interval word $v't^-$, which has the same image. Secondly, in v' we can change qt^- to t^-q for $q \in T^{\pm} - \{t^+\}$ and t^+q to qt^+

for $q \in T^{\pm} - \{t^-\}$; this change gives another interval word, which has the same image if t is internal or if q is part of an internal transition. This way, we can transform v' such that each t^+ is immediately followed by t^- for all internal transitions t; using the same rules, we can transform v' further such that no occurrence of some t^+ with internal t separates a start of a visible transition t_1 from the following end of t_1. Now Proposition 5.3.3 can be applied to construct from v' a step sequence with image $lab^*(w)$.   □

Corollary 5.4.8 *Interval-word equivalence implies step-language equivalence.*

Proof: By Proposition 5.4.7, those image interval words that are a concatenation of sequences $(a_1^+, e_1) \ldots (a_n^+, e_n)(a_1^-, e_1) \ldots (a_n^-, e_n)$ correspond to the image firing step sequences.   □

The idea of failure semantics is to describe one (partial) system run by the sequence of visible actions that occurred paired with a refusal set; a refusal set is a set of visible actions that cannot occur next. Image interval words are meant to replace image firing sequences in a situation where actions have a beginning and an end. Therefore, adapting failure semantics for this situation, we will use pairs consisting of an image interval word w and a refusal set X. In order to use the image firing rule to define the refusal set X, we cannot simply require $\neg M[a\rangle$ for all $a \in X$, where the ST-marking reached after w is (M, C); as already explained above, we also have to account for the internal transitions that are currently firing. Thus, we require in the next definition that $X \cap Ready = \emptyset$ for the ready set $Ready$ belonging to w. Otherwise this definition is a straightforward modification of the $\mathcal{F}$- and the $\mathcal{F}D$-semantics.

Definition 5.4.9 The *interval-word failure semantics* $iw\mathcal{F}(N)$ of a net N is defined by

$$iw\mathcal{F}(N) = \{(w, X) \mid w \text{ is an image interval word with ready set } Ready,$$
$$X \subseteq \Sigma, \text{ and } X \cap Ready = \emptyset\}.$$

We call an interval word or an image interval word *diverging*, if it reaches an ST-marking (M, C) such that M enables an infinite sequence of λ-transitions.

The *interval-word failure/divergence semantics* $iw\mathcal{F}D(N) = (iwF(N), iwD(N))$ of N is defined by

$$iwF(N) = \{(w, X) \mid w \in iwD(N), X \subseteq \Sigma\} \cup iw\mathcal{F}(N)$$
$$iwD(N) = \{w \mid w \text{ is an action interval sequence such that some prefix } v \text{ of } w$$
$$\text{is a diverging image interval word}\}.$$

First, we will compare $iw\mathcal{F}$- and $iw\mathcal{F}D$-equivalence with $\mathcal{F}$- and $\mathcal{F}D$-equivalence.

Proposition 5.4.10   *i) $iw\mathcal{F}$-equivalence implies $\mathcal{F}$-equivalence, $iw\mathcal{F}D$-equivalence implies $\mathcal{F}D$-equivalence.*

  ii) If nets are sequential and $\mathcal{F}$-equivalent, then they are $iw\mathcal{F}$-equivalent. If nets are sequential and $\mathcal{F}D$-equivalent, then they are $iw\mathcal{F}D$-equivalent.

Proof: i) This follows again immediately from Proposition 5.4.7.

ii) If a net N is sequential, then each terminated image interval word reaching an ST-marking $(M, \emptyset)$ corresponds by 5.4.7 to an image firing sequence reaching M. A non-terminated image interval word v is a terminated one followed by some (a^+, e); it can be obtained from the terminated image interval word $v(a^-, e)$ and, since the net is sequential, every action can be refused after v. The minimal elements of $iwD(N)$ (with respect to the prefix-relation) are diverging terminated image interval words corresponding to the minimal elements of $D(N)$. Thus, $iw\mathcal{F}(N)$ and $iw\mathcal{F}D(N)$ can be determined from $\mathcal{F}(N)$ and $\mathcal{F}D(N)$.   □

Congruence results based on interval words

Our next aim is to describe how the IW-, $iw\mathcal{F}$- and $iw\mathcal{F}D$-semantics of a refined net can be determined from the respective semantics of the unrefined net. Analogously to the definitions in the previous section, we define an interval word refinement function to formulate this result. Furthermore, we define when (a specific representative of) an action interval sequence is derived from (some representative of) an image interval word and such an interval word refinement function.

Definition 5.4.11 Let N be a net, ref a refinement function, and v_1 an image interval word of N over E. An *image interval word refinement function imiwref* for v_1 and ref assigns to each $e \in E$ a non-empty image interval word of $(N_{act(e)}, idle_{act(e)})$ over some set E_e; if e is terminated in v_1, then $imiwref(e)$ is complete, otherwise it is incomplete non-empty.
An action interval sequence v over $\bigcup_{e \in E}\{e\} \times E_e$ is *derived* from v_1 and $imiwref$ if

- for each $e \in E$, $imiwref(e)$ can be obtained from v by replacing each occurrence $(q, (e, e'))$ by (q, e') and deleting all $(q, (e_1, e'))$ with $e_1 \neq e$;
- deleting all occurrences (q, e) in v_1 for which $imiwref(e) = \lambda$ gives the same result as the following transformation of v:
  - replace, for each $e \in E$, the first occurrence of some $(q, (e, e'))$, $q \in \Sigma^\pm$, $e' \in E_e$, by $(act(e)^+, e)$
  - replace, for each $e \in E$, the last occurrence of some $(q, (e, e'))$, $q \in \Sigma^\pm$, $e' \in E_e$ by $(act(e)^-, e)$ provided that e is terminated in v_1
  - delete all other $(q, (e, e'))$.

First, we describe the image interval words of a refined net in quite a detailed and thus rather technical way.

Lemma 5.4.12 *Let N be a net and ref a refinement function. An action interval sequence v is an image interval word of $ref(N)$ reaching an ST-marking (M, C) if and only if it is derived from*

- *an image interval word v_1 of N reaching (M_1, C_1) and*

    – an image interval word refinement function *imiwref* for v_1 and *ref*, such that

        – for $t \in C_1$ with $lab(t) \neq \lambda$, the ST-marking (M_t, C_t) is reached after *imiwref*(e), where e is the non-terminated event of v_1 related to t and

        – (M, C) can be determined from (M_1, C_1) and (M_t, C_t), $t \in T_N$, according to Theorem 5.3.6.

        (For $t \in T_N - C_1$, we choose $(M_t, C_t) = (M_{lab(t)}, \emptyset)$; for $t \in C_1$ with $lab(t) = \lambda$, the ST-marking (M_t, C_t) is chosen such that M_t is 0 on the only place, $idle_\lambda$, and C_t contains the only transition, i.e. $C_t = T_\lambda$.)

Proof: '$\Rightarrow$' Let $v \in IW(ref(N))$ be given over some set $\tilde{E}$, and let w be an interval word underlying v with $(M_{ref(N)}, \emptyset)[w](M, C)$. Put $w_1 = proj^*(w)$, which is an interval word of N by Proposition 5.4.3, and let (M_1, C_1) be the ST-marking reached after w_1; let v_1 be the image of w_1 and assume that v_1 is given as an action interval sequence over some set E.

For all $t \in T_N$, we have by Proposition 5.4.3 that $proj_t^*(w)$ is an interval word of $N_{lab(t)}$, i.e. by Lemma 5.4.2 a concatenation of interval words of $(N_{lab(t)}, idle_{lab(t)})$. If t is visible, then each occurrence of t^+ in w_1 corresponds on the one hand to some $e \in E$, on the other hand to some occurrence $(t, t')^+$ in w where t' is a start transition of $(N_{lab(t)}, idle_{lab(t)})$. Thus, this occurrence of $(t, t')^+$ corresponds to an occurrence of t'^+ in $proj_t^*(w)$, which marks the beginning of one of the above mentioned interval words of $(N_{lab(t)}, idle_{lab(t)})$. Let *imiwref*$(e)$ be the image of this interval word. By construction, *imiwref*(e) is complete if e is terminated in v_1, otherwise it is incomplete non-empty; in the latter case let (M_t, C_t) be the ST-marking reached after *imiwref*(e) for the transition $t \in T$ related to e. By Proposition 5.4.3, the ST-marking (M, C) can be determined from (M_1, C_1) and (M_t, C_t). (Choose (M_t, C_t) as described at the end of the lemma if $t \notin C_1$ or if $t \in C_1$ is internal.) We have to show that v is derived from v_1 and *imiwref*.

First we check the condition on *imiwref*(e). Each occurrence $(q, \tilde{e})$ in v corresponds to some $(t, t')^+$ or $(t, t')^-$ in w, thus to some t'^+ or t'^- in $proj_t^*(w)$ and therefore to a unique of the above mentioned interval words of $(N_{lab(t)}, idle_{lab(t)})$ and a unique $e \in E$. Hence collecting all $(q, \tilde{e})$ from v belonging to a fixed $e \in E$ gives *imiwref*(e).

For the condition on v_1 we have to transform w such that v and (M, C) remain unchanged: Whenever $t \in T$ is visible, we move all $(t, t')^+$ and $(t, t')^-$ in w where t' is internal, such that they form sequences, each preceding immediately an occurrence $(t, t')^+$ where t' is visible. This is possible since end transitions are visible and only end transitions may be necessary to enable some $(t_0, t_0')^+$ with $t_0 \neq t$. – There is one exception, namely if some $(t, t')^+$ or $(t, t')^-$ with t' internal is not followed by some $(t, t')^+$ with t' visible. These occurrences remain where they are. The first of them leads to an occurrence t^+ in w_1 and $(lab(t)^+, e)$ in v_1, but *imiwref*$(e) = \lambda$ for this $e \in E$; hence this $(lab(t)^+, e)$ is deleted in v_1 in the condition we have to check.

Assuming that w is of the transformed form, we see that the first occurrence of some $(q, \tilde{e})$ where $\tilde{e}$ corresponds to a given $e \in E$ stems from some $(t, t')^+$ where t' is a start transition and t is visible, or at least it stems from an occurrence preceded by such a $(t, t')^+$. Such a $(t, t')^+$ gives rise to an occurrence of t^+ in w_1 and to the occurrence $(lab(t)^+, e)$ in v_1. If e is terminated in v_1, then the last occurrence $(q, \tilde{e})$ in v with $\tilde{e}$

corresponding to e stems from some $(t, t')^-$ in w where t' is an end transition – remember that end transitions are always visible. Thus, this $(t, t')^-$ corresponds to some t^- in w_1 and $(lab(t)^-, e)$ in v_1.

'$\Leftarrow$' Assume we are given $v_1 \in IW(N)$ over E, the function $imiwref$ for v_1 and ref, and firing sequences w_1 reaching (M_1, C_1) underlying v_1 and w_e underlying $imiwref(e)$ for $e \in E$, such that w_e reaches (M_e, C_e). Assume v is derived from v_1 and $imiwref$.

Each $e \in E$ corresponds via w_1 to some visible $t \in T$, each $e' \in E_e$ via w_e to some visible $t' \in T_{lab(t)}$. For each $e \in E$ and $e' \in E_e$, replace in v the occurrence $(a^+, (e, e'))$ by $(t, t')^+$ and $(a^-, (e, e'))$ (if it occurs) by $(t, t')^-$ (where $a = lab_{lab(t)}(t')$). Call the resulting sequence w'. We will obtain w from w' by inserting transition parts $(t, t')^+$ or $(t, t')^-$ where t' is internal. Any such w has image v; the only thing to check is that some $(q, (e_0, e'_0))$, $q \in \Sigma^\pm$, occuring in v after $(a^+, (e, e'))$ and before $(a^-, (e, e'))$ is not related to (t, t'), too. If e_0 is related to t and $e \neq e_0$, then we conclude from the derivability condition on v_1 that $(lab(t)^+, e)$ and $(lab(t)^+, e_0)$ appear in v_1 without being separated by $(lab(t)^-, e)$ or $(lab(t)^-, e_0)$; thus, there are two occurrences of t^+ in w_1 not being separated by t^-, a contradiction. Therefore, if $(q, (e_0, e'_0))$ is related to (t, t'), then $e_0 = e$. Now we conclude from the derivability condition on $imiwref(e)$, that in this image interval word (q, e'_0) occurs after (a^+, e') and before (a^-, e'), hence the transition corresponding to e'_0 cannot be t'.

First, we obtain w'' by inserting into w' occurrences $(t, t')^+$ where t is visible and t' is an internal start transition and occurrences $(t, t')^+$, $(t, t')^-$ with internal t such that $proj^*(w'') = w_1$. For this, observe that $proj^*(w)$ can be transformed to w_1 by insertion of some t^+ and t^-. Whenever t is internal, the corresponding t' is the only transition of T_λ; otherwise the missing transition part is a t^+ corresponding to some $e \in E$, and t' is the start transition of $(N_{lab(t)}, idle_{lab(t)})$ for which $imiwref(e)$ begins with t'^+. We also have to make sure that in the latter case $(t, t')^+$ is inserted before any $(t, t'')^+$ or $(t, t'')^-$ that correspond to some $(q, (e, e'))$ with $q \in \Sigma^\pm$, $e' \in E_e$; this is possible due to the derivability condition on v_1.

Now, for internal $t \in T$, $proj_t^*(w'')$ is an interval word of N_λ. For visible t, $proj_t^*(w'')$ can be viewed as the concatenation of sequences starting with some t'^+ where t' is a start transition and ending with the next t'^- where t' is an end transition (provided there is one). Each such sequence corresponds to some e and can be transformed to w_e by inserting internal transition parts due to the derivability condition on $imiwref(e)$; whenever t'^+ or t'^- has to be inserted, we insert $(t, t')^+$ or $(t, t')^-$ into w'' at the corresponding place. This way, we obtain w such that $proj^*(w) = w_1$, $proj_t^*(w)$ is an interval word of N_λ for internal $t \in T$, and $proj_t^*(w)$ is the concatenation of those w_e where e corresponds to t for visible $t \in T$. In the latter case, $proj_t^*(w)$ is an interval word of $N_{lab(t)}$ by Lemma 5.4.2; if furthermore $t \in C_1$, then t is not terminated in w_1, thus the last of the w_e is not complete and (M_e, C_e) is the marking (M_t, C_t) reached after $proj_t^*(w)$. Now the result follows from Proposition 5.4.3. $\square$

Next, we describe how the ready sets of the unrefined, the refined and the refinement nets are related.

Lemma 5.4.13 *Let (M, C) be an ST-marking of a refined net $ref(N)$, and let (M_1, C_1), (M_t, C_t), $t \in C_1$ with $lab(t) \neq \lambda$, be the related ST-markings of N and $ref(lab(t))$ according to Theorem 5.3.6. Let Ready, $Ready_1$ and $Ready_t$ be the sets associated with (M, C), (M_1, C_1) and (M_t, C_t) according to Definition 5.4.5. Then $Ready = \bigcup_{a \in Ready_1} init(ref(a))$ $\cup \bigcup_{t \in C_1, lab(t) \neq \lambda} Ready_t$.*

Proof: '$\subseteq$' Let $a \in Ready$, and let wt_0 be a shortest firing sequence with image a enabled under $M_0 = M + \Sigma_{t \in C, lab_{ref(N)}(t) = \lambda} W_{ref(N)}(t, .)$. By Proposition 5.4.7, we have that wt_0 corresponds to an interval word enabled under M_0 where each start of a transition is immediately followed by the end of that transition. Hence, we can use Theorem 5.3.6 to study wt_0. Remember that no end transition is internal, thus w does not contain an end transition except in the case that it corresponds to an internal transition of N. Let $t_0 = (t_1, t_1')$.

a) $t_1 \in C_1$.
In this case, w cannot contain a start transition of the form (t_1, t), and for all other transitions (t_1, t) the enabling depends only on the marking of the places (t_1, s). By choice of wt_0, it thus contains only transitions (t_1, t) and we conclude that $a \in Ready_{t_1}$.

b) $t_1 \notin C_1$.
In this case, wt_0 must contain a start transition of the form (t_1, t). As far as places (t_1, s) are concerned, this start transition is enabled under (M, C). Hence the first part of w (possibly empty) puts enough tokens on places $s \in S_N$ to enable this start transition. Consequently, this part (if not empty) contains some end transition; by the above remark, it must in fact correspond to a sequence of internal transitions of N, thus $lab_N(t_1) \in Ready_1$. Once the start transition (t_1, t) has fired, we argue as in a); therefore $a \in init(ref(lab_N(t_1)))$.

'$\supseteq$' a) Let $t \in C_1$ and $b \in Ready_t$.
If $t_1 \ldots t_n$ is a firing sequence with image a enabled under

$$M_t + \Sigma_{t' \in C_t, lab_{lab(t)}(t') = \lambda} W_{lab(t)}(t', .),$$

then $(t, t_1) \ldots (t, t_n)$ is enabled under M_0 as in '$\subseteq$' and we have $b \in Ready$.

b) Let $a \in Ready_1$ and $b \in init(ref(a))$.
A firing sequence of internal transitions of N enabling an a-labelled transition as required by $a \in Ready_1$ corresponds to a firing sequence of internal transitions of $ref(N)$ enabled under M_0 enabling all start transitions (t, t') for some t with $lab_N(t) = a$. Now we can fire $(t, t_1) \ldots (t, t_n)$, where $t_1 \ldots t_n \in FS(N_{lab(t)})$ has image b. Therefore $b \in Ready$. $\square$

With these very technical lemmata we can show that the three semantics we have based on image interval words are congruences. The first result is immediate from Lemma 5.4.12.

Theorem 5.4.14 *Let N be a net and ref a refinement function. An action interval sequence is an image interval word of $ref(N)$ if and only if it is derived from some image interval word v_1 of N and some image interval word refinement function for v_1 and ref.*

Interval-word equivalence is a congruence with respect to action refinement.

Theorem 5.4.15 *Let N be a net and ref a refinement function. Then $iw\mathcal{F}(ref(N))$ is the set of pairs (v, X) for which there are $(v_1, X_1) \in iw\mathcal{F}(N)$ and an image interval word refinement function imiwref for v_1 and ref such that:*

- *v is derived from v_1 and imiwref*

- *if e is non-terminated in v_1, then imiwref(e) is a non-empty incomplete image interval word of $(N_{act(e)}, idle_{act(e)})$ with ready set Ready$_e$ such that $X \cap$ Ready$_e = \emptyset$*

- *$X \cap \{a \in \Sigma \mid \exists b \in \Sigma - X_1,\ a \in init(ref(b))\} = \emptyset$.*

$iw\mathcal{F}$-equivalence is a congruence with respect to action refinement.

Proof: Follows from Lemmas 5.4.12 and 5.4.13. For the proof that each pair (v, X) constructed from (v_1, X_1) etc. really is in $iw\mathcal{F}(ref(N))$, observe that we can restrict attention to those (v_1, X_1) where $\Sigma - X_1$ is a ready set for v_1.   □

In this theorem we have not used the $iw\mathcal{F}$-semantics of the refinement nets; but the formulation shows that this would be possible provided that:

- we include pairs (λ, X) only if the refusal set corresponds to λ as a non-empty image interval word of the refinement net;

- we keep the information which image interval words are complete.

With these modifications, we could easily turn $iw\mathcal{F}$-equivalence into an equivalence of refinement nets and prove: if $ref(a)$ and $ref'(a)$ are equivalent for all $a \in \Sigma$, then $ref(N)$ and $ref'(N)$ are $iw\mathcal{F}$-equivalent for all nets N. Thus, we would get a congruence for action refinement, where action refinement is viewed as one operation with infinitely many arguments as discussed in the introduction of this chapter.

In the next theorem, which shows that $iw\mathcal{F}\mathcal{D}$-equivalence is a congruence with respect to action refinement, we restrict ourselves to refinement nets where not only all end transitions, but also all start transitions are visible. This is not a severe restriction since under some mild conditions $iw\mathcal{F}\mathcal{D}$-equivalence is a congruence for hiding; hence, action refinement with arbitrarily labelled start and end transitions can be performed by the restricted form of action refinement followed by some hiding, and it is therefore compatible with $iw\mathcal{F}\mathcal{D}$-equivalence. We need a lemma first.

Lemma 5.4.16 *Let N be a net, ref be a refinement function, $w_1 \in iwD(N)$, imiwref be an image interval word refinement function for w_1 and ref, and let w be derived from w_1 and imiwref. Then there exists a diverging image interval word v of $ref(N)$ that is a prefix of w.*

Proof: Let w_1 be given over the set E and w over $\bigcup_{e \in E}\{e\} \times E_e$ as described in Definition 5.4.11. Let v_1 be a diverging image interval word of N that is a prefix of w_1 and given over $E' \subseteq E$. Let v be that prefix of w for which the next $(q, (e, e'))$ in w is the first for which either $e \notin E'$ or $e \in E'$ is terminated in w_1 and not terminated in v_1 and $imiwref(e)$ ends with (q, e'). Define $imiwref'$ by restricting $imiwref$ to E' and, for $e \in E'$, letting $imiwref'(e)$ be that prefix of $imiwref(e)$ for which for all occurrences (q, e') there is an occurrence $(q, (e, e'))$ in v. Now one can check that v is derived from v_1 and $imiwref'$, and Lemma 5.4.12 shows that it is diverging due to an infinite sequence of λ-transitions corresponding to λ-transitions of N (– the enabling of these transitions of $ref(N)$ depends only on the marking of the places $s \in S_N$). $\qquad\square$

Theorem 5.4.17 *Let N be a net and ref a refinement function such that for all $a \in \Sigma$ all start transitions of $ref(a)$ are visible. Then $iwF(ref(N))$ is the set of pairs (v, X) for which $v \in iwD(ref(N))$ and $X \subseteq \Sigma$ or for which there are $(v_1, X_1) \in iwF(N)$ and an image interval word refinement function $imiwref$ for v_1 and ref such that:*

- *v is derived from v_1 and $imiwref$*

- *if e is non-terminated in v_1, then $imiwref(e)$ is a non-empty incomplete image interval word of $(N_{act(e)}, idle_{act(e)})$ with ready set $Ready_e$ such that $X \cap Ready_e = \emptyset$*

- *$X \cap \{a \in \Sigma \mid \exists b \in \Sigma - X_1, \, a \in init(ref(b))\} = \emptyset$.*

The set $iwD(ref(N))$ consists of action interval sequences w for which there exist a prefix v of w, $(v_1, \emptyset) \in iwF(N)$ and an image interval word refinement function $imiwref$ for v_1 and ref, such that v is derived from v_1 and $imiwref$, and $v_1 \in iwD(N)$ or some $imiwref(e)$ is diverging where e is non-terminated.

$iwFD$-equivalence is a congruence with respect to action refinement, if the refinement nets have visible start transitions.

Proof: First we consider $iwD(ref(N))$. If v is a prefix of $w \in iwD(ref(N))$ and for the ST-marking (M, C) reached after v an infinite sequence u of λ-transitions is enabled under M, then v is derived from some $v_1 \in IW(N)$, i.e. $(v_1, \emptyset) \in iwF(N)$, and some $imiwref$ according to Lemma 5.4.12; let (M_1, C_1) and (M_t, C_t), $t \in C_1$ with $lab(t) \neq \lambda$, be the corresponding ST-markings. The sequence u cannot contain any start or end transitions, except for those corresponding to λ-transitions of N. Now u may contain infinitely many λ-transitions derived from λ-transitions of N; in this case they give a sequence enabled under M_1 and therefore $v_1 \in iwD(N)$. Otherwise, it contains infinitely many λ-transitions that are neither start nor end transitions, i.e. have as first component some $t \in C_1$ with $lab(t) \neq \lambda$. Since C_1 is finite, we conclude that some $imiwref(e)$ is diverging where e is related to some $t \in C_1$, i.e. is non-terminated.

For the other inclusion, assume that v_1, $imiwref$ and a derived v are given. If $v_1 \in iwD(N)$, we are done by Lemma 5.4.16. Otherwise, v_1 is an image interval word of N and we are done by Lemma 5.4.12.

Now we consider the equality for $iwF(ref(N))$. If $(v, X) \in iwF(ref(N))$, then we may have $v \in iwD(ref(N))$, hence (v, X) is in the right-hand side, or we have $(v, X) \in iwF(ref(N))$ and (v, X) is in the right-hand side by Theorem 5.4.15. For the other inclusion, observe that $(v_1, X_1) \in iwF(N)$ implies $v_1 \in iwD(N)$, and then we are done by the iwD-part; for $(v_1, X_1) \in iwF(N)$, we are again done by Theorem 5.4.15. $\qquad\square$

Full abstractness results

In order to prove the announced full abstractness results, it remains to demonstrate that nets that are not interval-word-equivalent etc. really have to be distinguished if we want congruences for action refinement that respect language equivalence etc.

Theorem 5.4.18 *Interval-word-, $iw\mathcal{F}$- and $iw\mathcal{F}D$-equivalence are fully abstract with respect to action refinement and language-, $\mathcal{F}$- and $\mathcal{F}D$-equivalence. This also holds if only action refinement with finite refinement nets is considered.*

Proof: We only consider the $iw\mathcal{F}$-case, since the others are similar; as remarked above, we only have to show that all the distinctions made by the $iw\mathcal{F}$-equivalence are necessary. Thus, assume that we have nets N_1 and N_2 such that for all refinement functions ref – where all $ref(a)$, $a \in \Sigma$, are finite – we have $\mathcal{F}(ref(N_1)) = \mathcal{F}(ref(N_2))$. (In particular, we have $\mathcal{F}(N_1) = \mathcal{F}(N_2)$ in this case.) We have to show that $iw\mathcal{F}(N_1) = iw\mathcal{F}(N_2)$.

Let $(w_1, X_1) \in iw\mathcal{F}(N_1)$ and assume that w_1 is given over the set $E = \{1, \ldots, n\}$ with $n \in \mathbb{N}_0$. If $n = 0$, i.e. $w_1 = \lambda$, then $(w_1, X_1) \in \mathcal{F}(N_1) = \mathcal{F}(N_2)$ and $(w_1, X_1) \in iw\mathcal{F}(N_2)$; hence assume $n \neq 0$. Assume we have distinct actions $(a, e, j) \in \Sigma$ for $a \in \Sigma$, $e = 1, \ldots, n$, $j = 1, 2$. We define a refinement function ref by

$$
\begin{aligned}
S_a &= \{idle_a\} \cup \{s_e \mid e = 1, \ldots, n\}; \\
T_a &= \{t_{e,j} \mid e = 1, \ldots, n,\ j = 1, 2\}; \\
W_a(s, t) &= 1 \text{ if } (s = idle_a \text{ and } t = t_{e,1}) \text{ or } (s = s_e \text{ and } t = t_{e,2}) \text{ for some } e \\
W_a(t, s) &= 1 \text{ if } (s = idle_a \text{ and } t = t_{e,2}) \text{ or } (s = s_e \text{ and } t = t_{e,1}) \text{ for some } e \\
W_a &\text{ is } 0 \text{ otherwise} \\
M_a(idle_a) &= 1, \\
M_a(s_e) &= 0 \text{ for all } e \\
lab_a(t_{e,j}) &= (a, e, j)
\end{aligned}
$$

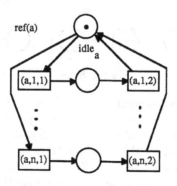

Figure 5.27

Figure 5.27 shows the refinement net $ref(a)$. One can view this refinement as replacing each action by a choice between n actions and splitting each of these into two. Thus, an image firing sequence of the refined net shows the beginnings and endings of actions from the unrefined net. The choices offered by the refinement nets allow to detect which ending belongs to which beginning. Without the choices, this detection would be difficult, if not impossible, in the presence of autoconcurrency.

Consequently, we now define an image interval word refinement function *imiwref* for w and *ref* such that every event from E corresponds to a different choice. A terminated event will correspond to both actions of this choice, a non-terminated only to the first action. Thus $imiwref(e) = ((act(e), e, 1)^+, 1)((act(e), e, 1)^-, 1)$ if $e \in E$ is non-terminated, $imiwref(e) = ((act(e), e, 1)^+, 1)((act(e), e, 1)^-, 1)((act(e), e, 2)^+, 2)((act(e), e, 2)^-, 2)$ otherwise. Observe that $imiwref(e)$ is complete or incomplete, whatever is required.

Now we obtain an action interval sequence w from w_1 by replacing each $(act(e)^+, e)$ by the sequence $((act(e), e, 1)^+, (e, 1))((act(e), e, 1)^-, (e, 1))$ and analogously each $(act(e)^-, e)$ by $((act(e), e, 2)^+, (e, 2))((act(e), e, 2)^-, (e, 2))$. Obviously this w is derived from w_1 and *imiwref*. Put $X = \{(a, e, 1) \mid a \in X_1,\ e \in E\}$. By Theorem 5.4.15, we have $(w, X) \in iw\mathcal{F}(ref(N_1))$.

Since w is terminated, we can assume that the ST-marking reached after w belonging to the refusal set X is of the form $(M, \emptyset)$. By Proposition 5.4.7, w corresponds to an image firing sequence $\tilde{w}$ of $ref(N_1)$ reaching M. Therefore $(\tilde{w}, X) \in \mathcal{F}(ref(N_1)) = \mathcal{F}(ref(N_2))$. By the same argument, we conclude that $(w, X) \in iw\mathcal{F}(ref(N_2))$. By Theorem 5.4.15 and the derivability condition on w_1, we conclude that for some X_2 we have $(w_1, X_2) \in iw\mathcal{F}(N_2)$ such that $X \cap \{(a, e, 1) \mid a \in \Sigma - X_2,\ e \in E\} = \emptyset$; hence $X_1 \subseteq X_2$ and $(w_1, X_1) \in iw\mathcal{F}(N_2)$.   □

In this proof we have not tried to minimize the size of refinement nets, but chosen them in order to make the proof as easy as possible. One could also choose more economic refinement nets; for example, if in w_1 each action from Σ starts at most once, it clearly suffices to use the refinement nets for $n = 1$, no matter what the size of E is.

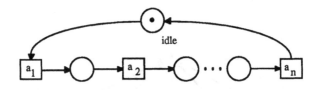

Figure 5.28

Probably this full abstractness result does not hold if we only consider arbitrary splittings, i.e. action refinements with nets as shown in Figure 5.28. The nets shown in Figure 5.29 are distinguished by interval-word-, $iw\mathcal{F}$- and $iw\mathcal{FD}$-equivalence, but it seems that, if they are both refined by the same arbitrary splittings, the resulting nets are language-, $\mathcal{F}$- and $\mathcal{FD}$-equivalent. (I thank Rob van Glabbeek for discussing this example with me.) It remains an open problem to determine the coarsest congruences with respect to splitting.

We conclude this section by showing the congruence result for $iw\mathcal{FD}$-equivalence and hiding that we have announced above. For this, we need a hiding operator on action

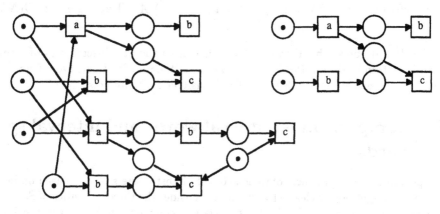

Figure 5.29

interval sequences. For $A \subseteq \Sigma$, we define $w \setminus A$ only for an action interval sequence w where all events e with $act(e) \in A$ are terminated; *hiding* A in such a sequence w gives the sequence $w \setminus A$, which is obtained from w by deleting all occurrences (a^+, e), (a^-, e) with $a \in A$. Obviously, if w is an image interval word of some net N, then $w \setminus A$ is an image interval word of $N \setminus A$ reaching the same ST-marking (M, C). The restriction on w ensures that each $t \in C$ is internal in N if and only if it is internal in $N \setminus A$; this is important for the ready and refusal sets associated with w and $w \setminus A$.

Theorem 5.4.19 *Let N be an image-finite net, $A \subseteq \Sigma$. Then*

$$\begin{aligned}
iwF(N \setminus A) \;=\; & \{(w, X) \mid w \in iwD(N \setminus A),\ X \subseteq \Sigma\} \\
\cup\; & \{(w \setminus A, X) \mid (w, X \cup A) \in iwF(N)\}
\end{aligned}$$

The set $iwD(N \setminus A)$ consists of those action interval sequences w for which there exists an action interval sequence v such that $v \setminus A$ is a prefix of w and :

– $v \in iwD(N)$

or

– for all $n \in \mathbb{N}_0$, there exists v_n with: $(v_n, \emptyset) \in iwF(N)$, $v_n = vu_n$, and u_n is a concatenation of n sequences $(a^+, e)(a^-, e)$ with $a \in A$.

Proof: We restrict the proof to the case where the precondition on N has to be applied, namely the '$\supseteq$'-part of the second equality for w from the second set. Let w, v and v_n, u_n, $n \in \mathbb{N}_0$, be given. We construct a directed graph whose vertices are those interval words of N whose images are prefixes of some v_n. There is an arc from one interval word u to another u' if $u' = uq$, $q \in T^{\pm}$. By the assumption on N we can apply König's Lemma and find some interval word u that can be elongated by an infinite sequence of parts of transitions with label in $A \cup \{\lambda\}$. We can terminate all a-labelled transitions of u and still find such an infinite sequence for the resulting interval word u'. Hence for the image u'' of u' we find that $u'' \setminus A$ is a diverging image interval word of $N \setminus A$. For some n the sequence

u'' is a prefix of v_n, i.e. $u'' \setminus A$ is a prefix of $v_n \setminus A = v \setminus A$. Therefore $w \in iwD(N \setminus A)$.
$\square$

As explained above, this theorem shows that often the $iw\mathcal{FD}$-semantics of a refined net can be determined from the $iw\mathcal{FD}$-semantics of the unrefined net, even if some start and end transitions are internal; especially this is true for finite nets.

5.5   Comparison of interval words and interval semi-words

In the previous section, we have obtained some full abstractness results based on interval words, but it might not be clear why we should consider the three semantics of Section 5.4 as partial order semantics. To clarify this, we will compare interval words and interval semiwords in this section. It will turn out that interval-word equivalence and interval-semiword equivalence coincide, thus we can finish the proofs left open in Section 5.3. One of the advantages of using interval words is that one can base on them a decision algorithm for interval-(semi)word equivalence, as we will also demonstrate in this section.

The comparison for linear-time semantics

The correspondence between interval words and interval semiwords is in fact quite natural, since the occurrence of some t^+ and that of the succeeding t^- in an interval word w mark the end points of an interval, i.e. a substring of w; see Figure 5.30.

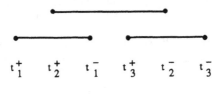

$$t_1^+ \qquad t_2^+ \qquad t_1^- \qquad t_3^+ \qquad t_2^- \qquad t_3^-$$

Figure 5.30

Definition 5.5.1 Let $w = q_1 \ldots q_n$ be an interval sequence of some net N, $q_i \in T^\pm$; let E be a set and $start : E \to \{1, \ldots, n\}$ be an injection such that q_i is a start of a transition if and only if $i = start(e)$ for some $e \in E$. Let $end : E \to \{1, \ldots, n\}$ be a partial function such that

- if $q_{start(e)} = t^+$ and the next occurrence of t^- in w is q_j, then $end(e) = j$
- if $q_{start(e)} = t^+$ is not followed by t^- in w, then $end(e)$ is undefined.

Let the labelled partial order $p = (E, <, l)$ be defined by ($e < e'$ iff $end(e) < start(e')$) and $q_{start(e)} = l(e)^+$. Then we say that w is an *interval representation* of the labelled interval order p.

Observe the close relationship between this definition and Definition 5.4.5, where we defined the image of an interval sequence. In the next lemma, we state that p as defined above is indeed a labelled interval order. Secondly, we note that each labelled interval order has some interval representation, and Figure 5.31 shows that there may be several such representations.

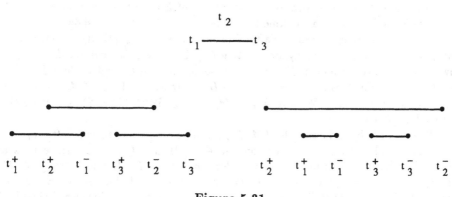

Figure 5.31

Lemma 5.5.2 *Each interval sequence of some net N is interval representation of some interval order labelled over T_N. Each interval order labelled over T_N has an interval representation that is terminated.*

Proof: The first part is obvious. For the second part, assume we are given a labelled interval order $(E, <, l)$ and intervals $[I_1(e), I_2(e)]$, $e \in E$, according to the definition of an interval order.

There are only finitely many reals $r_1 < r_2 < \cdots < r_n$ appearing as values of I_1 or I_2. Thus, whenever $I_1(e) = I_2(e') = r_i$, we can redefine $I_2(e')$ to take some value between r_i and r_{i+1} – or greater than r_n if $i = n$. The new intervals $[I_1(e), I_2(e)]$ still fit for $(E, <, l)$. Hence, we may assume that for all $e, e' \in E$ we have $I_1(e) \neq I_2(e')$. Now, if for some $E' \subseteq E$ we have $I_1(e) = r_i$ for all $e \in E'$, we can redefine the $I_1(e)$, $e \in E'$, to take different values between r_i and r_{i+1}; similarly for I_2.

Therefore, we may assume that all values $I_1(e)$, $e \in E$, and $I_2(e)$, $e \in E$, are different. We can map these values order-isomorphic onto some $\{1, \ldots, n\}$, $n \in \mathbb{N}_0$; hence we may assume that I_1, I_2 map E into $\{1, \ldots, n\}$ and for each $i \in \{1, \ldots, n\}$ there is a unique $(e, j) \in E \times \{1, 2\}$ such that $I_j(e) = i$.

Now define $w = q_1 \ldots q_n$ by letting $q_i = t^+$ if $i = I_1(e)$ and $l(e) = t$, and letting $q_i = t^-$ if $i = I_2(e)$ and $l(e) = t$.   □

Of course, the correspondence between interval sequences and labelled interval orders is of interest only because it is compatible with the firing rule.

Theorem 5.5.3 *Let N be a net, w an interval sequence of N and p an interval order labelled over T_N such that w is an interval representation of p. Then w is an interval word if and only if p is an interval semiword.*

Proof: Let $w = q_1 \ldots q_n$, $q_i \in T^{\pm}$ and $p = (E, <, l)$, and let $start$ and end be given as in Definition 5.5.1.

'$\Rightarrow$' Let B and $B \dot\cup C$ be downward-closed subsets of E as in the definition of a partial word. By Lemma 2.2.4, we may assume that C is a cut and we have to show that $l^*(C)$ is enabled under the marking M reached after B, i.e. $M = M_N + \sum_{e \in B} W(l(e), .) - W(., l(e))$. Put $m = \max_{e \in C} start(e)$ – we may assume that $C \neq \emptyset$. Since C is a cut and $B, B \cup C$ are downward-closed, every event in B is less than some event in C, and we find $B = \{e \in E \mid end(e) < m\}$. Furthermore $C = \{e \in E - B \mid start(e) \leq m\}$. Let (M_m, C_m) be the ST-marking after $q_1 \ldots q_m$. We have $M_m = M_N + \sum_{e \in B} W(l(e), .) - \sum_{e \in B \cup C} W(., l(e))$. Thus the claim follows.

'$\Leftarrow$' The proof is by induction; let (M_i, C_i) be the ST-marking reached after $q_1 \ldots q_i$, $i \in \{0, \ldots, n-1\}$. If $q_{i+1} = t_0^-$ for some $t_0 \in T$, then $(M_i, C_i)[q_{i+1}\rangle$ since w is an interval sequence. Thus assume $q_{i+1} = t_0^+$ for some $t_0 \in T$. Define $B = \{e \in E \mid end(e) \leq i\}$ and $C = \{e \in E - B \mid start(e) \leq i + 1\}$. If for $e, e' \in E$ we have $end(e) < start(e')$ and $end(e') \leq i$, then $end(e) \leq i$, hence B is downward-closed. If $end(e) < start(e') \leq i + 1$, we also have $end(e) \leq i$, hence $B \cup C$ is downward-closed and the elements of C are pairwise concurrent. Furthermore, $l^*(C) = C_i \cup \{t_0\}$ and $M_i = M_N + \sum_{e \in B}(W(l(e), .) - W(., l(e))) - \sum_{t \in C_i} W(., t)$. Since p is a partial word we have $\sum_{e \in C} W(., l(e)) \leq M_N + \sum_{e \in B}(W(l(e), .) - W(., l(e)))$, and we conclude $W(., t_0) \leq M_i$, i.e. $(M_i, C_i)[q_{i+1}\rangle$. Thus, w is an interval word. $\square$

Naturally, as a next step we translate the comparison of interval words and interval semiwords to the image level.

Definition 5.5.4 *Let w be an action interval sequence over some set E with its corresponding function act, and let $p = (E, <, l)$ be a partial order labelled over Σ. Then w is an interval representation of p, if for all $e, e' \in E$ we have*

- *$e < e'$ iff $(act(e)^-, e)$ occurs before $(act(e')^+, e')$ in w*

  and

- *$act(e) = l(e)$.*

Again, we see immediately that p must be a labelled interval order if it has an interval representation as an action interval sequence – the occurrences of $(act(e)^+, e)$ and $(act(e)^-, e)$ in w mark an interval for e. Obviously, p is uniquely determined by w.

Proposition 5.5.5 *Let N be a net.*

   i) *If the interval sequence w of N is an interval representation of the labelled partial order p, then the image of w is an interval representation of the image of p.*

ii) Let the action interval sequence v be an interval representation of the labelled partial order q and let w be an interval sequence of N. Then w has image v if and only if w is interval representation of some labelled partial order p with image q.

Proof: requires some considerations, but follows immediately from the definitions.   □

Corollary 5.5.6 *Let N be a net.*

$$IW(N) = \{w \mid w \text{ is interval representation of some } p \in ISW(N)\}$$
$$ISW(N) = \{p \mid \text{ some } w \in IW(N) \text{ is interval representation of } p\}$$

Interval-word equivalence and interval-semiword equivalence coincide.

Proof: The first equality follows from Theorem 5.5.3 and Propostion 5.5.5. The second equality follows from the first, since each interval order labelled over Σ has an interval representation and each action interval sequence is interval representation of a unique interval order labelled over Σ.   □

With this result, Theorem 5.4.18 immediately implies Theorem 5.3.18, whose proof was still missing. Thus, we have two different representations for the equivalence that is fully abstract with respect to action refinement and language equivalence.

The comparison for failure semantics

A natural question is whether we can also find a representation of $iw\mathcal{F}$- and $iw\mathcal{F}D$-equivalence based on interval semiwords. This is the case, and the two semantics have been presented in [Vog91b]; in a fairly obvious way, image interval semiwords are combined with a refusal set to obtain an alternative to the $iw\mathcal{F}$-semantics, the alternative to $iwD(N)$ consists of all image interval semiwords of N after which divergence can occur and their elongations, and the alternative to $iwF(N)$ is constructed from $iw\mathcal{F}(N)$ and $iwD(N)$ as usual.

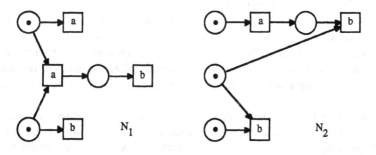

Figure 5.32

But there are two important subtleties. The first concerns the difference between terminated and non-terminated interval words. Consider the nets in Figure 5.32 (this example probably appeared first in [BDKP91]). These nets have the same action structures of processes, hence the same image semiwords and image interval semiwords.

In the beginning, they can refuse everything except a and b; after performing one of a and b they can refuse everything except the other; after performing a and b, concurrently or in some order, they can refuse every action. Therefore, if we define a partial order failure semantics $po\mathcal{F}$ by combining action structures of processes or image semiwords or image interval semiwords with refusal sets in a straightforward fashion (as suggested in [Pom88,PS92]), the nets of Figure 5.32 would be $po\mathcal{F}$-equivalent. But refining a by the sequence a_1a_2 only $ref(N_1)$ can refuse b after a_1, thus $ref(N_1)$ and $ref(N_2)$ are not $\mathcal{F}$-equivalent. In terms of the $iw\mathcal{F}$-semantics, the reason for this failure of the suggested $po\mathcal{F}$-semantics is that it ignores the non-terminated image interval words, and in fact $((a^+, e), \{b\}) \in iw\mathcal{F}(N_1) - iw\mathcal{F}(N_2)$.

The solution to this problem is to indicate in image interval semiwords just as in image interval words which events are terminated and which are non-terminated. In [Vog91b], a failure-type and a failure/divergence-type semantics are based on image interval semiwords *with termination set*; they are of the form $(E, <, l, ter)$, where $(E, <, l)$ is an image interval semiword and $ter \subseteq E$ is the set of terminated events. Naturally, if $e < e'$, i.e. if e' started after e finished, then e must be terminated. Therefore it is required that ter must contain all elements of E that are not maximal with respect to $<$.

The other subtlety concerns the question under which circumstances an image interval semiword $p = (E, <, l, ter)$ with termination set is an elongation of some other, $p' = (E', <', l', ter')$. This question has to be answered in order to define the alternative to $iwD(N)$. Naturally, one thinks of the prefix-relation; thus we require that $(E', <', l')$ is a prefix of $(E, <, l)$ and tentatively add the requirement $ter' \subseteq ter$, since an event that has terminated in p' should remain terminated in an elongation.

But consider the nets of Figure 5.33 and the indicated image interval semiword p where all events are terminated. Since a terminated a is a prefix of p and leads to divergence in both nets, we would include p in the alternative iwD-semantics of both nets. And in fact, more generally, the two nets of Figure 5.33 would have the same failure/divergence-type semantics, if we based this semantics on the prefix-relation. But if we refine a to a_1a_2, then a_1ba_2 is an image firing sequence of the refined nets; it shows that a and b can occur independently in the unrefined nets and thus corresponds to p; but $a_1ba_2 \in D(ref(N_1)) - D(ref(N_2))$.

Here, the solution is to consider p as an elongation of p', if p' is a *trunk* of p, i.e. a prefix where for all $e' \in ter'$ and $e \in E - E'$ we have $e' < e$ in p. The intuitive reason for the latter requirement is that the new events of p happen after p', thus they start after each event that is terminated in p. In fact, the trunk-relation is a direct translation of the prefix-relation of action interval sequences to image interval semiwords with termination set.

Based on image interval semiwords with termination set and the trunk-relation, two semantics are defined in [Vog91b] that induce the same equivalences as the $iw\mathcal{F}$- and the $iw\mathcal{F}D$-semantics. This can be seen from the full abstractness results that we have obtained here and that can be found in [Vog91b]. But also a direct translation between

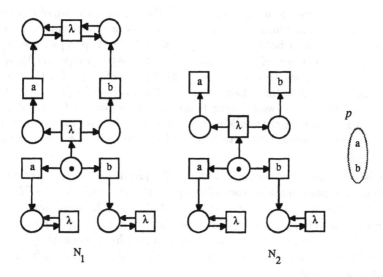

Figure 5.33

the semantics can be given along the lines of the translation between interval words and interval semiwords that we have given in this section.

Now a natural question is whether we can base a failure- or a failure/divergence-type semantics on image semiwords or action structures of processes if we use termination sets and the trunk-relation. I foresee no difficulties for the first case, i.e. a failure-type semantics based on image semiwords with termination set or on action structures of processes with termination set will probably induce a congruence with respect to action refinement. For the failure/divergence-case the situation is more difficult. If we define a *poFD*-semantics based on image semiwords or action structures of processes and use termination sets and the trunk-relation, then the two nets in Figure 5.34 are *poFD*-equivalent. It requires some considerations to check this. Just as an example consider the

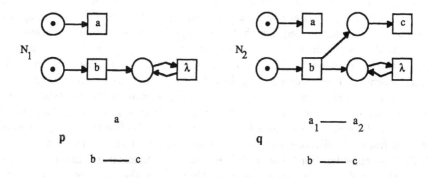

Figure 5.34

labelled partial order p shown in Figure 5.34. It is the action structure of a process only for N_2. But it has a trunk consisting of a non-terminated a concurrent to a terminated b; this trunk is diverging in both nets, thus $p \in poD(N_i)$ and $(p, X) \in poF(N_i)$ for all $X \in \Sigma$ and $i = 1, 2$. Now we refine a to $a_1 a_2$ and get $(q, \emptyset) \in poF(ref(N_2))$, where q is also depicted in Figure 5.34. Unfortunately, $(q, \emptyset) \notin poF(ref(N_1))$. On the one hand, q is not an image semiword or the action structure of a process of $ref(N_1)$. On the other hand, a diverging trunk of q would have to contain a terminated b, thus it must also contain a_2, which is not greater than b in q; hence a_1 would be terminated in this trunk, and therefore c, which cannot belong to this trunk, would have to be greater than a_1 in q. This is not the case.

Again it is important whether $iw\mathcal{F}$- and $iw\mathcal{FD}$-equivalence are congruences with respect to parallel composition; it is especially important, since $iw\mathcal{F}$- and $iw\mathcal{FD}$-semantics have been introduced to work with refinement and $\mathcal{F}$- and $\mathcal{FD}$-semantics, and these in turn were justified by referring to parallel composition. The following result can be shown directly, similarly to some proofs above, or it can be taken from [Vog91b]. (The necessary generalization from safe to self-concurrency-free nets is straightforward as far as the proof of this result is concerned.)

Theorem 5.5.7 $iw\mathcal{F}$- and $iw\mathcal{FD}$-equivalence are congruences with respect to parallel composition.

$iw\mathcal{F}$-equivalence, $iw\mathcal{FD}$-equivalence resp., are fully abstract with respect to deadlock-similarity, deadlock/divergence-similarity resp., parallel composition and action refinement. This also holds if all refinement nets are required to be finite or finite and safe, or if only finite nets are considered.

After our discussion of termination sets the question may arise why we had no need for these termination sets in our treatment of interval semiwords in this book. The answer is that we have only considered a linear-time semantics based on interval semiwords, i.e. we had no interest in the marking reached after an (image) interval semiword. Correspondingly, $IW(N)$ is already determined by the terminated image interval words for the following reason: an action interval sequence $v(a^-, e)$ is an image interval word if and only if v is.

When comparing the refinement of image interval words (Definition 5.4.11) and of image interval semiwords (Definition 5.3.11, Theorem 5.3.17), we observe that for the latter it is simply required that non-maximal events are refined by complete image interval semiwords, while the former requires that the image interval word refining an event e must be complete if e is terminated, and incomplete non-empty if e is not terminated. The reason for this difference is again that only in the word case we were concerned with the markings reached. The difference vanishes if we observe: instead of refining an image interval word $v(a^-, e)$ – here e is terminated and maximal in the corresponding image interval semiword – we can just as well refine v – here e is non-terminated and still maximal; and if we want to refine e by λ, we can instead refine the image interval word v' obtained from v by deleting (a^+, e), such that e does not appear in v' at all.

With these considerations, one can show that the refinement techniques for the semiword and the word case correspond to each other. Thus, the still missing proof of Theorem 5.3.17 follows from Theorem 5.4.14.

The respective advantages of interval semiwords and interval words

Having two representations for interval-(semi)word equivalence, we are faced with the question which of them is to be preferred. The main argument in favour of interval semiwords is that they express concurrency explicitly. Thus, it is immediately clear that the *ISW*-semantics is a partial order semantics distinguishing e.g. the nets of Figure 5.4. At least in my opinion, interval semiwords and their refinement are easier to grasp since they apply to our visual thinking, and in many cases they are conceptionally easier to work with. Consider, for example, the nets of Figure 5.35 that throw some light on the interplay of non-sequential behaviour and refusal sets. These nets have the same action structures of processes and the same $\mathcal{F}$-semantics, but only the second net can refuse c after performing a and b concurrently. Thus, considerations in terms of partial order semantics show easily that these nets are not $iw\mathcal{F}$-equivalent. Furthermore, image

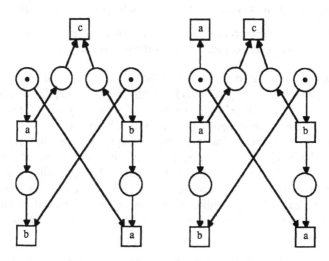

Figure 5.35

interval semiwords have (usually) several interval representations, i.e. they abstract from some irrelevant orderings found in image interval words.

On the other hand, interval words are based on the intuitive idea of splitting a transition into a beginning and an end. Their main advantage is that they are technically better to work with. In [Vog91b], I have tried as far as possible to argue directly with interval semiwords instead of using an interval representation of some sort; but I found myself unable to do this for those results that correspond to Theorems 5.4.17 and 5.4.18 of this book. Another example for the technical advantage of interval words is the result that will conclude this chapter; we will show that IW-equivalence is decidable for finite, bounded nets by considering $IW(N)$ as a regular set of sequences; similarly, we show the decidability of $iw\mathcal{F}$- and $iw\mathcal{F}D$-equivalence.

For this purpose, we can obviously not work with image interval words directly: if a net has arbitrarily long image interval words, we would need an infinite set of events, i.e. an infinite alphabet. To solve this problem we 'reuse' events: after some (a^-, e) has occurred, we can pair the next a^+ with the same e again; we only have to remember that (a^-, e) does not belong to just any (a^+, e) preceding it, but to the last one. Also, we can pair a^+ and b^+ with the same event e, since (a^-, e) belongs to some (a^+, e) and certainly not to some (b^+, e). We will use natural numbers as events.

Definition 5.5.8 An *indexed action interval sequence* w is an element of $(\Sigma^\pm \times \mathbb{N})'$ such that for all $a \in \Sigma$, $n \in \mathbb{N}$

i) (a^+, n) and (a^-, n) appear alternatingly in w beginning with (a^+, n)

ii) if $w = w_1(a^+, n)w_2$ then n is the minimal $i \in \mathbb{N}$ such that w_1 contains (a^+, i) as often as (a^-, i).

Such a sequence w *represents* an action interval sequence v over E if there is a function *index* : $E \to \mathbb{N}$ such that w is obtained from v by replacing $(act(e)^+, e)$ by $(act(e)^+, index(e))$ and $(act(e)^-, e)$ by $(act(e)^-, index(e))$ for all $e \in E$.

Lemma 5.5.9 *The representation-relation defined in 5.5.8 constitutes a bijection between indexed action interval sequences and action interval sequences.*

Proof: Given an action interval sequence v over E and some $e \in E$, the function value $index(e)$ is determined according to 5.5.8 ii) by the values $index(e')$ for those e' such that $act(e) = act(e')$ and $(act(e)^+, e')$ appears before $(act(e)^+, e)$ in v. Thus *index* and the representing sequence w can uniquely be determined by induction.

On the other hand, given an indexed interval sequence w the set E for a represented sequence v must be in bijective correspondence with the occurrences (a^+, i), $a \in \Sigma$, $i \in \mathbb{N}$, in w. An occurrence (a^-, i) must correspond to some (a^+, i) preceding it in order to make v an action interval sequence. Since (a^+, i) and (a^-, i) appear alternatingly, it must correspond to the last (a^+, i) preceding it. If this (a^+, i) is replaced by (a^+, e) to obtain v, then we must and may replace the (a^-, i) under consideration by (a^-, e). Thus, v is uniquely determined.   □

As a consequence of this lemma, we can compare the indexed action interval sequences of two nets instead of comparing their image interval words.

Theorem 5.5.10 *For finite, bounded nets IW-, iw$\mathcal{F}$- and iw$\mathcal{F}D$-equivalence are decidable.*

Proof: Let N be a finite m-bounded net. (The bound m can be determined from the coverability graph, see e.g. [Pet81,Rei85].) We will build a finite automaton A that recognizes the indexed action interval sequences representing the image interval words of N.

The states of A are all triples $(M, C, index)$ such that (M, C) is a reachable ST-marking of N and *index* is a function from $\{t \in C \mid lab(t) \neq \lambda\}$ to $\{1, \ldots, |T|\}$ satisfying $t, t' \in C \wedge lab(t) = lab(t') \neq \lambda \Rightarrow index(t) \neq index(t')$. (Usually, the reachable states are a small subset of the whole set of states we have just defined.) The initial state is $(M_N, \emptyset, \emptyset)$. All states are final.

The arcs of A are defined as follows:

- if $lab(t) = \lambda$ and $(M,C)[t^+\rangle(M',C')$ or $(M,C)[t^-\rangle(M',C')$, then there is a λ-labelled arc from each $(M,C,index)$ to $(M',C',index)$;

- if $lab(t) \in \Sigma$ and $(M,C)[t^+\rangle(M',C\cup\{t\})$, then there is a $(lab(t)^+, index'(t))$-labelled arc from each $(M,C,index)$ to $(M',C\cup\{t\},index')$ where $index'$ is the appropriate extension of $index$ defined by $index'(t) = \min\{i \in \mathbb{N} \mid \forall t' \in C : lab(t) = lab(t') \Rightarrow index(t') \neq i\}$;

- if $lab(t) \in \Sigma$ and $(M,C)[t^-\rangle(M,C-\{t\})$, then there is a $(lab(t)^-, index(t))$-labelled arc from each $(M,C,index)$ to $(M',C-\{t\},index')$ where $index'$ is the appropriate restriction of $index$.

By two induction proofs – one on the length of an accepting path, the other on the length of an interval word –, we see that A recognizes the desired language. Thus, the decidability of IW-equivalence follows from the decidability of the equality of regular languages.

Similarly to the proof of Theorem 3.2.8, the automaton A can be modified for $iw\mathcal{F}$-equivalence. The recognized sequences begin with an indexed action interval sequence representing an image interval word and close with a corresponding relevant refusal set; another modification also takes divergence into account. With these modifications, the other two decidability results follow. $\square$

Chapter 6

Action Refinement and Bisimulation

We continue the study of action refinement, which we have started in Chapter 5, and look for congruences of bisimulation type. Various combinations of partial order semantics and bisimulation do not give such a congruence. We show that the ST-idea can be used to lift these combinations in a uniform way to congruences, and in three cases out of four we supply further justification for the ST-idea by proving full abstractness.

A particularly interesting combination of partial order semantics and bisimulation is history-preserving bisimulation. Interleaving bisimulation for finite safe nets can be decided by exhaustive search, since there are only finitely many states to consider; this is not the case for history-preserving bisimulation, where states correspond to processes. We give a characterization of history-preserving bisimulation using so-called *ordered markings* and get decidability as a corollary.

6.1 Partial orders and bisimulation

In order to define an appropriate equivalence for systems as modelled by nets or e.g. CCS-terms, many researchers prefer bisimulation since bisimilar systems show the same behaviour in a very strong sense. In addition, it should be possible to change the level of abstraction in a system design, hence it is desirable to have a congruence with respect to action refinement as we have considered it in the last chapter. Here, bisimulation turns out to be not discriminating enough; it gives an interleaving semantics and is not a congruence for action refinement as already mentioned in Chapter 5. (The nets of Figure 5.4 are bisimilar, but after refining a to a_1a_2 they do not even have the same language).

It has been argued for quite a while that partial orders are useful when dealing with action refinement, and we have seen this in the previous chapter for several linear-time and failure-type semantics. Thus, if we want the discriminating power of bisimulation and a congruence for action refinement, it is natural to look for a combination of bisimulation and partial order semantics. As described in [BDKP91] and [GG89b], a very natural combination to look at was pomset bisimulation as it can be found in [BC87].

In interleaving bisimulation it is required that, whenever two bisimilar systems are in related states and one system evolves to another state by performing a sequence of actions, then the other must be able to evolve to a state that is related again by performing the same sequence of actions. In pomset bisimulation we require that the systems perform the same partial order of actions instead of a sequence; in terms of nets, they have to perform processes with the same action structure. Let us remark that pomset stands for partially ordered multiset, which is nothing else but a labelled partial order; but usually the word 'pomset' is associated with modelling causality, i.e. with processes in the case of nets.

In Section 5.5, we have seen that a failure-type semantics that simply combines e.g. action structures of processes with refusal sets would not give a congruence for action refinement. Thus, it is not so much of a surprise that pomset bisimulation is not such a congruence either. The nets in Figure 5.32 are pomset bisimilar, but when refining a to $a_1 a_2$ only $ref(N_2)$ can perform a_1 such that only a_2 is possible next.

What goes wrong in this example can also be expressed as: only N_2 can perform a such that the following b necessarily depends causally on a. This observation leads to what is called history-preserving bisimulation, which models the interplay of partial orders and branching in greater detail. It has been introduced in [RT88] under the name behaviour structure bisimulation and can also be found in [DDNM89]. When history-preserving bisimilar systems are in related states, they must not only be able to perform the same partial orders of actions; additionally, in both systems the new actions must be causally related in the same way to those actions that led to the states under consideration.

In [GG89b], it is shown that history-preserving bisimulation is a congruence with respect to action refinement for event structures without silent moves, and in [BDKP91] such a congruence result is shown for Petri nets (using the name fully concurrent bisimulation). The latter result considers general Petri nets, but involves some restrictions concerning self-concurrent firings and internal transitions. The problem with self-concurrency we have already discussed in the previous chapter; also without the restriction concerning internal transitions the result of [BDKP91] fails, and the result of [GG89b] does not hold for event structures with internal events either; see also [GW89a].

In the first part of this chapter, we will use event structures (more precisely: labelled prime event structures with binary conflicts, [NPW81]) with internal events as system models. Using these models instead of general Petri nets, we limit the technical complications. When dealing with event structures, it is most natural to refine actions by substructures that only describe a finite behaviour without choices; despite these limitations, it should be possible to generalize our results to nets. In Section 6.2, we introduce event structures and action refinement for event structures.

In Section 6.3, we will show that there is a uniform way of lifting interleaving bisimulation, pomset bisimulation and history-preserving bisimulation to bisimulations that are congruences with respect to action refinement – even if internal events are considered. This also works for partial-word bisimulation, which we introduce here as a bisimulation that considers the images of partial words just as pomset bisimulation considers action structures of processes.

Our approach is based on a proper treatment of those actions that have started, but have not finished yet. Thus, a system state contains information on the current actions, just as an ST-marking, and the important point is that a bisimulation relating two system states has to relate these current actions explicitly. Such a bisimulation has been defined for Petri nets under the name ST-bisimulation in [GV87]; in [Gla90c], it is shown that ST-bisimulation is a congruence with respect to action refinement for event structures without internal events. ST-bisimulation is the above mentioned refined version of interleaving bisimulation, hence we will speak in the following of interleaving ST-bisimulation and similarly of partial-word ST-bisimulation etc.

Our achievements in Section 6.3 are the following:

- We generalize the main result of [Gla90c], stating that interleaving ST-bisimulation gives a congruence for refinement, to event structures with silent moves.

- We show that history-preserving ST-bisimulation, which is a rather small variation of history-preserving bisimulation, gives a congruence for event structures with silent moves. We also prove that history-preserving ST-bisimilarity coincides with maximility-preserving bisimilarity which has independently been defined in [Dev90b].

- We answer a question raised in [GG89b]: history-preserving bisimulation is not the coarsest congruence (for event structures without silent moves) respecting pomset bisimulation, since pomset ST-bisimulation is a coarser congruence.

- We consider partial-word bisimulation; also this bisimulation can be lifted to a congruence.

- Finally, we demonstrate that the ST-versions are not arbitrary congruences, at least in three of our four cases. We can show that the ST-idea is, in the following sense, just right for dealing with action refinement: for image-finite event structures, interleaving / pomset / history-preserving ST-bisimulation is the coarsest congruence that respects interleaving / pomset / history-preserving bisimulation.

These results are no recommendation for history-preserving (ST-)bisimulation. If we want to have a congruence of bisimulation type, we can use interleaving ST-bisimulation. If we also want to take care of processes or partial words, we can use pomset-ST- or partial-word-ST-bisimulation. Still, I agree with the statement: 'If it is required to model the interplay of causality and branching in full detail, history-preserving bisimulation seems to be the coarsest suitable equivalence' [GG89b] – except that we should rather take history-preserving ST-bisimulation. But when dealing with history-preserving (ST-)bisimulation of nets, we meet the following problem.

As indicated above, when a system evolves from some state by performing some actions, then history-preserving bisimulation is concerned with how these actions are causally related to those actions that led to the state under consideration. To make this possible, a state of a net is not described by a marking in [BDKP91], but by the process that has been performed. This is somewhat counterintuitive, since it is more natural to describe a system by referring to the passive system elements, namely the places; and most of all, it has the unfortunate effect that we have to consider infinitely many states whenever the net allows some infinite behaviour. To some degree, this gives up the considerable advantage of Petri nets over event structures, the advantage that they can finitely describe an infinite behaviour. In particular, finite safe nets can describe an infinite behaviour, although they are themselves finite and have only finitely many markings.

Interleaving bisimulation for finite safe nets can be decided by exhaustive search; we just have to check all possible relations for the two finite sets of reachable markings. (Of course, more clever strategies exist as well, compare [PT87]). But how history-preserving bisimulation can be decided for finite safe nets is in not clear at all.

In Section 6.4, we will solve this problem by characterizing history-preserving bisimilarity by what we call OM-bisimilarity. Here, a system state is an *ordered marking*, a marking together with a pre-order on its tokens; this pre-order reflects the causal ordering of the transitions that produced the tokens.

To work out this idea clearly without obscuring it by technical details, we restrict our considerations to safe nets without internal transitions, where markings correspond to sets of places and history-preserving bisimulation coincides with its ST-version. A generalization to arbitrary nets seems possible; the transition system of ordered markings is in fact a refined version of a distributed transition system in the sense of [DM87], and distributed transition systems for general nets – as they are already defined in [DM87] – can probably be refined analogously.

A finite marking can be pre-ordered in only finitely many ways. Therefore, for a finite safe net, the transition system of ordered markings is finite; thus we can decide OM-bisimulation – or equivalently history-preserving bisimulation – for finite safe nets without internal transitions by exhaustive search.

The results of this chapter have been published in two extended abstracts as [Vog91c] and [Vog91d].

6.2 Event structures and action refinement

In this section, we introduce event structures (more precisely, prime event structures with binary conflict) and action refinement and various types of bisimulation for event structures. An *event structure* $\mathcal{E} = (E, <, \#, l)$ consists of a set E of *events*, a partial order $<$ on E, an irreflexive, symmetric relation $\#$ on E, and the *labelling* function $l : E \to \Sigma \cup \{\lambda\}$. An event e is an occurrence of the action $l(e)$, where e represents an internal action if $l(e) = \lambda$. In the latter case we also call e an *internal event*. The partial order $<$ is the *causality relation*, i.e. if $e' < e$ then e' is a necessary precondition for e to happen. We require that an event structure satisfies the *principle of finite cause*, i.e. for all $e \in E$ the set $\{e' \in E \mid e' < e\}$ is finite. The reason for this is that we assume that only finitely many events can occur in a finite amount of time, and therefore only events with finitely many causes can occur.

Finally, $\#$ is the *conflict relation*, and $e\#e'$ means that not both, e and e' can occur in the same system run. We require for an event structure that it satisfies the principle of *conflict heredity*, i.e. $e\#e' < e''$ implies $e\#e''$. The reason is: if a system run containing e cannot contain e', and e' is a necessary precondition for e'', then certainly a system run containing e cannot contain e''.

As already mentioned, event structures can be seen as special acyclic Petri nets. In such a net the events of the event structure are the transitions, an initially enabled event has a place with one token in its preset, an immediate conflict between events e and e' is modelled by a marked place in $^\bullet e \cap \,^\bullet e'$, while for an immediate predecessor e of e' we have an unmarked place in $e^\bullet \cap \,^\bullet e'$. Since such a net is acyclic, we do not have to distinguish between a transition and its possibly numerous occurrences, we can simply speak of events. Furthermore, a system run simply is a substructure of the event structure, whereas e.g. a process of a Petri net must be obtained by some 'unwinding' construction. Therefore, it is much easier to work with event structures. On the other hand, they have the disadvantage that they must be infinite in order to describe an infinite behaviour.

When depicting event structures (see Figure 6.1), we ignore the events and simply write down their labels, i.e. we are only interested in the isomorphism class of the event structure. We depict the immediate predecessor relation by lines, where $e < e'$ corresponds

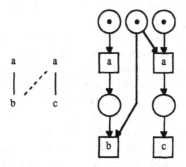

Figure 6.1   an event structure and a corresponding net

to a line going downward from $l(e)$ to $l(e')$. Furthermore, we only depict conflicts that are not implied by conflict heredity, and we depict them by dotted lines. Thus, the system shown in Figure 6.1 can perform a followed by b, and in this case neither the second a nor c can occur. Or it can perform a followed by c and another a independently of these. The corresponding Petri net is also shown in Figure 6.1.

If a system evolves to some state, then this state is defined by the set of events that have occurred. Such a set is called a configuration and it describes at the same time a system state and system run. A *configuration* C is a finite subset of E that is *conflict-free*, i.e. $e, e' \in C$ implies $\neg(e \# e')$, and *downward-closed*, i.e. $e < e' \in C$ implies $e \in C$. The set of configurations of $\mathcal{E}$ is denoted by $\mathcal{C}(\mathcal{E})$. The *initial state* of an event structure is the empty set. For $C, C' \in \mathcal{C}(\mathcal{E})$, the system can evolve from state C to C', denoted $C \to_\mathcal{E} C'$, if $C \subseteq C'$; if $\mathcal{E}$ is understood, we also write $C \to C'$.

If $C \in \mathcal{C}(\mathcal{E})$, then we denote by C or by some subset $C' \subseteq C$ not only the set itself, but also the labelled partial order it induces by restricting $<$ and l to C or C'. It will (hopefully) be clear from the context what is meant. Especially, we will often meet the situation that we speak of a *label-preserving* bijection f defined on some C'. In this case, we do not just refer to the set C', but require that for all $e \in C'$ we have $l(f(e)) = l(e)$. Similarly, if we say that the bijection f is a *homomorphism*, then f is label-preserving and respects the partial order, i.e. $e < e'$ implies $f(e) < f(e')$ for all $e, e' \in C'$; if we additionally have that $f(e) < f(e')$ implies $e < e'$ for all $e, e' \in C'$, then we call f an *isomorphism*. For a mapping f defined on some C' we denote by $f|_D$ the restriction of f to $C' \cap D$.

If $C' \subseteq E$, we will often be interested in the visible events of C', i.e. the set $vis_{\mathcal{E}}(C') = \{e \in C' \mid l(e) \in \Sigma\}$. For example, if $C \rightarrow_{\mathcal{E}} C'$, then the system evolves from state C to C' by performing the actions represented by the events in $vis_{\mathcal{E}}(C' - C)$. If p is the labelled partial order induced by $vis_{\mathcal{E}}(C' - C)$, then we write $C \xrightarrow{p}_{\mathcal{E}} C'$ or $C \xrightarrow{p} C'$. This labelled partial order corresponds for nets to an action structure of a process leading from one marking to another.

Let C, D be subsets of configurations (possibly of different event structures, and usually consisting of visible events only). Then C is called a *step* if we do not have $e < e'$ for any $e, e' \in C$, i.e. if C is a set of pairwise independent events. We already know when C – being a labelled partial order – is less sequential than D. Equivalently, we can say that C is *less sequential* than D if there exists a label-preserving bijection $f : C \rightarrow D$ such that f is a homomorphism.

Now we will describe what sort of action refinement we consider: we will refine actions to finite, conflict-free event structures. Both these restrictions are necessary to allow a natural description of the refined event structure. (For example, in order to allow conflicts in the refining structures, [GG90] define action refinement for flow event structures, which are a generalization of event structures.) We will also require that these event structures are non-empty. Replacing an action by an empty event structure cannot be explained by a change of abstraction level, and would also pose mathematical problems, see the concluding remarks of [Gla90c]. Note that such a forgetful refinement is different from *hiding*, where an observable action is changed into an internal action. Also, as in Chapter 5, we will not consider any refinement of internal actions.

An *action refinement ref* is a mapping that assigns to each $a \in \Sigma$ a non-empty, finite, conflict-free event structure $(E_a, <_a, \emptyset, l_a)$ (i.e. E_a is non-empty and finite). Additionally, we will always have $ref(\lambda) = (\{\lambda\}, \emptyset, \emptyset, \{(\lambda, \lambda)\})$; this means that internal actions cannot be refined.

When applying a refinement to an event structure $\mathcal{E}$, we will replace every $e \in E_{\mathcal{E}}$ by a disjoint copy of $ref(l(e))$, and the elements of this copy will inherit the precedences and the conflicts from e. This is analogous to the refinement of partial words etc. in Section 5.3.

Let *ref* be an action refinement, $\mathcal{E}$ an event structure. Then the *refined event structure* $ref(\mathcal{E})$ is defined by

- $E_{ref(\mathcal{E})} = \{(e, e') \mid e \in E,\ e' \in E_{ref(l_{\mathcal{E}}(e))}\}$

- $(e_1, e_1') <_{ref(\mathcal{E})} (e_2, e_2')$ iff $e_1 <_{\mathcal{E}} e_2$ or $(e_1 = e_2$ and $e_1' <_{ref(l_{\mathcal{E}}(e_1))} e_2')$

- $(e_1, e_1') \#_{ref(\mathcal{E})} (e_2, e_2')$ iff $e_1 \#_{\mathcal{E}} e_2$

- $l_{ref(\mathcal{E})}(e, e') = l_{ref(l_{\mathcal{E}}(e))}(e')$.

For a subset $\tilde{C}$ of $E_{ref(\mathcal{E})}$ we will write $pr_1(\tilde{C})$ for $\{e \in E_{\mathcal{E}} \mid$ there exists some $(e, e') \in \tilde{C}\}$.

This definition certainly describes the natural way to refine event structures, and it is easily checked that the refined event structure $r(\mathcal{E})$ is, indeed, an event structure again. This would fail if the event structures $ref(a)$ contained conflicts or infinitely many elements. Consider an event e preceding an event e'. If we replace e by conflicting events, then e' would be self-conflicting in the resulting structure by conflict-heredity; this is not what we would expect, and it violates the definition of an event structure. If we replace e by infinitely many events, then the resulting structure violates the principle of finite cause.

As already remarked in the previous chapter, we regard each refinement as a unary operator – here, on event structures. With this view, a congruence for action refinement is an equivalence of event structures such that, for every refinement, the refinement applied to equivalent event structures yields equivalent refined event structures. We do not consider the case of applying two different refinements ref and ref', where $ref(a)$ and $ref'(a)$ are equivalent for all $a \in \Sigma$.

Observe that relabelling and hiding are special cases of refinement. If we want to change the observable action a into an internal action, then we define a refinement ref such that $ref(a) = ref(\lambda)$ and, for all $b \in \Sigma - \{a\}$, $ref(b)$ is the one-element event structure whose only event is labelled b. If we want to relabel a into c, then we take the same refinement except that $ref(a)$ is the one-element event structure whose only event is labelled c.

Bisimulation

Now we give the definition of interleaving-, step-, pomset-, and partial-word-bisimulation for event structures. This definition is analogous to that of interleaving bisimulation for nets and its alternative formulation we have given in Proposition 2.3.3. Thus, in the interleaving case we are concerned with state changes where at most one visible action is performed, see Proposition 2.3.3. Consequently, a sequence of state changes corresponds to a sequence of actions, and concurrency is equated with arbitrary interleaving. The case of step bisimulation is analogous. Pomset- and partial-word-bisimulation are defined more in the spirit of the original definition of bisimulation; the state changes involve (parts of) system runs, described by partial orders of actions.

For pomset bisimulation, the notation $C \xrightarrow{p}_{\mathcal{E}} C'$ suitably describes a state change from C to C' with pomset p. A straightforward definition of partial-word bisimulation would change the definition of $C \xrightarrow{p}_{\mathcal{E}} C'$; for the partial-word case, $C \xrightarrow{p}_{\mathcal{E}} C'$ should mean that $vis_{\mathcal{E}}(C' - C)$ is less sequential than p. We will not use this straightforward approach; instead, we will say that a state change $C \xrightarrow{p}_{\mathcal{E}} C'$ in the system $\mathcal{E}$ can be matched by a state change $D \xrightarrow{q}_{\mathcal{F}} D'$, where q is less sequential than p. This form of definition is more suitable for a uniform treatment.

For the formal definition, observe the following: this definition includes four similar definitions; to stress the similarities, I have formulated it as one definition with a parameter α. Reading this definition requires some care in order to distinguish the four cases; but I hope the reader will have no difficulties in understanding the definition, especially if he or she spends as much time on it as (s)he would if it were approximately four times as long. Also observe, that this definition does not define one notion called α-bisimulation but four notions called interleaving bisimulation, step bisimulation etc. Similar remarks apply to the parameterized definitions and results that will follow.

Definition 6.2.1 The event structures $\mathcal{E}$, $\mathcal{F}$ are α-*bisimilar* with $\alpha \in \{$interleaving, step, pomset, partial-word$\}$ if there exists an α-*bisimulation*, i.e. a relation $\mathcal{B} \subseteq \mathcal{C}(\mathcal{E}) \times \mathcal{C}(\mathcal{F})$, such that:

i) $(\emptyset, \emptyset) \in \mathcal{B}$

ii) If $(C, D) \in \mathcal{B}$ and $C \xrightarrow{p}_\mathcal{E} C'$, where

  – p consists of at most one element if $\alpha = $ interleaving

  – p is a step if $\alpha = $ step,

  then there exist $D' \in \mathcal{C}(\mathcal{F})$ and a labelled partial order q such that

  – $D \xrightarrow{q}_\mathcal{F} D'$ and $(C', D') \in \mathcal{B}$

  – q is less sequential than p if $\alpha = $ partial-word and p, q are isomorphic otherwise.

iii) Vice versa, if $(C, D) \in \mathcal{B}$ and $D \xrightarrow{q}_\mathcal{F} D'$, where

  – q consists of at most one element if $\alpha = $ interleaving

  – q is a step if $\alpha = $ step,

  then there exists $C' \in \mathcal{C}(\mathcal{E})$ and a labelled partial order p such that

  – $C \xrightarrow{p}_\mathcal{E} C'$ and $(C', D') \in \mathcal{B}$

  – p is less sequential than q if $\alpha = $ partial-word and p, q are isomorphic otherwise.

Below, we will use another format for the definition of bisimulation. Roughly, in this format it is required that related states have equivalent pasts, in the sense that the same actions have been performed to reach them. In order to explain why this requirement does not change the equivalences on event structures, let us consider the partial-word case, which is probably the most unusual. Consider two partial-word bisimilar event structures $\mathcal{E}$ and $\mathcal{F}$ and two sequences $\emptyset = C_0 \subseteq C_1 ... \subseteq C_n$ in $\mathcal{C}(\mathcal{E})$ and $\emptyset = D_0 \subseteq D_1 ... \subseteq D_n$ in $\mathcal{C}(\mathcal{F})$. Suppose these sequences are matching computations: for each i, C_i and D_i are related by the bisimulation; from these states, one of the systems evolves to the next state and the other system simulates this; thus, the next states are related again and, by definition of a partial-word bisimulation, $vis_\mathcal{F}(D_i - D_{i+1})$ is less sequential than $vis_\mathcal{E}(C_i - C_{i+1})$ or the other way round. Hence, we can find a label-preserving bijection between these sets. Putting these bijections together, we find a label-preserving bijection from $vis_\mathcal{E}(C_n)$ onto $vis_\mathcal{F}(D_n)$, i.e. we find a match between the visible parts of C_n and D_n. Such a label-preserving bijection can always be found for related states, provided we are really interested in their relationship.

To explain what an uninteresting relationship is, consider two event structures that can both perform either a or b. A bisimulation might relate the state where the first structure has performed a with the state where the second structure has performed b, since both states have the same future – no future, essentially. But this relationship is of no importance, and eliminating this pair of states from the bisimulation still gives us a bisimulation.

In order to describe the equivalence of the pasts, we will add a third component to the elements of a bisimulation, and this component will be a label-preserving bijection between the visible parts of the other two components, which are still states of the event structures under consideration.

We choose this format because it fits better for our purposes; when we define the ST-variants of the bisimulations, we will have to match certain actions; for this we will need a corresponding function as a third component in an element of an ST-bisimulation. Furthermore, this format allows us a uniform treatment of history-preserving bisimulation.

In both, the pomset and the history-preserving case, the bisimilar systems must be able to perform the same partial orders; but in the latter case, the new actions must be causally related in both systems in the same way to the actions that led to the old state. As a consequence, the matching for the visible parts of related states, which we have discussed above, is an isomorphism. Consider the following example. Both event structures shown

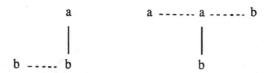

Figure 6.2   pomset- but not history-preserving-bisimilar, see [BDKP91]

in Figure 6.2 can perform the same pomsets – initially, and also after performing a or b; thus they are pomset bisimilar. But if the right-hand-side event structure performs the a in the middle, this can only be matched by the single a of the left-hand-side event structure; now both can perform b next. But in the right-hand-side event structure, this b will necessarily be causally dependent on a, whereas the left-hand-side event structure can perform a b that is independent of a. The resulting computations are not isomorphic; in other words, the relation of the new action b to the history of the computation is different in the two systems.

Definition 6.2.2 Let $\mathcal{E}$, $\mathcal{F}$ be event structures. Then $\mathcal{E}$ and $\mathcal{F}$ are *α-bisimilar*, denoted $\mathcal{E} \approx_\alpha \mathcal{F}$, with $\alpha \in \{i, s, p, pw, h\}$ if there exists an *α-bisimulation*, i.e. a relation $\mathcal{B} \subseteq \mathcal{C}(\mathcal{E}) \times \mathcal{C}(\mathcal{F}) \times \mathcal{P}(E_\mathcal{E} \times E_\mathcal{F})$, such that

i) $(\emptyset, \emptyset, \emptyset) \in \mathcal{B}$

ii) If $(C, D, f) \in \mathcal{B}$, then f is a label-preserving bijection from $vis_\mathcal{E}(C)$ onto $vis_\mathcal{F}(D)$ and if $\alpha = h$ then f is an isomorphism from $vis_\mathcal{E}(C)$ onto $vis_\mathcal{F}(D)$.

iii) If $(C, D, f) \in \mathcal{B}$ and $C \to_\mathcal{E} C'$, such that

  – $vis_\mathcal{E}(C' - C)$ has at most one element if $\alpha = i$

  – $vis_\mathcal{E}(C' - C)$ is a step if $\alpha = s$,

  then there are D', f' such that

- $D \to_{\mathcal{F}} D'$, $(C', D', f') \in \mathcal{B}$ and $f'|_C = f$
- f'^{-1} restricts to a homomorphism from $vis_{\mathcal{F}}(D' - D)$ onto $vis_{\mathcal{E}}(C' - C)$ if $\alpha = pw$ and to an isomorphism if $\alpha \in \{s, p\}$.

iv) Vice versa, if $(C, D, f) \in \mathcal{B}$ and $D \to_{\mathcal{F}} D'$, such that
- $vis_{\mathcal{F}}(D' - D)$ has at most one element if $\alpha = i$
- $vis_{\mathcal{F}}(D' - D)$ is a step if $\alpha = s$,

then there are C', f' such that
- $C \to_{\mathcal{E}} C'$, $(C', D', f') \in \mathcal{B}$ and $f'|_C = f$
- f' restricts to a homomorphism from $vis_{\mathcal{E}}(C' - C)$ onto $vis_{\mathcal{F}}(D' - D)$ if $\alpha = pw$ and to an isomorphism if $\alpha \in \{s, p\}$.

In the case $\alpha = h$ we speak of a *history-preserving bisimulation* and of *history-preserving bisimilar*.

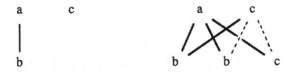

Figure 6.3   step- but not partial-word-bisimilar

As one example, let us consider pw-bisimulation for the event structures shown in Figure 6.3. From the initial state $\emptyset$, the left-hand-side structure can perform a followed by b and, independently of both, c; i.e. C' in Part iii) of the definition is the whole event set. Now we must find D' for the right-hand-side structure and a homomorphism from D' to C'. Thus, D' must consist of some a, b and c, and the only possible ordering would be between a and b. Such a D does not exist, and hence the structures are not pw-bisimilar.

For Part iii) of the above definition (and similarly for iv)) and the i-case, observe the following. By ii), f and f' are label-preserving bijections. Since f' extends f, the only element of $vis_{\mathcal{E}}(C' - C)$ must represent the same action as the only element of $vis_{\mathcal{F}}(D' - D)$. Thus, f'^{-1} restricts to an isomorphism also in the i-case. Since we have not required this explicitly, we can in fact drop the conditions that $vis_{\mathcal{E}}(C' - C)$ and $vis_{\mathcal{F}}(D' - D)$ have at most one element for $\alpha = i$ without changing the definition of an i-bisimulation. If $vis_{\mathcal{E}}(C' - C)$ contains several elements, then $\mathcal{E}$ can perform these actions in some order; $\mathcal{F}$ can simulate these actions one after the other, each time we extend f, and finally we arrive at some suitable D' and f'.

In the following proposition, we compare the first and the second definition, and we compare the different versions of bisimulation.

Proposition 6.2.3 *Let $\mathcal{E}$, $\mathcal{F}$ be event structures.*

i)   – $\mathcal{E} \approx_i \mathcal{F}$ if and only if $\mathcal{E}$ and $\mathcal{F}$ are interleaving bisimilar

      – $\mathcal{E} \approx_s \mathcal{F}$ if and only if $\mathcal{E}$ and $\mathcal{F}$ are step bisimilar

      – $\mathcal{E} \approx_p \mathcal{F}$ if and only if $\mathcal{E}$ and $\mathcal{F}$ are pomset bisimilar

      – $\mathcal{E} \approx_{pw} \mathcal{F}$ if and only if $\mathcal{E}$ and $\mathcal{F}$ are partial-word bisimilar

ii) $\mathcal{E} \approx_h \mathcal{F} \Rightarrow \mathcal{E} \approx_p \mathcal{F} \Rightarrow \mathcal{E} \approx_{pw} \mathcal{F} \Rightarrow \mathcal{E} \approx_s \mathcal{F} \Rightarrow \mathcal{E} \approx_i \mathcal{F}$ and none of these implications can be reversed in general, even for event stuctures without internal events.

Proof: i) '$\Rightarrow$' If $\mathcal{B}'$ is an α-bisimulation, $\alpha \in \{i, s, p, pw\}$, then we obtain a bisimulation $\mathcal{B}$ according to Definition 6.2.1 of the appropriate type by omitting the third component, i.e $\mathcal{B} = \{(C, D) | \exists f : (C, D, f) \in \mathcal{B}'\}$.

'$\Leftarrow$' Given an α-bisimulation $\mathcal{B}$ according to Definition 6.2.1, we define a corresponding bisimulation $\mathcal{B}'$ according to 6.2.2 as $\mathcal{B}' = \{(C, D, f) | (C, D) \in \mathcal{B}$ and f is a label-preserving bijection from $vis_{\mathcal{E}}(C)$ onto $vis_{\mathcal{F}}(D)\}$. Since p and q in 6.2.1 are just $vis_{\mathcal{E}}(C' - C)$ and $vis_{\mathcal{F}}(D' - D)$, and one of them is less sequential than the other, we can always extend f to a label-preserving bijection f' from $vis_{\mathcal{E}}(C')$ onto $vis_{\mathcal{F}}(D')$.

ii) The implications are not only true for the equivalences on event structures, they also hold for the bisimulations itself, i.e. every h-bisimulation is a p-bisimulation etc. In the following, we will not consider Part iv) since it is analogous to Part iii).

If $\mathcal{B}$ is an h-bisimulation, then in iii) f and f' are isomorphisms and f' extends f, thus f'^{-1} restricts to an isomorphism as required for a p-bisimulation. Hence, $\mathcal{B}$ is a p-bisimulation.

If $\mathcal{B}$ is a p-bisimulation, then in iii) f'^{-1} restricts to an isomorphism, thus also to a homorphism as required for a pw-bisimulation. Hence, $\mathcal{B}$ is a pw-bisimulation.

If $\mathcal{B}$ is a pw-bisimulation and $vis_{\mathcal{E}}(C' - C)$ in iii) is a step, then $vis_{\mathcal{F}}(D' - D)$ cannot be strictly less sequential since $vis_{\mathcal{E}}(C' - C)$ has an empty ordering. Thus, f'^{-1} restricts in fact to an isomorphism as required for an s-bisimulation. Hence, $\mathcal{B}$ is an s-bisimulation.

If $\mathcal{B}$ is an s-bisimulation, and $vis_{\mathcal{E}}(C' - C)$ in iii) has a single element, then $vis_{\mathcal{E}}(C' - C)$ is a special step. Hence, $\mathcal{B}$ is a i-bisimulation.

The second part follows from the counterexamples shown in Figure 6.2 to 6.5. Figure 6.4 shows the classical example that distinguishes interleaving from 'true' concurrency, compare Figure 2.2. For Figure 6.5 the essential point is the following: the least sequential 'complete' system run is the same for both systems, namely a and b independently; but only in the second system b may 'cause' a.   □

Figure 6.4   interleaving- but not step-bisimilar

Figure 6.5   partial-word- but not pomset-bisimilar

6.3   Congruence results for ST-bisimulations

In the previous chapter, we have seen that partial order semantics is useful to define
linear-time congruences for action refinement. We have also seen that a straightforward
combination of partial orders and the failure idea does not give a congruence. Similarly,
none of the bisimulations we have defined in Section 6.2 induce such a congruence, as
demonstrated by the example in Figure 6.6 [GW89a,BDKP91].

Figure 6.6   history-preserving bisimilar, but after refining a to $a_1 a_2$ not even interleaving
bisimilar

   If the right-hand-side system performs the right-hand-side a, then it decides at the
same time against performing b. This can be simulated by the left-hand-side system by
performing a followed by the internal action; in fact, the systems are history-preserving
bisimilar. If we refine a to the sequence $a_1 a_2$, then the right-hand-side system can start
the right-hand-side a by performing a_1; now it will never be able to do b. If the left-hand-
side system performs a_1, then it still has this ability after a_2. Thus, the refined systems
are not even interleaving bisimilar. History-preserving bisimulation induces a congruence
in the absence of silent moves [GG89b,BDKP91], but this is not true for the others, see
[GG89b].

   Interleaving ST-bisimulation equivalence was introduced in [GV87] and it was shown
to be a congruence w.r.t. refinement for event structures without silent moves in [Gla90c].
The ST-idea consists of a basic idea and a subtlety. The basic idea is the following: if
actions are not atomic, then a description of a system state should include the information
which actions are active, i.e. have started but have not finished yet. For Petri nets, this
means that a state description says how the tokens are distributed on the places (German:
Stellen) and which transitions (*Transitionen*) are active, hence the abbreviation ST. We
have met this basic idea already in the previous chapter.

   It might appear that there is the following easy way to distinguish active actions
from those that have already terminated: we simply split each observable action a into
a sequence $a_1 a_2$. In this splitted system, occurrence of a_1 indicates that in the original

system a has started and is active now; occurrence of a_2 indicates that in the original
system a has finished. If we compare systems by applying interleaving bisimilarity to their
splitted systems, then we speak of split bisimilarity. (This equivalence was advocated for
example in [AH89].)

That ST-bisimulation is different from split bisimulation is due to the subtlety mentioned above. The difference occurs only in case of autoconcurrency, where autoconcurrency means that in some state the system can perform a step consisting of two a's for
some observable action a. ST-bisimulation requires: if one system starts an action a and
the other system simulates this with the start of another action a, then these actions are
matched and we have to keep track of this match. Now, if the two systems are in states
where autoconcurrently two a's are active and one system finishes one of these a's, then
in split bisimulation the other system may simulate this by finishing any of the two active
a's. In ST-bisimulation, the other system can simulate this only by finishing the matching
active a. To describe this matching in a bisimulation, we need for each pair of related
states a label-preserving bijection between their active actions. This fits very well with
Definition 6.2.2, where we already have label-preserving bijections between related states.
Observe that the matching of active actions is implicitly concerned with connecting the
start and the end of an action; thus, it corresponds to the pointers we have used for action
interval sequences in Chapter 5.

This explains the intensional difference between split and ST-bisimulation. To demonstrate the extensional difference, i.e. to find an example where the induced equivalences
ST-bisimilarity and split bisimilarity differ, is much harder; such an example, the so-called
owl-example, can be found in [GV91].

In the following, we will show that the ST-idea enables us to lift various kinds of
bisimulation to finer ones that induce congruences w.r.t. refinement.

So far, we have used configurations to describe the state of a system. Now we have
to include which actions are active, just as we have done it in the ST-markings. Quite
naturally (but different from ST-markings), these active actions correspond to visible
events, since internal actions cannot be refined and, thus, can be regarded as atomic; furthermore, the events corresponding to active actions are maximal, since for an unfinished
event nothing can have happened afterwards. Following [Gla90c], we will describe a state
by a configuration and the set of finished events (the past):

Definition 6.3.1 An ST-*configuration* of $\mathcal{E}$ is a pair (C, P) of subsets of E such that
$P \subseteq C$, C is finite and conflict-free and $e' \in C$ with $l(e') = \lambda$ or $e' < e \in C$ implies
$e' \in P$. $\mathcal{S}(\mathcal{E})$ denotes the set of ST-configurations. If $\mathcal{E}$ is finite and conflict-free, we call
the ST-configuration $(E_{\mathcal{E}}, E_{\mathcal{E}})$ *complete*.

For ST-configurations (C, P) and (C', P'), we write $(C, P) \to_{\mathcal{E}} (C', P')$, or simply
$(C, P) \to (C', P')$, if $C \subseteq C'$ and $P \subseteq P'$.

If $(C, P) \in \mathcal{S}(\mathcal{E})$, then the events in $C - P$ represent the active actions; both, C
and P are configurations. Note that for every refinement *ref*, the event structure *ref*(λ)
has exactly two ST-configurations; for one, C and P are empty, for the other, C and P
contain both the only event. In contrast, an event structure consisting of a single visible
event has three ST-configurations.

The following lemma relates the ST-configurations $(\tilde{C}, \tilde{P})$ of a refined event structure to the ST-configurations (C, P) of the unrefined event structure, compare Lemma 4 of [Gla90c]. Basically, in order to construct $(\tilde{C}, \tilde{P})$, we replace every event e of C by a non-empty ST-configuration (C_e, P_e) of $ref(l_\mathcal{E}(e))$. A crucial point is that the events in P are the finished events, thus after refinement they correspond to complete ST-configurations. In particular, (C_e, P_e) is fixed for an internal event e.

Lemma 6.3.2 *Let $\mathcal{E}$ be an event structure, ref an action refinement.*

i) $(\tilde{C}, \tilde{P}) \in \mathcal{S}(ref(\mathcal{E}))$ *if and only if*
$\tilde{C} = \{(e, e') \mid e \in vis_\mathcal{E}(C), e' \in C_e\} \cup \{(e, \lambda) \mid e \in C, l_\mathcal{E}(e) = \lambda\}$ *and*
$\tilde{P} = \{(e, e') \mid e \in vis_\mathcal{E}(C), e' \in P_e\} \cup \{(e, \lambda) \mid e \in C, l_\mathcal{E}(e) = \lambda\}$ *where*

 - $(C, P) \in \mathcal{S}(\mathcal{E})$

 - $(C_e, P_e) \in \mathcal{S}(ref(l_\mathcal{E}(e)))$ *for every* $e \in vis_\mathcal{E}(C)$

 - $C_e \neq \emptyset$ *for every* $e \in vis_\mathcal{E}(C)$

 - *for every* $e \in vis_\mathcal{E}(C)$, $P_e = E_{ref(l_\mathcal{E}(e))}$ *if and only if* $e \in P$.

ii) *Each* $(\tilde{C}, \tilde{P}) \in \mathcal{S}(ref(\mathcal{E}))$ *uniquely determines the ST-configurations (C, P) and (C_e, P_e), $e \in vis_\mathcal{E}(C)$, according to i).*

iii) *Let* $(\tilde{C}, \tilde{P})$, $(\tilde{C}', \tilde{P}') \in \mathcal{S}(ref(\mathcal{E}))$ *determine the ST-configurations (C, P), $(C', P') \in \mathcal{S}(\mathcal{E})$, $(C_e, P_e) \in \mathcal{S}(ref(l_\mathcal{E}(e)))$ for $e \in vis_\mathcal{E}(C)$, and $(C'_e, P'_e) \in \mathcal{S}(ref(l_\mathcal{E}(e)))$ for $e \in vis_\mathcal{E}(C')$ according to i) and ii).*
Then we have $(\tilde{C}, \tilde{P}) \to_{ref(\mathcal{E})} (\tilde{C}', \tilde{P}')$ *if and only if* $(C, P) \to_\mathcal{E} (C', P')$ *and, for all $e \in vis_\mathcal{E}(C)$, $(C_e, P_e) \to_{ref(l(e))} (C'_e, P'_e)$.*

Proof: i) '$\Rightarrow$' Let $(\tilde{C}, \tilde{P}) \in \mathcal{S}(ref(\mathcal{E}))$. First, define $C = pr_1(\tilde{C})$, then $C_e = \{e' \mid (e, e') \in \tilde{C}\}$ and $P_e = \{e' \mid (e, e') \in \tilde{P}\}$ for $e \in vis_\mathcal{E}(C)$; finally put $P = \{e \in vis_\mathcal{E}(C) \mid P_e = E_{ref(l_\mathcal{E}(e))}\} \cup \{e \in C \mid l_\mathcal{E}(e) = \lambda\}$.

It is not hard to see that $C \in \mathcal{C}(\mathcal{E})$ since $\tilde{C} \in \mathcal{C}(ref(\mathcal{E}))$. Since $C = pr_1(\tilde{C})$ we have $C_e \neq \emptyset$ for all $e \in vis_\mathcal{E}(C)$, and one easily checks $(C_e, P_e) \in \mathcal{S}(ref(l_\mathcal{E}(e)))$. From the definition of P, we have for all $e \in vis_\mathcal{E}(C)$ that $P_e = E_{ref(l_\mathcal{E}(e))}$ if and only if $e \in P$. Also the equalities for $\tilde{C}$ and $\tilde{P}$ hold by definition.

Finally, we have to show that $(C, P) \in \mathcal{S}(\mathcal{E})$, and for this it remains to prove that P contains all internal and all non-maximal events of C. On the one hand, P contains all internal events by definition. On the other hand, if d is visible and $d < e \in C$, then for some e' we have $(e, e') \in \tilde{C}$; furthermore, for all $d' \in E_{ref(l_\mathcal{E}(d))}$ we have $(d, d') <_{ref(\mathcal{E})} (e, e')$, hence $(d, d') \in \tilde{P}$; thus, $P_d = E_{ref(l_\mathcal{E}(d))}$ and $d \in P$.

'$\Leftarrow$' Easy, but tedious.

ii) Since $C_e \neq \emptyset$ is required for $e \in vis_\mathcal{E}(C)$, it follows that C must be $pr_1(\tilde{C})$. Now the uniqueness of P and (C_e, P_e) follows easily.

iii) '$\Rightarrow$' Follows immediately from the construction in the proof of part i).

'$\Leftarrow$' immediate from the equalities for $\tilde{C}$ and $\tilde{P}$ in i). $\square$

Congruence results

Now we will define interleaving-ST-, partial-word-ST-, pomset-ST- and history-preserving-ST-bisimulation. (Since interleaving ST-bisimilar event structures are also step bisimilar, we will not deal with step ST-bisimulation here.)

Definition 6.3.3 Let $\mathcal{E}, \mathcal{F}$ be event structures. Then $\mathcal{E}$ and $\mathcal{F}$ are α-ST-*bisimilar* with $\alpha \in$ {interleaving, pomset, partial-word, history-preserving}, denoted by $\mathcal{E} \approx_{iST} \mathcal{F}$, $\mathcal{E} \approx_{pST} \mathcal{F}$, $\mathcal{E} \approx_{pwST} \mathcal{F}$ and $\mathcal{E} \approx_{hST} \mathcal{F}$, if there exists an α-ST-*bisimulation*, i.e. a relation $\mathcal{B} \subseteq \mathcal{S}(\mathcal{E}) \times \mathcal{S}(\mathcal{F}) \times \mathcal{P}(E_{\mathcal{E}} \times E_{\mathcal{F}})$ such that

i) $((\emptyset, \emptyset), (\emptyset, \emptyset), \emptyset) \in \mathcal{B}$

ii) If $((C, P), (D, Q), f) \in \mathcal{B}$, then f is a label-preserving bijection from $vis_{\mathcal{E}}(C)$ onto $vis_{\mathcal{F}}(D)$ such that $f(vis_{\mathcal{E}}(P)) = vis_{\mathcal{F}}(Q)$ and if $\alpha = h$ then $f : vis_{\mathcal{E}}(C) \to vis_{\mathcal{F}}(D)$ is an isomorphism.

iii) If $((C, P), (D, Q), f) \in \mathcal{B}$ and $(C, P) \to_{\mathcal{E}} (C', P')$, then there are (D', Q') and f' such that

 - $(D, Q) \to_{\mathcal{F}} (D', Q')$, $((C', P'), (D', Q'), f') \in \mathcal{B}$ and $f'|_C = f$
 - f'^{-1} restricts to a homomorphism from $vis_{\mathcal{F}}(D' - Q)$ onto $vis_{\mathcal{E}}(C' - P)$ if $\alpha = pw$ and to an isomorphism if $\alpha = p$.

iv) vice versa

Comparing this definition and Definition 6.2.2, one can see how each of the four types of bisimulation has been transformed in the same way to a corresponding ST-bisimulation. (Recall the remark after Definition 6.2.2 concerning the interleaving case.) Most important in ST-bisimulations is the following. If ST-configurations (C, P) and (D, Q) are related, then the active events are matched by f since $f(vis_{\mathcal{E}}(C)) = vis_{\mathcal{F}}(D)$ and $f(vis_{\mathcal{E}}(P)) = vis_{\mathcal{F}}(Q)$; this matching remains fixed if (C, P) evolves to (C', P') and (D, Q) to (D', Q') since f' extends f.

We have the following consequence. Suppose, $\mathcal{E}$ is in a state where two a-actions are active, i.e. $C - P$ contains two a-labelled events e and e', and $\mathcal{E}$ finishes one of them, say e. Then $C' = C$ and $P' = P \cup \{e\}$. Now $\mathcal{F}$ must finish the related a-action $f(e)$ (and not the other one!) in order to get into a bisimilar state again, since we must have $f' = f$ and $vis_{\mathcal{F}}(Q') = f(vis_{\mathcal{E}}(P')) = f(vis_{\mathcal{E}}(P)) \cup \{f(e)\} = vis_{\mathcal{F}}(Q) \cup \{f(e)\}$.

Proposition 6.3.4 *i) For $\alpha \in \{i, p, pw, h\}$ $\approx_{\alpha ST}$ is an equivalence.*
ii) Let $\mathcal{E}, \mathcal{F}$ be event structures, $\alpha \in \{i, p, pw, h\}$. Then $\mathcal{E} \approx_{\alpha ST} \mathcal{F}$ implies $\mathcal{E} \approx_{\alpha} \mathcal{F}$.

Proof: i) standard
  ii) Configurations $C \in \mathcal{C}(\mathcal{E})$ do not have any active events, thus they can be identified with ST-configurations (C, C). Now the result follows easily.   □

Now we will show the congruence results we aimed for.

Theorem 6.3.5 *Let $\mathcal{E}$ and $\mathcal{F}$ be event structures, ref be an action refinement and $\alpha \in \{i,$ pw, p, h\}. Then $\mathcal{E} \approx_{\alpha ST} \mathcal{F}$ implies $ref(\mathcal{E}) \approx_{\alpha ST} ref(\mathcal{F})$.*

Proof: Let $\mathcal{B} \subseteq \mathcal{S}(E_{\mathcal{E}}) \times \mathcal{S}(E_{\mathcal{F}}) \times \mathcal{P}(E_{\mathcal{E}} \times E_{\mathcal{F}})$ be an α-ST-bisimulation between $\mathcal{E}$ and $\mathcal{F}$. We will construct an α-ST-bisimulation $\tilde{\mathcal{B}}$ between $ref(\mathcal{E})$ and $ref(\mathcal{F})$ using Lemma 6.3.2. Given some $((C, P), (D, Q), f) \in \mathcal{B}$, Lemma 6.3.2 tells us how to refine (C, P) to an ST-configuration $(\tilde{C}, \tilde{P})$ of $ref(\mathcal{E})$; f relates C and D, and using this relation we can refine (D, Q) in the same way as (C, P); at the same time, we can translate f in the obvious way to the refined situation.

Define the relation $\tilde{\mathcal{B}}$ to consist of all

$$((\tilde{C}, \tilde{P}), (\tilde{D}, \tilde{Q}), \tilde{f}) \in \mathcal{S}(ref(\mathcal{E})) \times \mathcal{S}(ref(\mathcal{F})) \times \mathcal{P}(E_{ref(\mathcal{E})} \times E_{ref(\mathcal{F})})$$

such that there exist $((C, P), (D, Q), f) \in \mathcal{B}$ and for each $e \in vis_{\mathcal{E}}(C)$ some $(C_e, P_e) \in \mathcal{S}(ref(l_{\mathcal{E}}(e)))$ with

- for all $e \in vis_{\mathcal{E}}(C)$, $C_e \neq \emptyset$

- for all $e \in vis_{\mathcal{E}}(C)$, $P_e = E_{ref(l_{\mathcal{E}}(e))}$ if and only if $e \in P$

- $\tilde{C} = \{(e, e') \mid e \in vis_{\mathcal{E}}(C), e' \in C_e\} \cup \{(e, \lambda) \mid e \in C, l_{\mathcal{E}}(e) = \lambda\}$

- $\tilde{P} = \{(e, e') \mid e \in vis_{\mathcal{E}}(C), e' \in P_e\} \cup \{(e, \lambda) \mid e \in C, l_{\mathcal{E}}(e) = \lambda\}$

- $\tilde{D} = \{(f(e), e') \mid e \in vis_{\mathcal{E}}(C), e' \in C_e\} \cup \{(d, \lambda) \mid d \in D, l_{\mathcal{F}}(d) = \lambda\}$

- $\tilde{Q} = \{(f(e), e') \mid e \in vis_{\mathcal{E}}(C), e' \in P_e\} \cup \{(d, \lambda) \mid d \in D, l_{\mathcal{F}}(d) = \lambda\}$

- f' is the restriction to $vis_{ref(\mathcal{E})}(\tilde{C})$ of a mapping taking $(e, e') \in \tilde{C}$ to $(f(e), e')$ for $e \in vis_{\mathcal{E}}(C)$.

We have to show that $\tilde{\mathcal{B}}$ is an α-ST-bisimulation.

i) $((\emptyset, \emptyset), (\emptyset, \emptyset), \emptyset) \in \tilde{\mathcal{B}}$ since $((\emptyset, \emptyset), (\emptyset, \emptyset), \emptyset) \in \mathcal{B}$.

ii) For every $((C, P), (D, Q), f) \in \mathcal{B}$ and every appropriate family (C_e, P_e), $e \in vis_{\mathcal{E}}(C)$, $(\tilde{C}, \tilde{P})$ and $(\tilde{D}, \tilde{Q})$ are ST-configurations by Lemma 6.3.2 and by the fact that $f : vis_{\mathcal{E}}(C) \to vis_{\mathcal{F}}(D)$ is a label-preserving bijection with $f(vis_{\mathcal{E}}(P)) = vis_{\mathcal{F}}(Q)$. (The latter fact ensures that for all $d \in vis_{\mathcal{E}}(D)$, $P_d = E_{ref(l_{\mathcal{F}}(d))}$ if and only if $d \in Q$.) By construction, $\tilde{f}$ is a label-preserving bijection from $vis_{ref(\mathcal{E})}(\tilde{C})$ onto $vis_{ref(\mathcal{F})}(\tilde{D})$ with $\tilde{f}(vis_{ref(\mathcal{E})}(\tilde{P})) = vis_{ref(\mathcal{F})}(\tilde{Q})$.

In the case $\alpha = h$ we have for $(d, d'), (e, e') \in vis_{ref(\mathcal{E})}(\tilde{C})$:

$(d, d') <_{ref(\mathcal{E})} (e, e')$
$\Leftrightarrow \quad d, e \in vis_{\mathcal{E}}(C)$ and
$\qquad (d <_{\mathcal{E}} e$ or $d = e$ and $d' <_{ref(l_{\mathcal{E}}(e))} e')$
$\Leftrightarrow \quad d, e \in vis_{\mathcal{E}}(C)$ and
$\qquad (f(d) <_{\mathcal{F}} f(e)$ or $f(e) = f(d)$ and $d' <_{ref(l_{\mathcal{F}}(f(e)))} e')$
$\qquad$ (since f is an isomorphism)
$\Leftrightarrow \quad (f(d), d') <_{ref(\mathcal{F})} (f(e), e')$
$\Leftrightarrow \quad \tilde{f}(d, d') <_{ref(\mathcal{F})} \tilde{f}(e, e')$

Thus $\tilde{f}$ is an isomorphism in this case.

iii) Assume now that furthermore $(\tilde{C}, \tilde{P}) \to_{ref(\mathcal{E})} (\tilde{C}', \tilde{P}')$, i.e. $(\tilde{C}', \tilde{P}') \in \mathcal{S}(ref(\mathcal{E}))$, $\tilde{C} \subseteq \tilde{C}'$ and $\tilde{P} \subseteq \tilde{P}'$.

Take $(C', P') \in \mathcal{S}(\mathcal{E})$ and (C_e', P_e'), $e \in vis_{\mathcal{E}}(C')$, according to Lemma 6.3.2. This lemma gives $(C_e, P_e) \to_{ref(l(e))} (C_e', P_e')$ for $e \in vis_{\mathcal{E}}(C)$ and $(C, P) \to_{\mathcal{E}} (C', P')$; therefore – since $\mathcal{B}$ is an α-ST-bisimulation – there exist by definition D', Q', f' with the appropriate properties. Let

$$\tilde{D}' = \{(f'(e), e') \mid e' \in C_e', e \in vis_{\mathcal{E}}(C')\} \cup \{(d, \lambda) \mid d \in D', l_{\mathcal{F}}(d) = \lambda\},$$

$$\tilde{Q}' = \{(f'(e), e') \mid e' \in P_e', e \in vis_{\mathcal{E}}(C')\} \cup \{(d, \lambda) \mid d \in D', l_{\mathcal{F}}(d) = \lambda\},$$

and let $\tilde{f}'$ map each visible $(e, e') \in \tilde{C}'$ to $(f'(e), e') \in \tilde{D}'$.

Claim 1: $((\tilde{C}', \tilde{P}'), (\tilde{D}', \tilde{Q}'), \tilde{f}') \in \tilde{\mathcal{B}}$

This is immediate from the construction of $\tilde{\mathcal{B}}$, especially we have $(\tilde{D}', \tilde{Q}') \in \mathcal{S}(ref(\mathcal{F}))$, as observed above.

Claim 2: $\tilde{f}'|_{\tilde{C}} = \tilde{f}$

For $(e, e') \in vis_{\mathcal{E}}(\tilde{C})$ we have

$$
\begin{aligned}
\tilde{f}'(e, e') &= (f'(e), e') & \text{by construction} \\
&= (f(e), e') & \text{since } f'|_C = f \\
&= \tilde{f}(e, e') & \text{by construction.}
\end{aligned}
$$

Claim 3: $(\tilde{D}, \tilde{Q}) \to (\tilde{D}', \tilde{Q}')$

Let $(d, e') \in \tilde{D}$. If $d \in vis_{\mathcal{F}}(D)$, we find $e \in vis_{\mathcal{E}}(C)$ with $f(e) = d$ and $e' \in C_e$. Since $(C, P) \to_{\mathcal{E}} (C', P')$ we have $e \in vis_{\mathcal{E}}(C')$, since $(C_e, P_e) \to (C_e', P_e')$ we have $e' \in C_e'$. Thus $(d, e') = (f(e), e') = (f'(e), e') \in \tilde{D}'$. If $l_{\mathcal{F}}(d) = \lambda$ we get $d \in D'$, since $(D, Q) \to_{\mathcal{F}} (D', Q')$ by choice of (D', Q'). Thus $(d, e') = (d, \lambda) \in \tilde{D}'$. Analogously we have $\tilde{Q} \subseteq \tilde{Q}'$.

Case $\alpha = pw$ and $\alpha = p$.
First observe that $\tilde{f}'|_{\tilde{C}} = \tilde{f}$ and $\tilde{f}(vis_{ref(\mathcal{E})}(\tilde{P})) = vis_{ref(\mathcal{F})}(\tilde{Q})$, hence $\tilde{f}'$ maps $vis_{ref(\mathcal{E})}(\tilde{C}' - \tilde{P})$ onto $vis_{ref(\mathcal{F})}(\tilde{D}' - \tilde{Q})$. If $e \in P$ then $P_e = E_{ref(l_{\mathcal{E}}(e))}$, hence all elements (e, e') of $\tilde{C}$ are in $\tilde{P}$. Therefore $(e, e') \in vis_{ref(\mathcal{E})}(\tilde{C}' - \tilde{P})$ implies $e \in vis_{\mathcal{E}}(C' - P)$.

The rest of the proof is similar to the above proof of Part ii), case $\alpha = h$.

iv) is analogous to iii). $\square$

The definition of ST-bisimulation given above was very suitable for proving our congruence result. It relied on ST-configurations, which might appear as somewhat unusual. We will now show that we can give an alternative definition based on the usual configurations, but taking special care of maximal elements. The condition on maximal elements is rather technical and less intuitive than the ST-idea; nevertheless, I hope that this characterization will give some more insight into the ST-idea. We will restrict ourselves to interleaving-ST- and history-preserving-ST-bisimulations; similar results seem to be possible for partial-word-ST- and pomset-ST-bisimulation, but only if some additional requirements are included, which would make these results uncomfortably complicated.

To understand the following theorem, observe that a visible event of some configuration C might be maximal in $vis_{\mathcal{E}}(C)$ without being maximal in C. The reason would be that this event is succeeded by some internal event. The active events of an ST-configuration (C, P) are some of the visible maximal elements of C. The following theorem roughly says that it is enough to consider only those states where *all* the visible maximal events are active.

Theorem 6.3.6 *Event structures $\mathcal{E}$, $\mathcal{F}$ are α-ST-bisimilar for $\alpha = i, h$ if and only if there exists a relation $\mathcal{B} \subseteq \mathcal{C}(\mathcal{E}) \times \mathcal{C}(\mathcal{F}) \times \mathcal{P}(E_{\mathcal{E}} \times E_{\mathcal{F}})$ such that*

i) $(\emptyset, \emptyset, \emptyset) \in \mathcal{B}$

ii) *If $(C, D, f) \in \mathcal{B}$ then f is a label-preserving bijection from $vis_{\mathcal{E}}(C)$ onto $vis_{\mathcal{F}}(D)$ and for $\alpha = h$ an isomorphism.*

iii) *If $(C, D, f) \in \mathcal{B}$ and $C \to_{\mathcal{E}} C'$ then there are D', f' such that*

  (a) $D \to_{\mathcal{F}} D'$, $(C', D', f') \in \mathcal{B}$ and $f'|_C = f$,

  (b) *if $e \in vis_{\mathcal{E}}(C')$ is maximal in C', then $f'(e)$ is maximal in D' or $e \in C$ and $f(e)$ is not maximal in D.*

iv) *vice versa*

We can additionally restrict C' in iii) to the case that $vis_{\mathcal{E}}(C' - C)$ has at most one element, and D' in iv) to the case that $vis_{\mathcal{F}}(D' - D)$ has at most one element.

Proof: The proof is based on the idea that we consider only those ST-configurations (C, P) where $C - P$ consists of all visible maximal elements of C. The truth is a bit more complicated for the following reason. Consider two event structures, one consisting of a single a-labelled event e, the other of an a-labelled event e' followed by an internal event e''. These structures are interleaving ST-bisimilar. Both can start the a and get into related states where e and e' are active. Always in such a situation, both structures can finish the active events and get into related states; since this is always the case, we can ignore the resulting situation where maximal events are not active. But the second structure can also finish e' and perform e''. Now the first structure is forced to simulate this by finishing e; it reaches a state where e is maximal, but not active, and we cannot ignore this situation. Here, e is matched to e', which is not maximal; situations like this make the additional clause in iii) (b) necessary.

Given a relation $\mathcal{B}$ as in the theorem, we can construct an α-ST-bisimulation $\mathcal{B}' \subseteq \mathcal{S}(\mathcal{E}) \times \mathcal{S}(\mathcal{F}) \times \mathcal{P}(E_{\mathcal{E}} \times E_{\mathcal{F}})$ by $((C, P), (D, Q), f) \in \mathcal{B}'$ if and only if $(C, D, f) \in \mathcal{B}$ and $f(vis_{\mathcal{E}}(P)) = vis_{\mathcal{F}}(Q)$.

Definition 6.3.3 i) and ii) are obvious by the construction of $\mathcal{B}'$. Hence assume $((C, P), (D, Q), f) \in \mathcal{B}'$ and $(C, P) \to_{\mathcal{E}} (C', P')$. Thus $C \to_{\mathcal{E}} C'$, and by iii) above we can find D' and f'; we define $Q' = f'(vis_{\mathcal{E}}(P')) \cup \{d \in D' \mid l_{\mathcal{F}}(d) = \lambda\}$.

The elements of $D' - Q'$ are visible, hence $D' - Q' = f'(C' - P')$. If $e \in C' - P'$ then $e \in vis_{\mathcal{E}}(C')$ and e is maximal in C'. Since $e \notin P'$, we have $e \notin P$; hence $e \in C$ implies $f(e) \in D - Q$, i.e. $f(e)$ maximal in D. Therefore we conclude from iii) above that $f'(e)$

is maximal in D'. Thus $D' - Q'$ consists of elements which are maximal in D'. Now it is easy to see that $(D', Q') \in \mathcal{S}(\mathcal{F})$ and the remaining properties are immediate.

Vice versa, assume that an α-ST-bisimulation $\mathcal{B}'$ is given. We define $\mathcal{B} \subseteq \mathcal{C}(\mathcal{E}) \times \mathcal{C}(\mathcal{F}) \times \mathcal{P}(E_\mathcal{E} \times E_\mathcal{F})$ by:

$(C, D, f) \in \mathcal{B}$ if and only if for some Q we have $((C, P), (D, Q), f) \in \mathcal{B}'$ where P is defined by $C - P = \{e \in vis_\mathcal{E}(C) \mid e \text{ is maximal in } C \text{ and } f(e) \text{ maximal in } D\}$.

Conditions i) and ii) above are again obvious by the construction of $\mathcal{B}$. Hence assume $(C, D, f) \in \mathcal{B}$ and $C \rightarrow_\mathcal{E} C'$. We have $((C, P), (D, Q), f) \in \mathcal{B}'$ for some P and Q such that $C - P = \{e \in vis_\mathcal{E}(C) \mid e \text{ is maximal in } C \text{ and } f(e) \text{ maximal in } D\}$. We define P' by the equality $C' - P' = \{e \in vis_\mathcal{E}(C') \mid e \text{ is maximal in } C', \text{ and if } e \in C \text{ then } e \in C - P\}$.
Obviously $(C', P') \in \mathcal{S}(\mathcal{E})$ and $(C, P) \rightarrow_\mathcal{E} (C', P')$, hence by 6.3.3 iii) we can find some D', Q', f' such that $(D, Q) \rightarrow_\mathcal{F} (D', Q')$, $((C', P'), (D', Q'), f') \in \mathcal{B}$ and $f'|_C = f$.
If $e \in C' - P'$, then e is visible and maximal in C', furthermore $f'(e) \in D' - Q'$ is maximal in D'. On the other hand, if $e \in vis_\mathcal{E}(C')$ is maximal in C' and $f'(e)$ maximal in D', then $e \in C$ implies that e is maximal in C and $f(e)$ maximal in D, hence $e \in C - P$. Thus, we have $C' - P' = \{e \in vis_\mathcal{E}(C') \mid e \text{ is maximal in } C' \text{ and } f'(e') \text{ is maximal in } D'\}$. Hence $(C', D', f') \in \mathcal{B}$.
We directly have $D \rightarrow_\mathcal{F} D'$ and $f'|_C = f$. If some $e \in vis_\mathcal{E}(C')$ is maximal in C', but $f'(e)$ is not maximal in D', then by the above $e \notin C' - P'$. Thus, by choice of $C' - P'$, we must have $e \in C$ and $e \in P$. Since e is certainly maximal in C, we conclude that $f(e)$ is not maximal in D.
iv) is checked analogously.
The additional restriction can be dealt with as usual: If $C \rightarrow_\mathcal{E} C'$ we can find $C_0 = C, C_1, C_2, \ldots, C_n = C'$ such that $C_{i-1} \rightarrow_\mathcal{E} C_i$ and $C_i - C_{i-1}$ has at most one visible element, $i = 1, \ldots, n$. Then we find corresponding $D_0 = D, D_1, \ldots, D_n$ and $f_0 = f, f_1, \ldots, f_n$ and can define $D' = D_n$, $f' = f_n$ satisfying (a) and (b) of the theorem. $\square$

The characterization of Theorem 6.3.6 for history-preserving ST-bisimilarity shows that this equivalence coincides with maximality-preserving bisimilarity, which was defined and shown to be a congruence for action refinement of Petri nets in [Dev90b]. The characterization for interleaving ST-bisimilarity can be used to show that for systems without concurrency, i.e. where any two events are either causally related or in conflict, this equivalence coincides with delay bisimilarity; in delay bisimilarity, one must not simulate an observable event e by an equally labelled event followed by some internal events; this corresponds to (b) above: $e \in C' - C$ is maximal, thus $f(e)$ must be maximal, i.e. must not be followed by internal events. Delay bisimulation was first defined in [Mil81], the name was invented in [Gla90b]. Even stricter requirements for internal moves are enforced in branching bisimulation (see Chapter 2), which is shown to be a congruence w.r.t. refinement for systems without concurrency (but with silent moves) in [Gla90b].
Furthermore, we can deduce two interesting corollaries from Theorem 6.3.6: interleaving ST-bisimilarity is finer than step bisimilarity, and for systems without silent moves history-preserving ST-bisimilarity coincides with history-preserving bisimilarity.

Corollary 6.3.7 *If event structures $\mathcal{E}$, $\mathcal{F}$ are interleaving ST-bisimilar, then we have $\mathcal{E} \approx_s \mathcal{F}$.*

Proof: Take a relation $\mathcal{B}$ as in Theorem 6.3.6 for $\alpha = i$. This is a step bisimulation for the following reason: If $(C, D, f) \in \mathcal{B}$, $C \rightarrow_{\mathcal{E}} C'$ and $vis_{\mathcal{E}}(C' - C)$ is a step, then let C'' be the minimal left-closed subset of C' containing $C \cup vis_{\mathcal{E}}(C' - C)$. Then we have $C' \rightarrow_{\mathcal{E}} C'' \rightarrow_{\mathcal{E}} C'$ and can find D'', f'' and D', f' by 6.3.6 iii). The elements of $vis_{\mathcal{E}}(C'' - C)$ are maximal in C'', hence their images are maximal in D'', i.e. $vis_{\mathcal{F}}(D'' - D)$ is a step. Now $f' = f''$ is an isomorphism from $vis_{\mathcal{E}}(C' - C) = vis_{\mathcal{E}}(C'' - C)$ onto $vis_{\mathcal{F}}(D' - D) = vis_{\mathcal{F}}(D'' - D)$.   □

Corollary 6.3.8 *For event structures $\mathcal{E}$, $\mathcal{F}$ without internal events we have $\mathcal{E} \approx_h \mathcal{F}$ if and only if $\mathcal{E} \approx_{hST} \mathcal{F}$.*

Proof: '$\Rightarrow$' If we have $(C, D, f) \in \mathcal{B}$ for some h-bisimulation $\mathcal{B}$ then $vis_{\mathcal{E}}(C) = C$ and f, being an isomorphism, always maps maximal elements of C to maximal elements of D. Hence, the additional property (b) in 6.3.6 iii) and iv) holds automatically and $\mathcal{B}$ is a relation as in 6.3.6 for $\alpha = h$.
'$\Leftarrow$' Proposition 6.3.4.   □

With these corollaries we can show how the various bisimulation equivalences we have discussed so far are related.

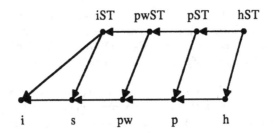

iST   pwST   pST   hST

i   s   pw   p   h

Figure 6.7   relations between the equivalences

Theorem 6.3.9 *Let $\approx_\beta$, $\approx_\gamma$ be two equivalences defined above. Then $\mathcal{E} \approx_\beta \mathcal{F}$ implies $\mathcal{E} \approx_\gamma \mathcal{F}$ for all event structures $\mathcal{E}$, $\mathcal{F}$ if and only if there is a directed path from β to γ in Figure 6.7. This also holds if only event structures without internal events are considered, except that $\approx_h$ equals $\approx_{hST}$ in this case.*

Proof: The horizontal implications follow directly from the definitions as in 6.2.3, the vertical implications follow from 6.3.4 and 6.3.7. Figure 6.6 gives an example of history-preserving bisimilar systems that are not interleaving ST-bisimilar, since after refining a to $a_1 a_2$ they are not interleaving bisimilar. Figure 6.2 gives an example of pomset bisimilar systems without internal events that are not interleaving ST-bisimilar (consider refining a to a_1a_2). Figure 6.5 gives an example of partial-word ST-bisimilar systems without internal events that are not pomset bisimilar.

Hence, in view of Proposition 6.2.3 and Corollary 6.3.8, the rest of the proof is provided by the counterexamples shown in Figures 6.8 and 6.9. (For Figure 6.8 consider the first system as a partial order that can be performed by this system; this or a less sequential partial order cannot be performed by the second system. In Figure 6.9 only the second system can perform a sequence of two a's such that the next action, b, is necessarily dependent on the first a.)   □

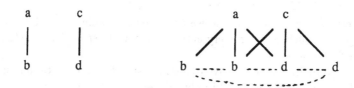

Figure 6.8   interleaving-ST-, but not partial-word-bisimilar

Figure 6.9   pomset-ST-, but not history-preserving-bisimilar

Full abstractness

So far, we have shown how we can uniformly construct from various notions of α-bisimulation corresponding notions of α-ST-bisimulation such that

- α-ST-bisimulation is a congruence w.r.t. action refinement,

- α-ST-bisimulation refines α-bisimulation.

In the rest of this section, we will show that for $\alpha \in \{i, p, h\}$ we have not only succeeded in constructing just any congruence, but that α-ST-bisimulation is the coarsest equivalence with the above properties, at least under some local finiteness condition on the event structures; i.e. α-ST-bisimulation is fully abstract for action refinement and α-bisimulation. Together with Rob van Glabbeek, I have observed this for interleaving ST-bisimulation for finite event structures without internal events, see [Gla90c, concluding remarks]. Here we will show this for interleaving-ST-, pomset-ST- and history-preserving-ST-bisimulation, and we will consider event structures with internal events, as long as they are image-finite, a notion we have already defined for nets.

Definition 6.3.10 An event structure $\mathcal{E}$ is *image-finite*, if for all $a \in \Sigma$ and $C \in \mathcal{C}(\mathcal{E})$ there are only finitely many $D \in \mathcal{C}(\mathcal{E})$ such that $C \to_{\mathcal{E}} D$ and $vis_{\mathcal{E}}(D-C)$ is either empty or it consists of one a-labelled element.

We have already argued in Section 3.3 that image-finiteness is a rather plausible restriction for systems one is really interested in. It is easy to obtain the following generalization by induction on the size of $vis_{\mathcal{E}}(D-C)$:

Proposition 6.3.11 *Let $\mathcal{E}$ be an image-finite event-structure and C, $D \in \mathcal{C}(\mathcal{E})$ with $C \to_{\mathcal{E}} D$. Then there are only finitely many $D' \in \mathcal{C}(\mathcal{E})$ such that $C \to_{\mathcal{E}} D'$ and there is a label-preserving bijection from $vis_{\mathcal{E}}(D-C)$ onto $vis_{\mathcal{E}}(D'-C)$.*

For the rest of this section assume we are given image-finite event structures $\mathcal{E}$ and $\mathcal{F}$ and $\alpha = \{i, p, h\}$ such that for every action refinement ref we have $ref(\mathcal{E}) \approx_{\alpha} ref(\mathcal{F})$. We will show that $\mathcal{E} \approx_{\alpha ST} \mathcal{F}$.

For this purpose, first of all, we have to distinguish active actions from finished actions. As explained above, this is easy: we simply split each a into a sequence $a_1 a_2$; now a start of a in the unrefined system corresponds to a_1 in the refined system, and the finish of a corresponds to a_2. But we also have to keep autoconcurrent actions apart, and this is much more difficult. If we were allowed to use conflicts in the structures $ref(a)$, we could refine a by a choice between the sequences $a_{11}a_{12}$ and $a_{21}a_{22}$, compare the proof of Theorem 5.4.18. Then, starting one a would correspond to a_{11}, starting a second a to a_{21}; now, the two autoconcurrent a's are distinguished since finishing the first a corresponds to a_{12} but not to a_{22}. Using more and more choices, we could distinguish more and more simultaneously active events with label a. Since such choices are not possible for us, we will use instead more and more parallel events in $ref(a)$, and somewhat surprisingly this allows us to distinguish more and more simultaneously active events with label a.

To work out this idea, let us define a sequence of action refinements ref_n, $n \in \mathbb{N}$. For $a \in \Sigma$ and $n \in \mathbb{N}$

$$ref_n(a) = (\{0, 1, \ldots, n\} \cup \{\infty\}, <_n, \emptyset, l_{n,a})$$

with $i <_n j$ if $i \neq j$ and $(i = 0$ or $j = \infty)$, and with $l_{n,a}(i) = (a, i) \in \Sigma$, see Figure 6.10.

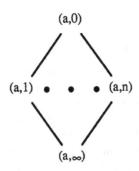

Figure 6.10   the labelled partial order $ref_n(a)$

Since Σ is infinite, we can assume that all (a, i) are distinct; since $l_{n+1,a}$ extends $l_{n,a}$, we will denote the labelling simply by l_a. As always, we have $ref_n(\lambda) = (\{\lambda\}, \emptyset, \emptyset, \{(\lambda, \lambda)\})$.

For α-bisimilar event structures $\mathcal{E}$ and $\mathcal{F}$, the union of α-bisimulations is obviously an α-bisimulation again, hence there exists a largest α-bisimulation between $\mathcal{E}$ and $\mathcal{F}$. Let $\mathcal{B}_n$ be the largest α-bisimulation between $ref_n(\mathcal{E})$ and $ref_n(\mathcal{F})$.

Our next aim is to construct a relation $\mathcal{R}_n \subseteq S(\mathcal{E}) \times S(\mathcal{F}) \times \mathcal{P}(E_\mathcal{E} \times E_\mathcal{F})$ from $\mathcal{B}_n$.

Let $(C, P) \in S(\mathcal{E})$, $n \in \mathbb{N}$. Using Lemma 6.3.2, we want to define $\tilde{C} \subseteq E_{ref_n(\mathcal{E})}$ such that active events of (C, P) with the same label can be distinguished after refinement. This can be done only if there are at most n of them. If this is the case, we assign injectively to each of them a number between 1 and n; if e has number j then $C_e = P_e = \{0, j\}$, see Figure 6.11. If we have more than n active events of (C, P) with the same label, then we distinguish n of them; for the others, we choose $C_e = P_e = \{0\}$.

- For each $a \in \Sigma$ such that there are at least n a-labelled elements in $C - P$, let $event_a : \{1, \ldots, n\} \to \{e \in C - P \mid l_\mathcal{E}(e) = a\}$ be an injection. Define $C_{event_a(i)} = P_{event_a(i)} = \{0, i\}$, for $i = 1, \ldots, n$, and $C_e = P_e = \{0\}$ for $e \in C - P - event_a\{1, \ldots, n\}$ with $l_\mathcal{E}(e) = a$.

- For each $a \in \Sigma$ such that there are less than n a-labelled elements in $C - P$, let $in_a : \{e \in C - P \mid l_\mathcal{E}(e) = a\} \to \{1, \ldots, n\}$ be an injection, and put $C_e = P_e = \{0, in_a(e)\}$ for $e \in C - P$ with $l_\mathcal{E}(e) = a$.

- For $e \in P$, let $C_e = P_e = E_{ref(l_\mathcal{E}(e))}$.

Put $\tilde{C} = \bigcup_{e \in C}\{e\} \times C_e$. By Lemma 6.3.2, $\tilde{C} \in \mathcal{C}(ref_n(\mathcal{E}))$.

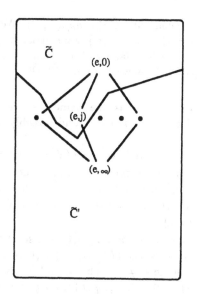
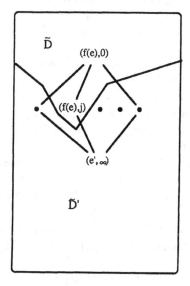

Figure 6.11

Define a mapping $con_{\mathcal{E},n} : \mathcal{S}(\mathcal{E}) \to \mathcal{P}(\mathcal{C}(ref_n(\mathcal{E})))$ that assigns to each (C,P) the set of all configurations $\tilde{C}$ that can be constructed from (C,P) as above. Analogously, define $con_{\mathcal{F},n} : \mathcal{S}(\mathcal{F}) \to \mathcal{P}(\mathcal{C}(ref_n(\mathcal{F})))$.

Using the mappings $con_{\mathcal{E},n}$ and $con_{\mathcal{F},n}$, we define a relation $\mathcal{R}_n \subseteq \mathcal{S}(\mathcal{E}) \times \mathcal{S}(\mathcal{F}) \times \mathcal{P}(E_{\mathcal{E}} \times E_{\mathcal{F}})$ by (see Figure 6.11)

$((C,P),(D,Q),f) \in \mathcal{R}_n$ if and only if

there exist $\tilde{C} \in con_{\mathcal{E},n}(C,P)$, $\tilde{D} \in con_{\mathcal{F},n}(D,Q)$ and $\tilde{f}$ such that $(\tilde{C}, \tilde{D}, \tilde{f}) \in \mathcal{B}_n$

and $f : vis_{\mathcal{E}}(C) \to vis_{\mathcal{F}}(D)$ satisfies $\tilde{f}(e,i) = (f(e),i)$ for all $(e,i) \in vis_{ref_n(\mathcal{E})}(\tilde{C})$.

The relations $\mathcal{R}_n$ are not α-ST-bisimulations, but we will construct an α-ST-bisimulation $\mathcal{R}$ from them, namely $\mathcal{R} =$

$$\{((C,P),(D,Q),f) \mid \text{ there are infinitely many } n \text{ with } ((C,P),(D,Q),f) \in \mathcal{R}_n\}.$$

We have $((\emptyset,\emptyset),(\emptyset,\emptyset),\emptyset) \in \mathcal{R}$ since $(\emptyset,\emptyset,\emptyset) \in \mathcal{B}_n$ for all $n \in \mathbb{N}$.

If $((C,P),(D,Q),f) \in \mathcal{R}$ and for some $n \in \mathbb{N}$ $(\tilde{C},\tilde{D},\tilde{f})$ is the corresponding element of $\mathcal{B}_n$, then the partial order induced by $\{(e,i) \in \tilde{C} \mid i = 0\}$ is isomorphic to $vis_{\mathcal{E}}(C)$ via $(e,0) \to e$. Hence, from the corresponding properties for $\tilde{f}$ we conclude that f is a label-preserving bijection from $vis_{\mathcal{E}}(C)$ onto $vis_{\mathcal{F}}(D)$, and that f is an isomorphism if $\alpha = h$. We have $e \in vis_{\mathcal{E}}(P)$ if and only if $(e,\infty) \in vis_{ref_n(\mathcal{E})}(\tilde{C})$ if and only if $\tilde{f}(e,\infty) = (f(e),\infty) \in vis_{ref_n(\mathcal{F})}(\tilde{D})$ if and only if $f(e) \in vis_{\mathcal{F}}(Q)$. Hence $f(vis_{\mathcal{E}}(P)) = vis_{\mathcal{F}}(Q)$.

Therefore, in order to prove that $\mathcal{R}$ is an α-ST-bisimulation we have to prove that 6.3.3 iii) holds. (6.3.3 iv) is analogous then.)

Hence, assume $((C,P),(D,Q),f) \in \mathcal{R}$ and $(C,P) \to_{\mathcal{E}} (C',P')$, where in the case $\alpha = i$ we can restrict ourselves to the case that $\mid vis_{\mathcal{E}}(C' - C) \mid + \mid vis_{\mathcal{E}}(P' - P) \mid \leq 1$. Consider $n > \mid vis_{\mathcal{E}}(C' - P)\mid$ such that some $(\tilde{C},\tilde{D},\tilde{f}) \in \mathcal{B}_n$ corresponds to $((C,P), (D,Q), f)$ according to the definition of $\mathcal{R}_n$.

For each $a \in \Sigma$, we have used an injection in_a to construct $\tilde{C} \in con_{\mathcal{E},n}(C,P)$. We can extend this to an injection $in_a : \{e \in C' - P \mid l_{\mathcal{E}}(e) = a\} \to \{1, \dots, n\}$ and restrict it again to $\{e \in C' - P' \mid l_{\mathcal{E}}(e) = a\}$. From this function (or rather: family of functions), we can construct $\tilde{C}' \in con_{\mathcal{E},n}(C',P')$ such that $\tilde{C} \to_{ref_n(\mathcal{E})} \tilde{C}'$. Now we can find $(\tilde{C}',\tilde{D}',\tilde{f}') \in \mathcal{B}_n$ according to 6.2.2 iii).

Let us first consider the case $\alpha = i$. If $vis_{\mathcal{E}}(P' - P)$ has one element e, we have obtained $\tilde{C}'$ by adding (e,∞) and $n - 1$ other elements (e,i) to $\tilde{C}$ (and possibly some invisible elements), see Figure 6.11.

Since $\tilde{f}'$ maps $vis_{ref_n(\mathcal{E})}(\tilde{C}' - \tilde{C})$ to $vis_{ref_n(\mathcal{F})}(\tilde{D}' - \tilde{D})$ we have some (e',∞) and $n - 1$ other visible elements in $\tilde{D}' - \tilde{D}$; these must all have the same first component, since $(e',\infty) \in \tilde{D}'$ implies $(e',i) \in \tilde{D}'$ for $i = 0, \dots, n$, and in fact $e' = f(e)$. Therefore $\tilde{D}' \in con_{\mathcal{F},n}(D',Q')$ for some (D',Q') with $(D,Q) \to_{\mathcal{F}} (D',Q')$; for all $(e,i) \in vis_{ref_n(\mathcal{E})}(\tilde{C}')$ we have $\tilde{f}'(e,i) = (f(e),i)$; therefore $((C',P'),(D',Q'),f) \in \mathcal{R}_n$.

If $vis_{\mathcal{E}}(C' - C)$ has one element e, we have obtained $\tilde{C}'$ by adding $(e,0)$ and some (e,j) to $\tilde{C}$ (and possibly some invisible elements). Thus we have $vis_{\mathcal{F}}(\tilde{D}' - \tilde{D}) = \{(e',0),(e'',j)\}$. If $e' \neq e''$, then we have $(e'',0), (e'',k) \in \tilde{D}$ for some $k \neq j$. This would imply that $\tilde{D}' \to_{ref_n(\mathcal{F})} \tilde{D}''$ with $vis_{ref_n(\mathcal{F})}(\tilde{D}'' - \tilde{D}') = \{(e'',i) \mid i \in \{1, \dots, n\} \cup \{\infty\}, i \neq j, i \neq k\}$, but $\tilde{C}'$ has still the form shown in Figure 6.11, i.e. for every (e_0,∞) not in $\tilde{C}'$ there is at most one $m \in \{1, \ldots, n\}$ with (e_0, m) in $\tilde{C}'$; hence, it is impossible to find $\tilde{C}''$ with $\tilde{C}' \to_{ref_n(\mathcal{E})} \tilde{C}''$ such that the labels of the elements of $vis_{ref_n(\mathcal{E})}(\tilde{C}'' - \tilde{C}')$ form exactly the set $\{(l_{\mathcal{F}}(e''), i) \mid i \in \{1, \ldots, n\} \cup \{\infty\}, i \neq j, i \neq k\}$. This would be a contradiction to $\mathcal{B}_n$ being an i-bisimulation, hence we conclude $e'' = e'$. Thus we can extend f to f' by $f'(e) = e'$; now we can find (D', Q') such that $(D, Q) \to_{\mathcal{F}} (D', Q')$, $\tilde{D} \in con_{\mathcal{F},n}(D', Q')$ and $f' : vis_{\mathcal{E}}(C') \to vis_{\mathcal{F}}(D')$ such that for all $(e, i) \in vis_{ref_n(\mathcal{E})}(\tilde{C}')$ we have $\tilde{f}'(e, i) = (f'(e), i)$. Thus $((C', P'), (D', Q'), f') \in \mathcal{R}_n$, and $f'|_C = f$.

If $vis_{\mathcal{E}}((C' - C) \cup (P' - P)) = \emptyset$ we can find (D', Q') and f' with these properties easily.

Now we consider the cases where $\alpha \in \{p, h\}$. The mapping $\tilde{f}'$ restricts to an isomorphism $vis_{ref_n(\mathcal{E})}(\tilde{C}' - \tilde{C}) \to vis_{ref_n(\mathcal{F})}(\tilde{D}' - \tilde{D})$. Consider for some $e \in vis_{\mathcal{E}}(C' - P)$ the elements (e, i) of $\tilde{C}' - \tilde{C}$ with $i \in \{0, \ldots, n\} \cup \{\infty\}$ and the immediate predecessor relation. With this relation these elements form a connected graph, thus they are mapped to an isomorphic graph in $vis_{ref_n(\mathcal{F})}(\tilde{D}' - \tilde{D})$, i.e. we find some e' with $\tilde{f}'(e, i) = (e', i)$ for all i. If $(e, 0) \notin \tilde{C}' - \tilde{C}$, then $(e, in_{l(e)}(e)) \notin \tilde{C}' - \tilde{C}$, too; hence we find $e' = f(e)$. Otherwise we can extend f to f' by putting $f'(e) = e'$.

Thus, we see that $\tilde{D}' \in con_{\mathcal{F},n}(D', Q')$ for some $(D', Q') \in \mathcal{S}(\mathcal{F})$ with $(D, Q) \to_{\mathcal{F}} (D', Q')$, we have $f' : vis_{\mathcal{E}}(C') \to vis_{\mathcal{F}}(D')$ with $f'|_C = f$, and for all $(e, i) \in vis_{ref_n(\mathcal{E})}(\tilde{C}')$ we have $\tilde{f}'(e, i) = (f'(e), i)$. Hence $((C', P'), (D', Q'), f) \in \mathcal{R}_n$. Furthermore, since the elements (e, i), (e', j) with $e \neq e'$ inherit their ordering from e, e', we conclude that f' is an isomorphism from $vis_{\mathcal{E}}(C' - P)$ onto $vis_{\mathcal{F}}(D' - Q)$ since $\tilde{f}'$ is an isomorphism.

For all cases of α, we have just seen that for infinitely many n we have found D_n, Q_n, f_n such that $(D, Q) \to_{\mathcal{F}} (D_n, Q_n)$, $((C', P'), (D_n, Q_n), f_n) \in \mathcal{R}_n$, $f_n|_C = f$ and, if $\alpha = p$, then f_n is an isomorphism from $vis_{\mathcal{E}}(C' - P)$ onto $vis_{\mathcal{F}}(D_n - Q)$. (Thus, $\mathcal{R}_n$ satisfies the desired properties for the element $((C, P), (D, Q), f)$ of $\mathcal{R}$ and the ST-configuration (C', P') we have chosen; but $\mathcal{R}_n$ might violate the desired properties for other choices.)

By Proposition 6.3.11, infinitely many of the (D_n, Q_n) coincide with one (D', Q'). Since there are only finitely many mappings from $vis_{\mathcal{E}}(C')$ to $vis_{\mathcal{F}}(D')$, for infinitely many n with $(D_n, Q_n) = (D', Q')$ the mapping f_n coincides with one f'. We conclude that $((C', P'), (D', Q'), f') \in \mathcal{R}$ and that $\mathcal{R}$ is indeed an α-ST-bisimulation. This proves:

Theorem 6.3.12 *Image finite event structures $\mathcal{E}$, $\mathcal{F}$ are interleaving / pomset / history-preserving ST-bisimilar if $r(\mathcal{E})$ and $r(\mathcal{F})$ are interleaving / pomset / history-preserving bisimilar for all refinements r.*

Corollary 6.3.13 *For image-finite event structures, interleaving / pomset / history-preserving ST-bisimulation are fully abstract with respect to interleaving / pomset / history-preserving bisimulation and action refinement.*

For simplicity, we have obtained our results using event structures as system models; I expect that all the results also hold for more general models like flow event structures and Petri nets (at least for Petri nets without selfconcurrency, see [BDKP91]). For history-preserving ST-bisimulation, this has independently been shown in [Dev90b] already, where for Petri nets it is proven that maximality-preserving bisimulation induces a congruence w.r.t. refinement; our characterization of history-preserving ST-bisimilarity in

Theorem 6.3.6 shows that this equivalence agrees with maximality-preserving bisimilarity. As a reaction to our above results, Raymond Devillers has provided the generalization to Petri nets also for interleaving ST-bisimulation (including the coarsest congruence results), see [Dev91].

In [Dev91], it is also shown that simple splitting is enough to show the coarsest congruence result for history-preserving ST-bisimulation; thus, in this case a restriction to image finite systems is not necessary. The coarsest congruence result of [Dev91] for interleaving ST-bisimulation is based on refining with structures that contain conflicts, namely those structures we have used to prove Theorem 5.4.18; with conflicts, it is much easier to distinguish autoconcurrent actions. We have managed to distinguish them 'by parallel actions'; but I have failed to do this for the case of partial-word ST-bisimulation. A coarsest congruence result for this case should be easy to obtain, if refining systems with conflicts are used. It is somewhat intriguing whether such a result is also possible in our restricted setting; if not, this could mean: one can work with a simpler equivalence like split partial-word bisimulation and still get a congruence for refinement – provided one only wants to refine an action to one run consisting of several subactions, which might also be concurrent. Another result for a restricted form of refinement has recently been obtained in a process algebra setting in [GL91]: interleaving ST-bisimulation gives the coarsest congruence for arbitrary splitting, i.e. for refinements where each $ref(a)$ is a sequence.

Another recent result for refinement in process algebras can be found in [AH91]: for a rather rich finite process algebra, interleaving ST-bisimulation is a congruence for action refinement. This result is especially interesting, since action refinement in this paper differs essentially from our action refinement. When we refine concurrent events, the inserted event structures inherit conflicts and causal relationships and, thus, each event in one of them is concurrent with each event in the other; in particular, we do not have any form of synchronization between the two structures. In [AH91], a natural definition of action refinement for CCS has the effect that such a synchronization is possible. So far, realistic examples or case studies for the application of action refinement are missing – which is a severe shortcoming. Thus, only the future can show which form of action refinement is really useful; the result of [AH91] is promising in so far as it shows the robustness of the ST-idea not only w.r.t. changes in the type of basic bisimulation but also w.r.t. changes in the specific form of action refinement.

6.4 History-preserving and OM-bisimulation

In the natural translation of history-preserving bisimulation from event structures to nets, the states of a net are described by its processes, and the isomorphism relating two states is an isomorphism between the action structures of the corresponding processes. The aim of this section is to give a characterization of history-preserving bisimulation as OM-bisimulation. For OM-bisimulation, a state is described by an ordered marking, i.e. a marking together with a pre-order on its tokens; this seems to be more natural than to take processes, since this way states are related to the passive net elements, the places, just as in the case of ordinary markings.

In this section, we will only consider safe nets that have no internal transitions and are T-restricted. Thus, the labelling of a net is always a function into Σ, and we assume all nets to be of arc weight 1. For these nets, history-preserving bisimulation is a congruence with respect to action refinement by a result of [BDKP91]. (A slightly restricted version of action refinement is used in that paper, where in a refinement net all places except *idle* are initially unmarked.) Recall that self-concurrency is a problem when dealing with action refinement (see Section 5.2) – but T-restricted safe nets are free of self-concurrency –, and that internal actions can make the congruence result fail, compare the previous sections. Technically, safe nets have the advantage that we can (and will) consider the reachable markings as sets of places. Thus, we can view a token as the place it lies on, and this makes it much easier to deal with pre-orders on tokens.

Since we are working with T-restricted nets, we have for every process π of a net N that $\min \pi$ and $\max \pi$ consist of conditions only; l_π is injective on $\min \pi$ and $\max \pi$ for safe nets, $l_\pi(\min \pi)$ equals M_N (regarded as subset of S_N), and $l_\pi(\max \pi)$ is a reachable marking by Theorem 2.2.1. If some $t \in T$ is enabled under $l_\pi(\max \pi)$, we can extend π to a process π' by adding a new event e with label t and for each $s \in t^\bullet$ a new condition b with label s, such that ${}^\bullet e = \{b \in \max \pi \mid l_\pi(b) \in {}^\bullet t\}$ and $e^\bullet$ is the set of new conditions. We write $\pi \xrightarrow{t} \pi'$ in this case. In the following, we will not work with the strict partial order F^+ induced by π, but with the weak partial order F^*, the reflexive, transitive closure of F.

In view of Theorem 6.3.6, we take the following definition of history-preserving bisimulation for safe nets, see also Proposition 5.5 of [BDKP91].

Definition 6.4.1 Let N_1, N_2 be safe nets. A set $\mathcal{B}$ of triples (π_1, π_2, f) is a *history-preserving bisimulation* for N_1, N_2 if

i) $(\pi_0(N_1), \pi_0(N_2), \emptyset) \in \mathcal{B}$

ii) If $(\pi_1, \pi_2, f) \in \mathcal{B}$ then π_1 is a process of N_1, π_2 a process of N_2 and f is an isomorphism from $ac(\pi_1)$ onto $ac(\pi_2)$.

iii) If $(\pi_1, \pi_2, f) \in \mathcal{B}$ and $\pi_1 \xrightarrow{t} \pi_1'$ for some $t \in T_1$, then there exist t', π_2', f' with $\pi_2 \xrightarrow{t'} \pi_2'$, $(\pi_1', \pi_2', f') \in \mathcal{B}$ and $f'|_{ac(\pi_1)} = f$.

iv) Vice versa, i.e. if $(\pi_1, \pi_2, f) \in \mathcal{B}$ and $\pi_2 \xrightarrow{t'} \pi_2'$ for some $t' \in T_2$, then there exist t, π_1', f' with $\pi_1 \xrightarrow{t} \pi_1'$, $(\pi_1', \pi_2', f') \in \mathcal{B}$ and $f'|_{ac(\pi_1)} = f$.

If there exists a history-preserving bisimulation for nets N_1, N_2, then they are called *history-preserving bisimilar*.

As already explained, this definition regards processes as system states. Conditions iii) and iv) guarantee, that if one system can perform some action then this can be matched by the other system. The conditions on f and f' ensure that the new transitions t and t' have the same label, i.e. represent the same action, and that this action extends the processes π_1 and π_2 (or rather their action structures) in the same way. This definition considers only the extension of a process by one action. It can be shown that equivalently we may just as well consider extensions by arbitrarily many actions, see [BDKP91] and compare Theorem 6.3.6; in this case, one system performs a partial order of actions which can be matched in the other system by the same partial order due to the conditions on f and f'.

Next, we introduce a new description of system states, ordered markings. They consist of a safe marking together with a pre-order. This pre-order reflects precedences in the generation of tokens, where the precedence is not strict, i.e. if tokens s, s' are generated together we have $s \leq s'$ and $s' \leq s$. We define a corresponding occurrence rule, where for the follower ordered marking we have: if s precedes some s'' in the old ordered marking and s'' is used to produce a new token s', then s must precede s' in the new ordered marking.

Definition 6.4.2 An *ordered marking* of a net N is a pair $(M, \leq)$ such that $M \subseteq S$ and $\leq$ is a pre-order (i.e. a reflexive and transitive relation) on M. The *initial ordered marking* $initOM(N)$ is $(M_N, M_N \times M_N)$.

Let $t \in T$ and let M and M' be safe markings with $M[t\rangle M'$. Then, for an ordered marking $(M, \leq)$, we say that t is *enabled* under $(M, \leq)$, denoted $(M, \leq)[t\rangle$. If t occurs, it yields the *follower ordered marking* $(M', \leq')$, denoted $(M, \leq)[t\rangle(M', \leq')$, where $\leq'$ is defined by: for all $s, s' \in M'$ we have $s \leq' s'$ if and only if

- $s, s' \in M - {}^\bullet t$ and $s \leq s'$ or

- $s \in M - {}^\bullet t$, $s' \in t^\bullet$ and there exists $s'' \in {}^\bullet t$ with $s \leq s''$ or

- $s, s' \in t^\bullet$.

As usual, this is extended to sequences, and an ordered marking $(M, \leq)$ is *reachable* if for some $w \in T^*$ we have $initOM(N)[w\rangle(M, \leq)$. The set of reachable ordered markings is denoted by $OM(N)$.

It is easy to check that in this definition $(M', \leq')$ is well-defined, i.e. that $\leq'$ is indeed a pre-order on M'.

Remark: If we wanted to apply the usual bisimulation idea in a straightforward way, we could define a transition system as follows: the vertices are the reachable ordered markings, $initOM(N)$ is the root, and there is an a-labelled arc from $(M, \leq)$ to $(M', \leq')$ whenever $(M, \leq)[t\rangle(M', \leq')$ and $lab(t) = a$ (– we could write $(M, \leq)[a\rangle)(M', \leq')$ in this case). But it is important in the following that if $(M, \leq)[t\rangle(M', \leq')$ then we know which tokens of M t consumes, which tokens of M' t produces, and which tokens of M remain idle and appear in M' again. This information would be lost, if we would just write $(M, \leq)[lab(t)\rangle)(M', \leq')$ and forget about the transition. Instead, we will in fact refine the notion of a distributed transition system [DM87], although we will not define the transition system explicitly.

It seems possible to generalize our approach to non-safe nets. But then an ordered marking would have to be a pre-ordered set labelled with places of the net, since we could have several tokens on one place, i.e. several elements of an ordered marking with the same label. To describe the transition of one ordered marking to another, we would therefore have to specify which of the equally labelled elements are consumed and which stay idle.

This would make the connection to distributed transition systems much clearer, but it seems to be technically quite involved.   □

Next, we define how an ordered marking is generated by a process. This definition already indicates how we will use tokens to represent transitions of a process: a token s is less or equal than another token s' if the transition generating s is less or equal than the one generating s'. Therefore, the pre-order will enable us to extend an isomorphism as required in 6.4.1 iii) and iv) without looking at the process, but just by considering the generated ordered marking.

Definition 6.4.3 Let π be a process of a safe net N, $(M, \leq)$ be an ordered marking. Then we write $initOM(N)[\pi](M, \leq)$ if l is a bijection from $\max \pi$ to M such that for all $b, b' \in \max \pi$ we have $l(b) \leq l(b')$ if and only if either $b \in \min \pi$ or ${}^\bullet b \neq \emptyset \neq {}^\bullet b'$ and ${}^\bullet b \, F^* \, {}^\bullet b'$.

Here it is important that we use F^* instead of F^+. If s and s' are generated by the same transition, then $s \leq s'$ and $s' \leq s$. But if they are generated by different concurrent transitions, then s and s' are incomparable.

We have the following facts: for every process π, the mapping l is injective on $\max \pi$; $l(\max \pi)$ is a reachable marking; F^* is a partial order. Using these facts, it is easy to see that for every process π the ordered marking $(M, \leq)$ exists and is uniquely determined by this definition.

The following lemma exhibits an important relationship between the last two definitions.

Lemma 6.4.4 *Let π be a process of a safe net N, $t \in T$, and $(M, \leq)$, $(M', \leq')$ be ordered markings such that $initOM(N)[\pi](M, \leq)$. Then $(M, \leq)[t](M', \leq')$ if and only if there exists a process π' such that $\pi \xrightarrow{t} \pi'$ and $initOM(N)[\pi'](M', \leq')$.*

Proof: We have $(M, \leq)[t]$ iff $M[t]$ iff $\pi \xrightarrow{t} \pi'$ for some process π'. In this case, we obtain π from π' by adding a new event e with $l'(e) = t$ and for each $s \in t^\bullet$ a new condition b with $l'(b) = s$; furthermore, we define ${}^\bullet e$ to consist of all $b \in \max \pi$ with $l(b) \in {}^\bullet t$ and $e^\bullet$ to consist of all new conditions. It is standard (and not difficult to see), that π' is a process of N, and l' is a bijection from $\max \pi'$ onto M', where M' is defined by $M[t]M'$. Define $(M', \leq')$ by $(M, \leq)[t](M', \leq')$ and $(M', \leq'')$ by $initOM(N)[\pi'](M', \leq'')$. We have to show that $\leq''$ equals $\leq'$.

Hence let $s, s' \in M'$ and $b, b' \in \max \pi'$ with $l'(b) = s$, $l'(b') = s'$. We have $s, s' \in t^\bullet$ iff $b, b' \notin \max \pi$; in this case $s \leq' s'$ by Definition 6.4.2 and, since ${}^\bullet b = e = {}^\bullet b'$, $s \leq'' s'$ by Definition 6.4.3. We have $s, s' \in M - {}^\bullet t$ iff $b, b' \in \max \pi$; in this case $s \leq' s'$ iff $s \leq s'$ by 6.4.2 iff $s \leq'' s'$ by two applications of 6.4.3 to π and π'.

If finally $s \in M - {}^\bullet t$ and $s' \in t^\bullet$, then $s \leq' s'$ implies that there exists some $s'' \in {}^\bullet t$ with $s \leq s''$. Thus we have $b \in \max \pi \cap \max \pi'$, $b' \in \max \pi'$ and $b'' \in \max \pi$ such that $l(b'') = s''$, $b'' \in {}^\bullet e$, $b' \in e^\bullet$ and ($b \in \min \pi = \min \pi'$ or ${}^\bullet b \, F^* \, {}^\bullet b''$). Since ${}^\bullet b \, F^* \, {}^\bullet b''$ implies ${}^\bullet b \, F'^* \, {}^\bullet b'' \, F'^* \, {}^\bullet b'$, we get $s \leq'' s'$. Vice versa, assume $s \leq'' s'$. If $b \notin \max \pi$ and $b' \in \max \pi$, then 6.4.3 implies $e \, F'^* \, {}^\bullet b'$, a contradiction. Hence it remains the case that $b \in \max \pi$,

$b' \in e^{\bullet}$ and $(b \in \min \pi = \min \pi'$ or $^{\bullet}b\ F'^{*}\ e)$. If $^{\bullet}b\ F'^{*}\ e$, let $b'' \in {}^{\bullet}e$ with $^{\bullet}b\ F'^{*}\ b''$; if $b \in \min \pi$, choose an arbitrary $b'' \in {}^{\bullet}e$. (Since N is T-restricted, $^{\bullet}e \neq \emptyset$.) We have $b'' \in \max \pi$ and $(b \in \min \pi$ or $^{\bullet}b\ F^{*}\ ^{\bullet}b'')$, hence $s \leq l(b'') \in {}^{\bullet}t$. Therefore $s \leq' s'$.   □

Now we define our new bisimulation. As we will see in Proposition 6.4.7 below, we may assume for any element $((M_1, \leq_1), (M_2, \leq_2), \beta)$ of such an OM-bisimulation, that β is a relation between M_1 and M_2 that respects the pre-orders.

Definition 6.4.5 Let N_1, N_2 be safe nets. A relation $\mathcal{B} \subseteq OM(N_1) \times OM(N_2) \times \mathcal{P}(S_1 \times S_2)$ is an *OM-bisimulation* for N_1, N_2 if

i) $(initOM(N_1), initOM(N_2), M_{N_1} \times M_{N_2}) \in \mathcal{B}$

ii) If $((M_1, \leq_1), (M_2, \leq_2), \beta) \in \mathcal{B}$ and $(M_1, \leq_1)[t_1\rangle(M_1', \leq_1')$ then there exist $t_2 \in T_2$ and $(M_2', \leq_2')$ such that for β' defined by

  – for all $s_1 \in M_1'$, $s_2 \in M_2'$ we have $(s_1, s_2) \in \beta'$ if and only if

    – $s_1 \in M_1 - {}^{\bullet}t_1$, $s_2 \in M_2 - {}^{\bullet}t_2$, $(s_1, s_2) \in \beta$

    or

    – $s_1 \in t_1^{\bullet}$, $s_2 \in t_2^{\bullet}$

  we have

    – $(M_2, \leq_2)[t_2\rangle(M_2', \leq_2')$, $((M_1', \leq_1'), (M_2', \leq_2'), \beta') \in \mathcal{B}$ and $l_1(t_1) = l_2(t_2)$

    – for all $s_1 \in {}^{\bullet}t_1$ there are $s_1' \in {}^{\bullet}t_1$, $s_2' \in {}^{\bullet}t_2$ such that $s_1 \leq_1 s_1'$ and $(s_1', s_2') \in \beta$, and

      for all $s_2 \in {}^{\bullet}t_2$ there are $s_2' \in {}^{\bullet}t_2$, $s_1' \in {}^{\bullet}t_1$ such that $s_2 \leq_2 s_2'$ and $(s_1', s_2') \in \beta$.

iii) vice versa

If such a relation exists, then N_1, N_2 are called *OM-bisimilar*.

The basic idea of an OM-bisimulation is that in ii) the transition t_2 has to match t_1 not only by representing the same action, but also by consuming and producing related tokens. But note that, since tokens only interest us as far as precedences in their generation are concerned, we do not require that the tokens consumed/produced by t_1 are in a bijective correspondence to those consumed/produced by t_2. We do not even require that a token s_1 consumed by t_1 is itself related to some s_2 consumed by t_2; it is enough that s_1 precedes some s_1' that is consumed by t_1 and related to some token consumed by t_2.

What we have in mind with this definition is to relate an OM-bisimulation $\mathcal{B}$ to a history-preserving bisimulation $\mathcal{R}$ in the following way. If we have $(\pi_1, \pi_2, f) \in \mathcal{R}$, and the processes π_1, π_2 generate $(M_1, \leq_1)$, $(M_2, \leq_2)$, then we have $((M_1, \leq_1), (M_2, \leq_2), \beta) \in \mathcal{B}$ such that tokens s, s' are related by β if the transition generating s is mapped to the transition generating s' by f. The definition of β' in 6.4.5 ii) ensures that in the new element $((M_1', \leq_1'), (M_2', \leq_2'), \beta')$ of $\mathcal{B}$ this relation stores the correct information. The last part of 6.4.5 ii) ensures that we can extend f to an isomorphism between the extensions of π_1 by t_1 and π_2 by t_2.

Proposition 6.4.7 below gives some more insight into the nature of an OM-bisimulation.

Definition 6.4.6 Let N_1, N_2 be safe nets and $\mathcal{B}$ be an *OM*-bisimulation for N_1 and N_2. We call $((M'_1, \leq'_1), (M'_2, \leq'_2), \beta') \in \mathcal{B}$ *reachable*, if

i) $((M'_1, \leq'_1), (M'_2, \leq'_2), \beta') = (initOM(N_1), initOM(N_2), M_{N_1} \times M_{N_2})$

or

ii) there exist a reachable $((M_1, \leq_1), (M_2, \leq_2), \beta) \in \mathcal{B}$ and $t_1 \in T_1$, $t_2 \in T_2$ such that the properties of Definition 6.4.5 ii) are satisfied.

Proposition 6.4.7 *Let N_1, N_2 be safe nets and $\mathcal{B}$ be an OM-bisimulation for N_1 and N_2.*

i) $\{((M_1, \leq_1), (M_2, \leq_2), \beta) \in \mathcal{B} \mid ((M_1, \leq_1), (M_2, \leq_2), \beta)$ is reachable$\}$ is an OM-bi-simulation for N_1 and N_2.

ii) If $((M_1, \leq_1), (M_2, \leq_2), \beta) \in \mathcal{B}$ is reachable, then $\beta \subseteq M_1 \times M_2$ and whenever $(s_1, s_2), (s'_1, s'_2) \in \beta$ then $s_1 \leq_1 s'_1$ if and only if $s_2 \leq_2 s'_2$.

Proof: i) by induction.

ii) By symmetry it is enough to show one implication, which we will do by induction, and the claim is obvious for $(initOM(N_1), initOM(N_2), M_{N_1} \times M_{N_2})$. Now assume that $((M'_1, \leq'_1), (M'_2, \leq'_2), \beta')$ is constructed from a reachable $((M_1, \leq_1), (M_2, \leq_2), \beta) \in \mathcal{B}$, t_1 and t_2 according to Definition 6.4.5 ii). Obviously $\beta' \subseteq M'_1 \times M'_2$. Let $(s_1, s_2), (s'_1, s'_2) \in \beta'$ and $s_1 \leq'_1 s'_1$. If $s_1 \in t_1^{\bullet}$ and $s'_1 \in t_1^{\bullet}$, then $s_2 \leq'_2 s'_2$ since $s_2, s'_2 \in t_2^{\bullet}$. If $s_1 \in M_1 - {}^{\bullet}t_1$ and $s'_1 \in M_1 - {}^{\bullet}t_1$ then $s_2, s'_2 \in M_2 - {}^{\bullet}t_2$, and we are done by induction.

Finally, assume $s_1 \in M_1 - {}^{\bullet}t_1$ and $s'_1 \in t_1^{\bullet}$; hence, by the definition of β' in 6.4.5 ii), we have $s_2 \in M_2 - {}^{\bullet}t_2$, $(s_1, s_2) \in \beta$ and $s'_2 \in t_2^{\bullet}$. Since $s_1 \leq'_1 s'_1$, there exists $s''_1 \in {}^{\bullet}t_1$ with $s_1 \leq_1 s''_1$. By Definition 6.4.5 ii) there exist $s'''_1 \in {}^{\bullet}t_1$ and $s'''_2 \in {}^{\bullet}t_2$ such that $s''_1 \leq_1 s'''_1$ and $(s'''_1, s'''_2) \in \beta$. Thus we have $s_1 \leq_1 s'''_1$, hence by induction $s_2 \leq_2 s'''_2$. Since $s_2 \in M_2 - {}^{\bullet}t_2$ and $s'_2 \in t_2^{\bullet}$, this implies $s_2 \leq'_2 s'_2$. □

Hopefully, this proposition makes the definition of an *OM*-bisimulation appear more natural: we may assume that for an element $((M_1, \leq_1), (M_2, \leq_2), \beta)$ of an *OM*-bisimula-tion β is a relation between M_1 and M_2 that respects the pre-orders. We have not included this requirement in Definition 6.4.5, since it would only complicate the definition further.

The characterization

Now we will show the main result of this section.

Theorem 6.4.8 *Let N_1, N_2 be safe nets. Then N_1, N_2 are history-preserving bisimilar if and only if they are OM-bisimilar.*

Proof: '$\Rightarrow$' Let $\mathcal{R}$ be a history-preserving bisimulation for N_1, N_2. Then we define an *OM*-bisimulation $\mathcal{B}$ as follows:

$((M_1, \leq_1), (M_2, \leq_2), \beta) \in \mathcal{B}$ if and only if there is $(\pi_1, \pi_2, f) \in \mathcal{R}$ such that

- $initOM(N_1)[\pi_1](M_1, \leq_1)$ and $initOM(N_2)[\pi_2](M_2, \leq_2)$

- for all $s_1 \in S_1$ and $s_2 \in S_2$, we have $(s_1, s_2) \in \beta$ iff $s_1 \in M_1$, $s_2 \in M_2$ and there are $b_1 \in \max \pi_1$ and $b_2 \in \max \pi_2$ such that $l_1(b_1) = s_1$, $l_2(b_2) = s_2$ and

  - $b_1 \in \min \pi_1$ and $b_2 \in \min \pi_2$ or

  - ${}^\bullet b_1 \neq \emptyset$ and $f({}^\bullet b_1) = {}^\bullet b_2$.

We have to show that $\mathcal{B}$ is an OM-bisimulation.

Obviously, we have $(initOM(N_1), initOM(N_2), M_{N_1} \times M_{N_2}) \in \beta$. Hence assume we are given $((M_1, \leq_1), (M_2, \leq_2), \beta) \in \mathcal{B}$ with a corresponding $(\pi_1, \pi_2, f) \in \mathcal{R}$ and we have $(M_1, \leq_1)[t_1](M_1', \leq_1')$. By Lemma 6.4.4, we find some π_1' such that $\pi_1 \xrightarrow{t_1} \pi_1'$ and $initOM(N_1)[\pi_1'](M_1', \leq_1')$. Applying the definition of $\mathcal{R}$ gives us π_2', t_2, f' such that $\pi_2 \xrightarrow{t_2} \pi_2'$, $(\pi_1', \pi_2', f') \in \mathcal{R}$ and $f'|_{ac(\pi_1)} = f$. Define $(M_2', \leq_2')$ and β' such that $initOM(N_2)[\pi_2'](M_2', \leq_2')$ and $((M_1', \leq_1'), (M_2', \leq_2'), \beta') \in \mathcal{B}$ according to the definition of $\mathcal{B}$ applied to $(\pi_1', \pi_2', f') \in \mathcal{R}$. Lemma 6.4.4 shows that $(M_2, \leq_2)[t_2](M_2', \leq_2')$, and since f' maps the additional event e_1 of π_1' to the additional event e_2 of π_2', we have $lab_1(t_1) = lab_2(t_2)$. This gives us the first part of Part ii) of Definition 6.4.5.

We have to check that the above β' coincides with the β' defined in 6.4.5 ii); let $s_1 \in M_1'$, $s_2 \in M_2'$. By the above definition of β', we have $(s_1, s_2) \in \beta'$ if and only if

(∗) there are $b_1 \in \max \pi_1'$ and $b_2 \in \max \pi_2'$ such that $l_1'(b_1) = s_1$, $l_2'(b_2) = s_2$ and either $b_1 \in \min \pi_1'$ and $b_2 \in \min \pi_2'$ or ${}^\bullet b_1 \neq \emptyset$ and $f'({}^\bullet b_1) = {}^\bullet b_2$.

Assume $(s_1, s_2) \in \beta'$. If $s_1 \in t_1^\bullet$ or $s_2 \in t_2^\bullet$, then (∗) implies ${}^\bullet b_1 = e_1$ or ${}^\bullet b_2 = e_2$, thus ${}^\bullet b_1 \neq \emptyset$ and $f'({}^\bullet b_1) = {}^\bullet b_2$, hence ${}^\bullet b_1 = e_1$ and ${}^\bullet b_2 = e_2$, and therefore $s_1 \in t_1^\bullet$ and $s_2 \in t_2^\bullet$. If on the other hand, $s_1 \notin t_1^\bullet$ and $s_2 \notin t_2^\bullet$, then (∗) implies $b_1 \in \max \pi_1$ and $b_2 \in \max \pi_2$, thus $s_1 \in M_1$, $s_2 \in M_2$, and the properties of (∗) hold for π_1 and π_2. Hence $(s_1, s_2) \in \beta$. Now $b_1 \in \max \pi_1'$ and $b_2 \in \max \pi_2'$ imply $b_1 \notin {}^\bullet e_1$ and $b_2 \notin {}^\bullet e_2$; since $b_1 \in \max \pi_1$ and $b_2 \in \max \pi_2$, we conclude that $s_1 \notin {}^\bullet t_1$ and $s_2 \notin {}^\bullet t_2$. Therefore $s_1 \in M_1 - {}^\bullet t_1$ and $s_2 \in M_2 - {}^\bullet t_2$.

For the reverse implication, $s_1 \in t_1^\bullet$ and $s_2 \in t_2^\bullet$ imply (∗) with $b_1 \in e_1^\bullet$ and $b_2 \in e_2^\bullet$. Now assume $s_1 \in M_1 - {}^\bullet t_1$, $s_2 \in M_2 - {}^\bullet t_2$ and $(s_1, s_2) \in \beta$. The latter gives us (∗) with $b_1 \in \max \pi_1$, $b_2 \in \max \pi_2$ etc., while $s_1 \notin {}^\bullet t_1$ and $s_2 \notin {}^\bullet t_2$ imply $b_1 \in \max \pi_1'$ and $b_2 \in \max \pi_2'$, and (∗) follows.

For the last part of ii), take $s_1 \in {}^\bullet t_1$ and let $b_1 \in {}^\bullet e_1$ with $l_1(b_1) = s_1$. There are two cases:

a) e_1 is a minimal element of $ac(\pi_1')$. Then $b_1 \in \min \pi_1 \cap \max \pi_1$. Since N_2 is T-restricted, we have ${}^\bullet t_2 \neq \emptyset$, hence let $s_2' \in {}^\bullet t_2$, $b_2' \in {}^\bullet e_2$ with $l_2(b_2') = s_2'$, and $s_1' = s_1$. Since f' is an isomorphism, e_2 is a minimal element of $ac(\pi_2')$, and we have $b_2' \in \min \pi_2 \cap \max \pi_2$. Thus, we have $s_1 \leq_1 s_1'$ and $(s_1', s_2') \in \beta$ by definition of $\mathcal{B}$.

b) e_1 is not a minimal element of $ac(\pi_1')$. If $b_1 \in \min \pi_1$, let e_1' be any immediate predecessor of e_1 in $ac(\pi_1')$; if $b_1 \notin \min \pi_1$, let $e_1'' = {}^\bullet b_1$ and let e_1' be an immediate predecessor of e_1 in $ac(\pi_1')$ with $e_1'' F_1'^* e_1'$. Let $e_2' = f(e_1')$. There exist $b_1' \in {}^\bullet e_1 \cap e_1'^\bullet$ and $b_2' \in {}^\bullet e_2 \cap e_2'^\bullet$, since e_2' is an immediate predecessor of e_2 in $ac(\pi_2')$. Let $s_1' = l_1(b_1')$ and

$s_2' = l_2(b_2')$. By choice of e_1' we have $s_1 \leq_1 s_1'$; since $b_1' \in \max \pi_1$ and $b_2' \in \max \pi_2$, we have $(s_1', s_2') \in \beta$ by definition of $\mathcal{B}$.

For $s_2 \in {}^\bullet t_2$ we proceed analogously, which finishes the last part.

'$\Leftarrow$' Let $\mathcal{B}$ be an *OM*-bisimulation for N_1, N_2. Then we define $\mathcal{R}$ and a function $o : \mathcal{R} \to \mathcal{P}(\mathcal{B})$ as follows

a) $(\pi_0(N_1), \pi_0(N_2), \emptyset) \in \mathcal{R}$ and $o(\pi_0(N_1), \pi_0(N_2), \emptyset) = \{(initOM(N_1), initOM(N_2), M_{N_1} \times M_{N_2})\}$

b) Let $(\pi_1, \pi_2, f) \in \mathcal{R}$ and $((M_1, \leq_1), (M_2, \leq_2), \beta) \in o(\pi_1, \pi_2, f)$. Let $(M_1, \leq_1)$, $(M_2, \leq_2)$, β, t_1, t_2, β' etc. satisfy the properties of Definition 6.4.5 ii), and let $\pi_1 \xrightarrow{t_1} \pi_1'$, $\pi_2 \xrightarrow{t_2} \pi_2'$. Then $(\pi_1', \pi_2', f') \in \mathcal{R}$ where $f'|_{ac(\pi_1)} = f$ and $f'(e_1) = e_2$ for the additional events e_1 of π_1' and e_2 of π_2'. Furthermore $((M_1', \leq_1'), (M_2', \leq_2'), \beta') \in o(\pi_1', \pi_2', f')$.

We show some claims first, where Claim 1 is obvious:

Claim 1: For all $(\pi_1, \pi_2, f) \in \mathcal{R}$ we have $o(\pi_1, \pi_2, f) \neq \emptyset$.

Claim 2: Whenever we have $((M_1, \leq_1), (M_2, \leq_2), \beta) \in o(\pi_1, \pi_2, f)$ for some $(\pi_1, \pi_2, f) \in \mathcal{R}$, then $initOM(N_1)[\pi_1](M_1, \leq_1)$ and $initOM(N_2[\pi_2](M_2, \leq_2)$.

Proof of Claim 2: By induction, the claim being obvious for $\pi_1 = \pi_0(N_1)$, $\pi_2 = \pi_0(N_2)$. Hence let $(\pi_1', \pi_2', f') \in \mathcal{R}$ be constructed from (π_1, π_2, f) as in b). Then $initOM(N_1)[\pi_1]$ $(M_1, \leq_1)$ by induction, furthermore $(M_1, \leq_1)[t_1](M_1', \leq_1')$, thus $initOM(N_1)[\pi_1'](M', \leq_1')$ by Lemma 6.4.4. The second part is analogous.

Claim 3: Let $(\pi_1, \pi_2, f) \in \mathcal{R}$ and $((M_1, \leq_1), (M_2, \leq_2), \beta) \in o(\pi_1, \pi_2, f)$. Let $b_1 \in \max \pi_1$, $b_2 \in \max \pi_2$ such that $(l_1(b_1), l_2(b_2)) \in \beta$. Then

- ${}^\bullet b_1 \neq \emptyset$ implies $f({}^\bullet b_1) = {}^\bullet b_2$

- ${}^\bullet b_2 \neq \emptyset$ implies $f^{-1}({}^\bullet b_2) = {}^\bullet b_1$.

Proof of Claim 3: By induction, and for $(\pi_0(N_1), \pi_0(N_2), \emptyset)$ there is nothing to show. Let $(\pi_1', \pi_2', f') \in \mathcal{R}$ be constructed from (π_1, π_2, f) according to b). If ${}^\bullet b_1 = e_1$, then $l_1'(b_1) \in t_1^\bullet$. By the definition of β' in 6.4.5 ii), we must have $l_2'(b_2) \in t_2^\bullet$, hence ${}^\bullet b_2 = e_2$ and $f'({}^\bullet b_1) = {}^\bullet b_2$. The same follows if ${}^\bullet b_2 = e_2$. If ${}^\bullet b_1 \neq e_1$ and ${}^\bullet b_2 \neq e_2$, then $b_1 \in \max \pi_1$, $b_2 \in \max \pi_2$ and $l_1(b_1) \in M_1 - {}^\bullet t_1$ and $l_2(b_2) \in M_2 - {}^\bullet t_2$ with Claim 2; the definition of β' in 6.4.5 ii) gives $(l_1(b_1), l_2(b_2)) \in \beta$. Therefore, by induction, ${}^\bullet b_1 \neq \emptyset$ implies $f({}^\bullet b_1) = {}^\bullet b_2$, i.e. $f'({}^\bullet b_1) = {}^\bullet b_2$, and ${}^\bullet b_2 \neq \emptyset$ implies $f'^{-1}({}^\bullet b_2) = {}^\bullet b_1$.

Remark: It is also possible to show that in fact for $b_1 \in \max \pi_1$ and $b_2 \in \max \pi_2$ we have $(l_1(b_1), l_2(b_2)) \in \beta$ if and only if either ${}^\bullet b_1 = \emptyset = {}^\bullet b_2$ or ${}^\bullet b_1 \neq \emptyset \neq {}^\bullet b_2$ and $f({}^\bullet b_1) = {}^\bullet b_2$. Thus, o really is a function from $\mathcal{R}$ to $\mathcal{B}$. But this is not important for this proof. □

Proof of Theorem 6.4.8 continued:

We have to show that $\mathcal{R}$ satisfies Definition 6.4.1; i) is obvious.

For ii) assume that some $(\pi_1', \pi_2', f') \in \mathcal{R}$ is constructed from $(\pi_1, \pi_2, f) \in \mathcal{R}$ etc. We have to show that f' is an isomorphism from $ac(\pi_1')$ to $ac(\pi_2')$. Since by induction f is an isomorphism from $ac(\pi_1)$ to $ac(\pi_2)$, f' is by construction a label-preserving bijection from $ac(\pi_1')$ to $ac(\pi_2')$. Since e_1, e_2 are maximal in $ac(\pi_1')$, $ac(\pi_2')$, we only have to check:

- If e'_1 is an immediate predecessor of e_1 in $ac(\pi'_1)$, then $f(e'_1)$ is a predecessor of e_2 in $ac(\pi'_2)$.

- If e'_2 is an immediate predecessor of e_2 in $ac(\pi'_2)$, then $f^{-1}(e'_2)$ is a predecessor of e_1 in $ac(\pi'_1)$.

Hence let e'_1 be given. Since π'_1 is a process, we find $b_1 \in {}^{\bullet}e_1 \cap e'_1{}^{\bullet}$. We have $l_1(b_1) \in {}^{\bullet}t_1$, hence by the last part of Definition 6.4.5 ii) there are $s'_1 \in {}^{\bullet}t_1$, $s'_2 \in {}^{\bullet}t_2$ such that $l_1(b_1) \leq_1 s'_1$ and $(s'_1, s'_2) \in \beta$. First, we find $b'_1 \in \max \pi_1$ and $b'_2 \in \max \pi_2$ such that $l_1(b'_1) = s'_1$ and $l_2(b'_2) = s'_2$. By Claim 2 we have ${}^{\bullet}b_1\, F^{*}_1\, {}^{\bullet}b'_1$, and by Claim 3 we have $f({}^{\bullet}b'_1) = {}^{\bullet}b'_2$. Furthermore $b'_2{}^{\bullet} = e_2$. Therefore $f(e'_1) = f({}^{\bullet}b_1)\, F^{*}_2\, f({}^{\bullet}b'_1) = {}^{\bullet}b'_2\, F^{*}_2\, e_2$. The property for an immediate predecessor e'_2 of e_2 is shown analogously.

Finally, we will show Definition 6.4.1 iii) (iv) is analogous). Let $(\pi_1, \pi_2, f) \in \mathcal{R}$, $t_1 \in T_1$ and π'_1 with $\pi_1 \xrightarrow{t_1} \pi'_1$ be given. By Claim 1 we find $((M_1, \leq_1), (M_2, \leq_2), \beta) \in o(\pi_1, \pi_2, f)$. Define $(M'_1, \leq'_1)$ by $initOM(N_1)[\pi'_1](M'_1, \leq'_1)$. By Claim 2 and Lemma 6.4.4, we have $(M_1, \leq_1)[t_1](M'_1, \leq'_1)$. Therefore, we can find $t_2 \in T_2$, $(M'_2, \leq'_2)$ and β' such that the properties of Definition 6.4.5 ii) hold. We are done if we can show that $\pi_2 \xrightarrow{t_2} \pi'_2$ for some π'_2. But this is clear since $initOM(N_2)[\pi_2](M_2, \leq_2)$ by Claim 2, and $(M_2, \leq_2)[t_2]$ by choice of t_2.   □

Corollary 6.4.9 *It is decidable, whether finite safe nets are history-preserving bisimilar.*

Proof: By Theorem 6.4.8 we have to check whether there exists an *OM*-bisimulation $\mathcal{B}$ for the given nets N_1 and N_2. $OM(N_1)$, $OM(N_2)$ and $S_1 \times S_2$ are finite, thus there are only finitely many objects that are possible elements of $\mathcal{B}$. Therefore, we can check by exhaustive search whether one of the finitely many possible sets $\mathcal{B}$ is an *OM*-bisimulation.

By Proposition 6.4.7 i), we can restrict our search to those relations $\mathcal{B}$ that have reachable elements only.   □

It must be remarked that in general the complexity of this decision procedure is prohibitive. Already the number of reachable markings of a safe net can be exponential in the size of the net.

A topic for future research is to generalize *OM*-bisimulation to nets that are not necessarily safe; this should be possible, but might be technically quite involved, compare [DDNM89]. Furthermore, internal transitions should be incorporated; here the results of the previous sections and [Dev90b] should be helpful.

Chapter 7

Partial Order Semantics for Nets with Capacities

So far, we have only considered nets without capacities and their semantics, especially their partial order semantics. In this chapter, which reports joint work with Robert Gold [GV90], we will continue the comparative study of partial order semantics for Petri nets *with* capacities initiated in [HRT89]. A number of properties are defined in [HRT89] which one might want a semantics to share:

On the one hand, the semantics should support the modular construction of nets, i.e. it should be possible to determine the semantics of a composed net from the semantics of the components. The composition operator ∥ on nets, which we have defined in Section 3.1, and a corresponding operator ∥ on sets of labelled partial orders are considered; a semantics Sem is called compositional if we have $Sem(N_1 ∥ N_2) = Sem(N_1) ∥ Sem(N_2)$.

On the other hand, certain net modifications, which are related to capacities, should not change the semantics of the net. If this is the case, the semantics is called complementation-invariant and capacity-oriented.

We will continue this study. In Section 7.1 it is shown that there is only one compositional semantics for 1-bounded nets, thereby improving a result of [HRT89]. In Section 7.2 we give a characterization of complementation-invariance and a sufficient condition for capacity-orientation. We introduce the *Flow* semantics, a new semantics which shares all the above mentioned properties. Using some rather pathological partial order semantics, we show that a new criterion is necessary to exclude these from our considerations; therefore, we introduce a property called uniformity. Then we show a closure-property for each semantics that is well-behaved, i.e. fulfills all the properties mentioned so far. We give a characterization of capacity-orientation and show that *Flow* is the unique minimal well-behaved semantics. With a certain view of causality, this makes *Flow* the only well-behaved semantics that models causality.

Finally, we discuss in Section 7.3 whether compositionality as it is defined in [HRT89] really is such a desirable property: in [Ber87], S-modification is introduced, and one would expect that such a modification would preserve the semantics of a net. We show that this is not the case for any compositional semantics. Therefore, we consider another operator introduced in [Gra81] and show that there is a unique partial order semantics which is compositional with respect to this operator and fulfills all our quality criteria.

7.1   Compositionality

This chapter gives only a first look at possible partial order semantics for nets with capacities, and we will restrict our considerations to linear-time semantics. Also, in this chapter all nets will be of a restricted form only, especially they will be unlabelled.

General assumption In this chapter all nets are finite, unlabelled, and of arc weight 1; we also forbid loops and isolated transitions to avoid technical complications.

Unlabelled nets do not have an explicit labelling function; nevertheless, we can use notions that are defined for labelled nets by taking the labelling to be the identity. Thus e.g. firing sequences and image firing sequences coincide in this chapter; another consequence of this view is that isomorphic nets must have the same transitions while places may be renamed.

On the one hand, the nets in this chapter are quite restricted. On the other hand, they are more general: the nets in this chapter have a capacity function, which assigns to each place the maximal number of tokens that are allowed on this place; if the number of tokens is unlimited, then the assigned value is infinity, denoted by ω. We calculate with ω as usual.

Thus a net N is a tuple (S, T, W, K, M_N), where S, T, W and M_N have the usual meaning, K is a function $S \rightarrow \mathbb{N}_0 \cup \{\omega\}$, called the *capacity function*, and we require that $\forall s \in S : M_N(s) \leq K(s)$. The last condition means that the initial marking does not violate the capacity constraint given by K. Analogously, we have to supplement the firing rule; a step μ is *enabled* under a marking M, denoted by $M[\mu\rangle$, if not only $\sum_{t \in \mu} \mu(t) \cdot W(s, t) \leq M(s)$ for all $s \in S$, but also $\sum_{t \in \mu} \mu(t) \cdot W(t, s) + M(s) \leq K(s)$ for all $s \in S$. With this notion of enabling the firing of steps and transitions, follower marking, reachable markings etc. are defined as before. Observe especially that the definition of a partial word makes use of the enabledness of steps; thus, partial words of a net with capacities are defined as before – except that enabledness of a step refers to the supplemented definition given above. Since the nets in this chapter are pure, i.e. without loops, we can give a stronger version of Proposition 2.2.5; [Kie88] contains a proof for nets without capacities, which stays virtually the same for our setting.

Proposition 7.1.1 *A labelled partial order p over T_N is enabled under some marking M of a net N if and only if for every linearization $w \in T^*$ of p we have $M[w\rangle$.*

Especially, we will have a look at the class of 1-bounded nets, which contains safe nets (i.e. 1-bounded nets where all capacities are ω) and EN-systems (i.e. nets where all capacities are 1): N is called *1-bounded* if $\forall M \in [M_N\rangle \forall s \in S_N : M(s) \leq 1$.

We have already met various partial order semantics for nets. To give a framework for this chapter, we require that every semantics we deal with assigns to each net a set of partial words. Thus – as announced above – we will only deal with linear-time semantics. We also require that at least the firing sequences are 'covered' in some sense.

Definition 7.1.2 *A semantics Sem of nets (i.e. a function with domain the class of all nets) is called admissible if for all nets N we have:*

i) $Sem(N) \subseteq PW(N)$

ii) Every firing sequence is a linearization of some $p \in Sem(N)$.

We call *Sem consistent*, if additionally

iii) every step firing sequence is a step linearization of some $p \in Sem(N)$.

If i) and ii), or i), ii) and iii) only hold for a class $\mathcal{C}$ of nets, then *Sem* is admissible or consistent for $\mathcal{C}$. *Sem* is called *prefix-closed*, if $Sem(N)$ is prefix-closed for all nets N.

If a semantics is consistent (a notion introduced in [HRT89]), then it represents in some sense the full parallelism of each net. Thus, admissability is a minimal requirement for a semantics, consistency is a desirable stronger property. Prefix-closure is a natural condition for any linear-time semantics; the intuition for such a semantics is to be the set of all (not necessarily complete) system runs, and the prefix of a system run is a begin of this run and thus a run itself. Following [HRT89], we consider the operator ∥ for the composition of nets, where synchronization is over the intersection of the alphabets of the two nets, see Definition 3.1.1. Nets are unlabelled in this chapter, i.e. the implicit labelling function is the identity and thus injective; hence, the composition of nets is obtained by taking their disjoint union and merging the common transitions.

As we have already seen in Proposition 3.1.2, this composition is commutative and associative and has the empty net as neutral element. Hence, we can generalize the operator to finite families $(N_i)_{i \in I}$ of nets; in this case we write the result as $\|_{i \in I} N_i$. With this operator, unlabelled nets can uniquely be decomposed into their atomic subnets. (See [HRT89]; for nets without capacities see e.g. [Maz88].)

Definition 7.1.3 A net N is *atomic* if it has exactly one place. We write such a net as (I, O, k, n), where the name of the place s is suppressed, $I = {}^\bullet s, O = s^\bullet, k = K(s), n = M_N(s)$. The elements of I are called *input-transition*, those of O *output-transitions*. For a net N and a place $s \in S$, the *atomic subnet* $N(s)$ of N induced by s is $({}^\bullet s, s^\bullet, K(s), M_N(s))$.

Proposition 7.1.4 *Each net N is a unique composition of atomic nets, namely $N = \|_{s \in S_N} N(s)$.*

For the proof, observe that the nets in this paper do not have isolated transitions. If we want to construct nets in a modular fashion, we must be able to determine the semantics of a composed net from the semantics of the components – just as in the preceding chapters. Here we consider compositionality with respect to the net operator ∥ as an important property of a semantics:

Definition 7.1.5 Let *op* be an operator that assigns to each pair consisting of a set of partial words labelled over some Z_1 and a set of partial words labelled over some Z_2 a set of partial words labelled over $Z_1 \cup Z_2$; let *Sem* be an admissible semantics of nets. Then *Sem* is called *compositional with respect to op*, if for all nets N_1, N_2 we have $Sem(N_1 \| N_2) = op(Sem(N_1), Sem(N_2))$.

Since the net composition is commutative and associative, it follows that an operator as in Definition 7.1.5 must be commutative and associative, at least when applied to sets $Sem(N)$, N a net. Furthermore, for any admissible semantics Sem the semantics of the empty net is the empty set, thus the empty set is a neutral element, as far as we are interested. Hence, when applying op to a finite family $(Sem(N_i))_{i \in I}$, we can write the result as $op_{i \in I} Sem(N_i)$. As observed e.g. in [Gra81] and [Maz84], the semantics of a net can be determined from the semantics of the atomic subnets if the semantics is compositional.

Proposition 7.1.6 *Let Sem be an admissible semantics, and op be an operator as in 7.1.5 which is commutative and associative and has the empty set as neutral element. Then Sem is compositional with respect to op if and only if for all nets N*

$$Sem(N) = op_{s \in S_N} Sem(N(s)).$$

Next we define the operator $\|$ for labelled partial orders, which corresponds to the net composition operator $\|$ in much the same way as the operator $\|_A$ on labelled partial orders corresponds to the net composition operator $\|_A$. But there is a subtlety, and therefore we have not given this definition in Chaper 3 already. The operator $\|$ synchronizes over the intersection of the alphabets of the two operands. In order to maintain the correspondence between $\|$ on nets and on labelled partial orders, it is important that the synchronization set is the same when we apply $\|$ to nets N_1 and N_2 or when we apply $\|$ to a partial word of N_1 and a partial word of N_2. Thus, the alphabet of a labelled partial order must not be the set of labels actually appearing, but a partial order p labelled over some Z must have Z as alphabet, which is denoted by $\alpha(p)$. Consequently we must work with qualified pomsets in the sense of [Maz88], which have the alphabet as a separate component. But strictly speaking, the alphabet Z is the codomain of the labelling l and thus part of the labelled partial order already. Hence, we will not add the alphabet of a labelled partial order as an explicit fourth component, and we trust that this does not lead to any confusion.

Observe that, as a consequence of our considerations, a partial order labelled over Z_1 is always different from a partial order labelled over Z_2 for $Z_1 \neq Z_2$.

A partial word of a net N is labelled over T; since the nets in this chapter have the identity as labelling, such a partial word coincides with its image except that the image is labelled over Σ. In view of this small difference, we will use $PW(N)$ in this chapter to denote the set of partial words of N, i.e. $PW(N)$ is a set of partial orders labelled over T.

Since in Definition 3.1.5 we have only considered the parallel composition of partial orders labelled over Σ, we will spell out the definition of $\|$ below, although basicly we have $p_1 \| p_2 = p_1 \|_{\alpha(p_1) \cap \alpha(p_2)} p_2$. We also change the definition of the projection id_Z^* of a labelled partial order to some set Z slightly: in this chapter, when applying id_Z^* we have Z as the alphabet of the result.

Definition 7.1.7 Let p_1, p_2 be labelled partial orders over sets Z_1 and Z_2. If $p_1 = (X_1, <_1, l_1)$ and $p_2 = (X_2, <_2, l_2)$ are representatives such that

- $\{x \in X_1 \mid l_1(x) \in Z_1 \cap Z_2\} = X_1 \cap X_2 = \{x \in X_2 \mid l_2(x) = Z_1 \cap Z_2\}$

- $l_1(x) = l_2(x)$ for all $x \in X_1 \cap X_2$
- $(<_1 \cup <_2)^+$ is a partial order on $X_1 \cup X_2$,

then $p = (X_1 \cup X_2, (<_1 \cup <_2)^+, l)$ is a *synchronization* of p_1 and p_2, where l is that function into $Z_1 \cup Z_2$ for which $l_1 = l|_{X_1 \times Z_1}$ (i.e. l_1 is $l|_{X_1}$ with the codomain changed to Z_1) and $l_2 = l|_{X_2 \times Z_2}$.

The set of all synchronizations of p_1 and p_2 is denoted by $p_1 \parallel p_2$. If L_i is a set of partial orders labelled over Z_i, i=1,2, then the synchronization of L_1 and L_2 is defined as $L_1 \parallel L_2 = \{p \in p_1 \parallel p_2 \mid p_1 \in L_1, p_2 \in L_2\}$.

The synchronization of two labelled partial orders is not unique. Consider the partial orders p, q labelled over $\{a, b\}$ in Figure 7.1: we have $p \parallel p = \{p, q\}$. This is very similar to the example shown in Figure 3.5.

Figure 7.1

We have the following easy observations:

Lemma 7.1.8 *Let p_i be a partial order labelled over Z_i, $i = 1, 2$, let $p \in p_1 \parallel p_2$.*

i) $p_1 \preceq id^*_{Z_1}(p)$ and $p_2 \preceq id^*_{Z_2}(p)$; especially, if $Z_1 = Z_2$ then $p_1 \preceq p$ and $p_2 \preceq p$.

ii) $p \in id^*_{Z_1}(p) \parallel p_2$ and $p \in p_1 \parallel id^*_{Z_2}(p)$

iii) If $p'_1 \preceq p_1$, then there exists p' with $p' \preceq p$ and $p' \in p'_1 \parallel p_2$.

As observed in [Maz88] for nets without capacities, it is not hard to see that $\parallel$ is commutative and associative. Applied to labelled partial orders, it has the empty word over the empty alphabet as neutral element. Applied to sets of labelled partial orders, it has the set consisting of the empty word over the empty alphabet as neutral element. Hence, for a finite family $(L_i)_{i \in I}$ we may write $\parallel_{i \in I} L_i$. If a semantics is compositional with respect to $\parallel$, we will simply say that it is *compositional*.

We recall the following result from [Gra81], which partly corresponds to Theorem 3.1.4 ii):

Proposition 7.1.9 *Let N_1, N_2 be nets, $p_1 \in PW(N_1), p_2 \in PW(N_2), p \in p_1 \parallel p_2$. Then $p \in PW(N_1 \parallel N_2)$.*

Proof: In [Gra81], this is shown for nets with a final marking and without capacities, and for a slightly different operator for partial words, but the proof stays virtually the same.   □

The following theorem of [Maz88] and [HRT89] shows that in order to define a compositional semantics it is enough to define it on atomic nets. Then it can uniquely be extended to a semantics of nets in general, and this extension preserves the desirable properties of consistency and prefix-closure.

Theorem 7.1.10 *Let Sem be an admissible semantics of atomic nets, then Sem can uniquely be extended to a compositional semantics of nets by*

$$Sem(N) = \|_{s \in S} Sem(N(s)).$$

If the given semantics is consistent for atomic nets, then the extended semantics is consistent for all nets. If the given semantics is prefix-closed, then the extended semantics is prefix-closed, too.

Before applying this theorem, let us note the following lemma:

Lemma 7.1.11 *Let $(E, <, l)$ be a partial word that is enabled in an atomic net (I, O, k, n), let $e \in E$ be an immediate predecessor of $e' \in E$, such that both $l(e)$ and $l(e')$ are in I or both $l(e)$ and $l(e')$ are in O. Then the partial word $(E, < -\{(e, e')\}, l)$ is enabled in (I, O, k, n).*

Proof: Straightforward, but tedious.   □

Definition 7.1.12   i) The semantics PWA is defined by $PWA(N) = \|_{s \in S} PW(N(s))$, for a net N. (*A* stands for atomic.)

ii) For an atomic net N, define $Proc(N)$ as the prefix-closure of the set of least sequential partial words of N. Then the semantics $Proc$ of nets is defined by $Proc(N) = \|_{s \in S} Proc(N(s))$.

iii) Let (I, O, k, n) be an atomic net. Define $Flow(I, O, k, n) = \{(E, <, l) \in PW(I, O, k, n) \mid \forall e, e' \in E : e$ is an immediate predecessor of $e' \Rightarrow (l(e) \in I \Leftrightarrow l(e') \in O)\}$. Then the semantics $Flow$ is defined by $Flow(N) = \|_{s \in S} Flow(N(s))$.

Definition 7.1.12 is an application of the scheme presented in Theorem 7.1.10. PWA (called Pr_2 in [HRT89]) represents (a modification of) the idea that concurrency is different from arbitrary interleaving, but includes (for atomic nets) arbitrary interleaving. Therefore PWA of an atomic net is seq-closed – it simply is the partial-word semantics. But note that PWA is *not* seq-closed on nets in general, since direct dependences can only exist between transitions that belong to a common atomic subnet. PWA is the maximal compositional, admissible semantics of nets.

On the other hand, $Proc$ (called Pr_1 in [HRT89]) represents the idea that, at least for atomic nets, we should only consider a minimal set of dependences; this way it is hoped to model causality; but see the examples below. Furthermore, the semantics is made prefix-closed by force. For nets with infinite capacities, $Proc$ is just the set of event structures of processes (see [HRT89]), i.e. Definition 7.1.12 extends our definition of $Proc$ given in Section 2.2.

The third semantics, *Flow*, is new. It represents a combination of the above ideas: by Lemma 7.1.11, direct causality in the sense of the *Proc*-semantics can only hold between an input- and an output-transition. With this in mind, *Flow* includes some sequentializations, but not all. Note that in the *Proc*-semantics a direct causal dependence can only exist *from* an input- *to* an output-transition. But in the presence of finite capacities, an output-transition might create space needed by an input-transition; thus, intuitively, we can also have a direct causal dependence *from* an output- *to* an input-transition. The *Flow*-semantics will play an important rôle in our considerations.

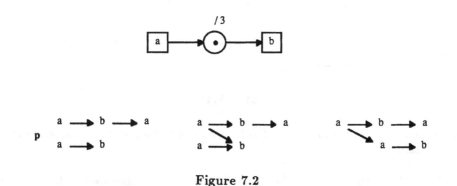

Figure 7.2

Figure 7.2 shows three partial words of the atomic net $N = (\{a\}, \{b\}, 3, 1)$, where the first is in $Proc(N)$, the second is in $Flow(N) - Proc(N)$, and the third is in $PWA(N) - Flow(N)$.

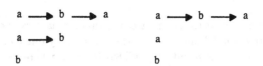

Figure 7.3

Proc has been defined in such a way that it is prefix-closed, which seems to be a very natural property. Unfortunately, this has the consequence that elements of *Proc* do *not* always have a minimal set of dependences for atomic nets. Figure 7.3 shows a minimally ordered element of $Proc(N)$ which has p from Figure 7.2 as a prefix; therefore p really is an element of $Proc(N)$, although it is more sequential than the second element of $Proc(N)$ shown in Figure 7.3.

Even if we decide to do without prefix-closure and restrict ourselves to infinite capacities, a compositional semantics could not consist of minimally ordered partial words only: Figure 7.4 shows a net N and three partial words; the first is a minimally ordered element of $PW(N(s))$, the second is a minimally ordered element of $PW(N(s'))$; the third is the composition of the first two, but it is not minimally ordered in $PW(N)$.

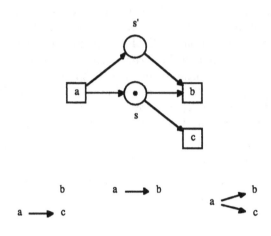

Figure 7.4

Proposition 7.1.13 *The semantics PWA, Proc and Flow are consistent, prefix-closed and compositional.*

Proof: *PW* is consistent by definition and prefix-closure follows easily from the definition. Therefore, *PWA* is consistent and prefix-closed on atomic nets. From this and the definition, it follows directly that *Proc* is consistent and prefix-closed for atomic nets. For an atomic net (I, O, k, n), we have by Lemma 7.1.11 that $Proc(I, O, k, n) \subseteq Flow(I, O, k, n)$, thus *Flow* is consistent for atomic nets; the prefix-closure of *Flow* for atomic nets follows from the definition. Hence, the claim follows from Theorem 7.1.10. □

Next we will obtain a result that compares the three semantics above. We will close this section by showing that for 1-bounded nets there is exactly one compositional semantics. This extends a corresponding result of [HRT89] for nets where all capacities are 1. We note a lemma first:

Lemma 7.1.14 *Let N, N_1, N_2 be nets, $N = N_1 \parallel N_2$, $p \in PW(N)$.*

   i) $id^*_{T_1}(p) \in PW(N_1)$

   ii) *If N is 1-bounded and N_1 an atomic subnet of N, then $id^*_{T_1}(p) \in L(N_1)$.*

Proposition 7.1.15   i) *Let Sem_1 and Sem_2 be compositional semantics of nets such that for all atomic nets N we have $Sem_1(N) \subseteq Sem_2(N)$. Then for all nets N we have $Sem_1(N) \subseteq Sem_2(N)$.*

   ii) *For all nets N, we have $Proc(N) \subseteq Flow(N) \subseteq PWA(N)$.*

Proof: i) obvious
      ii) follows from 7.1.11 and i).   □

Theorem 7.1.16 *Let Sem be a compositional semantics of nets. Then for all 1-bounded nets N we have*

$$Sem(N) = PWA(N).$$

Proof: '$\subseteq$' By assumption, we have $Sem(N) \subseteq PW(N)$ for all atomic nets N. Now we apply Proposition 7.1.15 i).
'$\supseteq$' Let $p \in PWA(N)$. By Lemma 7.1.14, we have $id^*_{\bullet s \cup s \bullet}(p) \in L(N(s))$ for all atomic nets $N(s), s \in S$. Since Sem is admissible, there exist $p(s) \in Sem(N(s))$ for all $s \in S$ such that $p(s) \preceq id^*_{\bullet s \cup s \bullet}(p)$. By Lemma 7.1.8 ii), we have $p \in \|_{s \in S} id^*_{\bullet s \cup s \bullet}(p)$, and thus by 7.1.8 iii) we find some $p' \in Sem(N)$ with $p' \in \|_{s \in S} p(s)$ and $p' \preceq p$. If e is an immediate predecessor of e' in p, then $l(e), l(e') \in {}^\bullet s \cup s^\bullet$ for some $s \in S$ by definition of $\|$. Since $p' \in PW(N)$, we have $id^*_{\bullet s \cup s \bullet}(p') \in L(N)$ by 7.1.14. Thus e and e' must be ordered in p', hence $p' \preceq p$ implies $e <' e'$ in p' and we conclude $p = p' \in Sem(N)$. $\square$

This result shows in particular that the interval semiword semantics, which is favoured by the results of Chapter 5, is not compositional for safe nets. This is a consequence of the specific notion of 'compositional' we use in this chapter; compare also Section 7.3.

7.2   Complementation-invariance and capacity-orientation

Consistency or at least admissibility, prefix-closure, and compositionality are desirable properties for a semantics of nets. But many net semantics can be defined that are consistent, prefix-closed and compositional, and many of them are not very sensible. Following the approach of [HRT89], we define two more properties – capacity-orientation and complementation-invariance –, which one might want a semantics to share. Both properties state that certain natural and quite moderate net transformations do not change the semantics of a net.

Definition 7.2.1 Let N be a net and $s \in S$. The *actual capacity* of s, denoted by $AC(s)$, is the maximal number of tokens s may hold in any reachable marking of N, i.e.

$$AC(s) := sup\{M(s) \mid M \in [M_N\rangle\}.$$

The place s is called *contact-free* if

$$\forall M \in [M_N\rangle \forall t \in T : ((\forall s' \in S : M(s') \geq W(s', t)) \Rightarrow (M(s) + W(t, s) \leq K(s))).$$

If the capacity of a place s is greater than $AC(s)$, then s is contact-free. In any case, if s is contact-free, then it seems that $K(s)$ does not influence the behaviour of the net, i.e. we would expect that changing $K(s)$ to any number not less than $AC(s)$ does not change the semantics of the net.

Definition 7.2.2 Let N be a net.

i) Let $\bar{S} \subseteq S$ be a set of contact-free places. A net N' is called *capacity-modification* of N, if

  - $S' = S, T' = T, W' = W, M_{N'} = M_N,$
  - $K'(s) = K(s)$ for $s \in S - \bar{S},$
  - $K'(s) \geq AC(s)$ for $s \in \bar{S}.$

ii) Let $\bar{S} \subseteq S$ be a set of places with finite capacity and $\tilde{S}$ a copy of $\bar{S}$ disjoint from S, i.e. there exists a bijection $\tilde{\ }: \bar{S} \to \tilde{S}$. A net N' is called *complementation* of N if

  - $S' = S \cup \tilde{S}, T' = T,$
  - $W'(x,y) = W(x,y)$ for $x,y \in S \cup T,$
    $W'(\tilde{s},t) = W(t,s), W'(t,\tilde{s}) = W(s,t)$ for $s \in \bar{S}, t \in T,$
  - $K'(s) = K(s)$ for $s \in S, K'(\tilde{s}) = K(s)$ for $s \in \bar{S},$
  - $M_{N'}(s) = M_N(s)$ for $s \in S, M_{N'}(\tilde{s}) = K(s) - M_N(s)$ for $s \in \bar{S}.$

iii) An admissible semantics Sem of nets is called *capacity-oriented (complementation-invariant)* if for all nets N and all capacity-modifications (complementations) N' of N we have $Sem(N) = Sem(N').$

Let us note that every capacity-modification can be seen as a series of modifications that change the capacity of a single place only. Furthermore, if N' is a capacity-modification of N where only the capacity of some place s is increased, then s is contact-free in N' and has the same actual capacity as in N. Therefore N is also a capacity-modification of N', and to check whether a given semantics is capacity-oriented it is enough to check all capacity-modifications where the capacity of a single place is decreased. (Warning: if N' is a capacity-modification of N where the capacity of a place s is decreased to $AC(s)$, then s is possibly not contact-free anymore!)

Complementation is often used to make nets contact-free: all places with finite capacity are complemented, and then one changes the capacities of the complemented places and of the complements to ω. This way one wants to reduce the study of nets with capacities to the study of nets without capacities. It is easy to see that for a capacity-oriented, complementation-invariant semantics this approach is possible:

Proposition 7.2.3   *i) Let N be a net and N' a complementation of N, i.e. there exist $\bar{S}, \tilde{S}$ as in Definition 7.2.2 such that $S' = S \cup \tilde{S}$. Then every $s \in \bar{S} \cup \tilde{S}$ is contact-free in N'.*

  ii) Let Sem be a complementation-invariant, capacity-oriented semantics of nets, let the net N' be obtained from a net N by complementing all places with finite capacities first and setting all capacities equal to ω afterwards. Then $Sem(N) = Sem(N')$.

For compositional semantics we can characterize complementation-invariance as follows:

Theorem 7.2.4 *Let Sem be a compositional semantics of nets. Then the following are equivalent:*

i) Sem is complementation-invariant.

ii) For all atomic nets (I, O, k, n) with $k \in \mathbb{N}$, we have $Sem(I, O, k, n) = Sem(I, O, k, n) \parallel Sem(O, I, k, k - n)$.

iii) For all atomic nets (I, O, k, n) with $k \in \mathbb{N}$, we have $Sem(I, O, k, n) = Sem(O, I, k, k - n)$ and $Sem(I, O, k, n) = Sem(I, O, k, n) \parallel Sem(I, O, k, n)$.

Proof: ii) is the restriction of i) to atomic nets, ii) and iii) are easily seen to be equivalent. Now let us assume ii) and suppose we are given a net N and a complementation N' as in Definition 7.2.2 ii). Then $Sem(N) = \parallel_{s \in S} Sem(N(s)) = (\parallel_{s \in \bar{S}} Sem(N(s))) \parallel (\parallel_{s \in \bar{S}} (Sem(N(s)) \parallel Sem(N(\bar{s})))) = Sem(N')$. □

Since for partial words over the same alphabet synchronization corresponds to a kind of superposition (see Lemma 7.1.8), we can say that complementation-invariance means that the semantics of each atomic net is closed under superposition and equals the semantics of the complement net.

Corollary 7.2.5   *i) PWA is complementation-invariant.*

ii) Proc is not complementation-invariant.

iii) Flow is complementation-invariant.

Proof: i) was already observed in [HRT89]. It follows quite easily from Theorem 7.2.4 iii). (For the second equality, observe that for all partial words p we have $p \in p \parallel p$; furthermore, use Lemma 7.1.8 i) and seq-closure of PW.)

ii) Consider $N = (\{a\}, \{b\}, 2, 0)$, its complementation $\bar{N}$ and the partial words p, q shown in Figure 7.1. Since $p \in Proc(N)$ we have $q \in Proc(\bar{N})$, but $q \notin Proc(N)$.

iii) Again we apply Theorem 7.2.4 iii): the first equality follows from $PW(I, O, k, n) = PW(O, I, k, k - n)$. One inclusion of the second equality is always true; for the other, consider $p_1, p_2 \in Flow(I, O, k, n)$ and $p \in p_1 \parallel p_2$, and observe that e is an immediate predecessor of e' in p only if it is an immediate predecessor of e' in p_1 or p_2. □

With regard to capacity-orientation, we first observe that directly from the definitions, especially the definition of contact-freeness, we have that L is capacity-oriented, and this implies by Proposition 7.1.1 that PW is, too.

Proposition 7.2.6 *L and PW are capacity-oriented.*

Now we will show a sufficient criterion for capacity-orientation. For its formulation we need the concept of flow-sequentialization. This is a sequentialization that can be obtained by adding precedences that concern an input- *and* an output-transition, and not two input- or two output-transitions.

Definition 7.2.7 Let (I, O, k, n) be an atomic net, $p, q \in PW(I, O, k, n)$. Then $p = (E, <, l)$ is a *flow-sequentialization* of q, if $p \succeq q$ and for all $e, e', \in E$ such that e is an immediate predecessor of e' in p we have $e <_q e'$ or $(l(e) \in I \Leftrightarrow l(e') \in O)$. We call p the *flow-reduction* of q if p is the maximal $p \in Flow(I, O, k, n)$ such that $p \preceq q$.

For $Sem(I, O, k, n) \subseteq PW(I, O, k, n)$ we write $FlowseqSem(I, O, k, n) = \{p \mid p$ is a flow-sequentialization of some $q \in Sem(I, O, k, n)\}$.

$Sem(I, O, k, n)$ is *flowseq-closed* if $FlowseqSem(I, O, k, n) = Sem(I, O, k, n)$. A semantics Sem is *flowseq-closed* if $Sem(N)$ is flowseq-closed for all atomic nets N.

Note that repeated application of Lemma 7.1.11 to some $q \in PW(I, O, k, n)$ yields a flow-reduction, and it is easy to see that it is unique.

Obviously, PWA and $Flow$ are flowseq-closed, while $Proc$ is not. We note the following connection between flow-sequentialization and the $Flow$-semantics.

Proposition 7.2.8 Let $p, q \in PW(I, O, k, n)$. Then p is a flow-sequentialization of q if and only if there is some $r \in Flow(I, O, k, n)$ with $p \in q \parallel r$.

Proof: '$\Rightarrow$' Let r be the flow-reduction of p.   '$\Leftarrow$' follows from the definitions.   □

Now we are ready to obtain the announced sufficient criterion for capacity-orientation.

Theorem 7.2.9 Let Sem be a compositional semantics of nets such that for all atomic nets (I, O, k, n) and $k' \geq k$, $k' \in \mathbb{N}_0 \cup \{\omega\}$ we have

  i) $FlowseqSem(I, O, k', n) \cap PW(I, O, k, n) \subseteq Sem(I, O, k, n)$

  ii) $Sem(I, O, k, n) \subseteq Sem(I, O, k', n)$.

Then Sem is capacity-oriented.

Proof: Let N' be a net and N be a capacity-modification of N' where the only change is that $K(s) = k < k' = K'(s)$ for a single place s. Let $N = \bar{N} \parallel N(s)$, $N(s) = (I, O, k, n)$. By compositionality and ii), we have immediately that $Sem(N) \subseteq Sem(N')$.

For the other inclusion, let $p \in Sem(N')$ and choose $q \in Sem(\bar{N})$ and $q' \in Sem(I, O, k', n)$ such that $p \in q \parallel q'$. For $p_1 = id^{\bullet}_{sUs\bullet}(p)$, we have by Lemma 7.1.8 that $p_1 \succeq q'$ and $p \in q \parallel p_1$. Since PW is capacity-oriented we have $p \in PW(N)$, hence by Lemma 7.1.14 $p_1 \in PW(I, O, k, n)$. We can choose representatives of p, q and q' such that $E_p = E_q \cup E_{q'}$, $l_p = l_q \cup l_{q'}$ and $<_p = (<_q \cup <_{q'})^+$. Then $p_1 = (E_{q'}, <_1, l_{q'})$ where $<_1$ is the appropriate restriction of $<_p$. We now reduce $<_1$ by repeated application of Lemma 7.1.11 to $<_2$ such that $<_2 \supseteq <_{q'}$ and any further application of 7.1.11 would violate this inclusion. Then $p_2 = (E_{q'}, <_2, l_{q'}) \in PW(I, O, k, n)$ by Lemma 7.1.11, and p_2 is by construction a flow-sequentialization of q'. Thus, by i), $p_2 \in Sem(I, O, k, n)$. Furthermore $<_p = (<_q \cup <_{q'})^+ = (<_q \cup <_1)^+$ and $<_1 \supseteq <_2 \supseteq <_{q'}$, hence $<_p = (<_q \cup <_2)^+$ and $p \in q \parallel p_2$. Therefore $p \in Sem(N)$.   □

Corollary 7.2.10 PWA and $Flow$ are capacity-oriented.

This corollary answers a question raised in [HRT89] whether *PWA* is the only semantics which is consistent, prefix-closed, compositional, complementation-invariant and capacity-oriented. The discussion in [HRT89] might lead to the question whether such a semantics is necessarily seq-closed on atomic nets: we see that the answer to this question is negative, since *Flow* is not seq-closed on atomic nets.

Uniformity

Theorems 7.2.4 and 7.2.9 give a characterization of complementation-invariance and a sufficient condition for capacity-orientation which only refer to atomic nets. This makes these results so useful. Now we are able to construct a number of semantics for which we can easily see that they share all the above mentioned properties. Especially, we will construct such semantics Sem_1, Sem_2 and Sem_3 which are somewhat pathological. They show that we should add another property to our list of desirable properties, and we will define such a property below.

Define for an atomic net (I, O, k, n)

$$Sem_i(I, O, k, n) = \{p \in PW(I, O, k, n) \mid p \in Flow(I, O, k, n) \text{ or } prop_i(I, O, p)\},$$

where $prop_i$ is a property defined as follows:

- Fix some t: $prop_1(I, O, p) \Leftrightarrow t \in I \cup O$

- $prop_2(I, O, p) \Leftrightarrow \mid I \mid = \mid O \mid$

- $prop_3(I, O, p) \Leftrightarrow$ all minimal elements of p have the same label

Applying Theorem 7.1.10, we can extend these functions to compositional, prefix-closed and consistent semantics $Sem_i, i = 1, 2, 3$. From Theorem 7.2.4 and Corollary 7.2.5, we conclude that all three are complementation-invariant; for showing $Sem_3(I, O, k, n) \parallel Sem_3(I, O, k, n) \subseteq Sem_3(I, O, k, n)$ observe: if $r \in p \parallel q$, $p, q \in Sem_3(I, O, k, n)$, and all minimal elements of p have the same label, then by Lemma 7.1.8 i) all minimal elements of r have the same label. Finally, Theorem 7.2.9 and Corollary 7.2.10 show that Sem_1, Sem_2 and Sem_3 are capacity-oriented.

These semantics share all the desirable properties mentioned so far, but they are quite pathological:

- Sem_1 cares about the specific name of a transition; instead, we would expect that changing the name of a transition results in a corresponding name change in all partial words of the semantics, but in nothing else.

- Considering Sem_2, we have the phenomenon that adding some transitions to an atomic net might add or disallow some partial words which do not use the new transitions at all. I think that this is undesirable.

- Sem_3 violates our expectation that a semantics does not distinguish between different input- and different output-transitions. (This expectation is somewhat disputable since it is also violated by semiwords [Sta81] and Mazurkiewicz-traces [Maz87]; it should be noted that these are not consistent either.)

The next definition formalizes the above expectations which were violated by Sem_2 and Sem_3. We close this section by showing that Sem_1 and similar semantics are automatically excluded then.

Definition 7.2.11 Let Sem be a compositional semantics. Sem is called *uniform* if for all atomic nets (I, O, k, n):

i) If $I \subseteq I', O \subseteq O'$, then

$$Sem(I, O, k, n) = \{id^*_{I \cup O}(p) \mid p = (E, <, l) \in Sem(I', O', k, n) \ \wedge \ l(E) \subseteq I \cup O\}.$$

ii) If $X = I$ or $X = O$, $(E, <, l) \in Sem(I, O, k, n)$ and $e \in E$ with $l(e) \in X$, then $(E, <, l') \in Sem(I, O, k, n)$ for $l'|_{E-\{e\}} = l|_{E-\{e\}}$ and $l'(e) \in X$.

Part i) of this definition says that, if we delete some transitions of an atomic net, then the semantics of the modified net should be the set of those partial words in the semantics of the original net that do not use the deleted transitions. Part ii) says that a semantics should be closed under substituting in any of its elements one occurrence of an input-transition by an occurrence of another input-transition, and similarly for output-transitions.

Proposition 7.2.12 *PWA, Proc and Flow are uniform.*

Theorem 7.2.13 *For fixed i and o, let Sem be an admissible semantics of atomic nets $(\{i\}, \{o\}, k, n)$, $k \in \mathbb{N} \cup \{\omega\}$, $n \in \mathbb{N}_0$. Then Sem can uniquely be extended to a uniform semantics of nets. This extension preserves consistency and prefix-closure.*

Proof: Let us assume that $i \notin O$ and $o \notin I$, and let us determine $Sem(I, O, k, n)$. (The general case is similar but more involved.) By Definition 7.2.11 i), we must have $Sem(\{i\}, \{o\}, k, n) = \{id^*_{\{i,o\}}(p) \mid p = (E, <, l) \in Sem(I \cup \{i\}, O \cup \{o\}, k, n) \wedge l(E) \subseteq \{i, o\}\}$. Now we can obtain $Sem(\{I \cup \{i\}, O \cup \{o\}, k, n)$ by relabelling in every $p \in Sem(\{i\}, \{o\}, k, n)$ every i-labelled element with an arbitrary transition from $I \cup \{i\}$, every o-labelled element with an arbitrary transition from $O \cup \{o\}$, and changing the alphabet of p to $I \cup O \cup \{i, o\}$. Finally $Sem(I, O, k, n) = \{id^*_{I \cup O}(p) \mid p = (E, <, l) \in Sem(I \cup \{i\}, O \cup \{o\}, k, n) \wedge l(E) \subseteq I \cup O\}$. This extends Sem uniquely to a uniform semantics of atomic nets in a consistency- and prefix-closure-preserving fashion. The result follows from Theorem 7.1.10.   □

Corollary 7.2.14 *Let Sem be a uniform semantics of nets. Let N and N' be weakly isomorphic nets, i.e. we have a bijection $\phi : S \cup T \to S' \cup T'$ such that:*

- *$\phi(S) = S', \phi(T) = T',$*
- *$W_N(x, y) = W_{N'}(\phi(x), \phi(y))$ for all $x, y \in S \cup T$,*
- *$K_N(s) = K_{N'}(\phi(s))$ for all $s \in S$,*
- *$M_N(s) = M_{N'}(\phi(s))$ for all $s \in S$.*

Then $Sem(N') = \{(E, <, \phi \circ l) \mid (E, <, l) \in Sem(N)\}$.

Proof: As remarked at the beginning of Section 7.1, we distinguish nets only up to renaming of places. Thus, for atomic nets the result follows from the construction in the proof of Theorem 7.2.13, and carries over to all nets by compositionality.   □

Our next aim is to give a characterization of capacity-orientation. We will also show that a capacity-oriented, complementation-invariant and compositional semantics must be flowseq-closed. These results can be used to find out whether a given semantics fulfills all the nice properties mentioned above; they may also help to construct a semantics with these and some other properties. In particular, we can exhibit the unique minimal semantics fulfilling all the above requirements. First, we will give a necessary condition for capacity-orientation. In the proof we will need the following notation: if $p = (E, <, l)$ is a labelled partial order (or, especially, a sequence or a sequence of multisets), then $\#(p, Z) := |\{e \in E \mid l(e) \in Z\}|$.

Proposition 7.2.15 *If a uniform semantics Sem is capacity-oriented, then for all atomic nets (I, O, k, n), $k' > k$ and $p \in PW(O, I, \omega, k - n)$ we have $Sem(I, O, k, n) \parallel \{p\} = Sem(I, O, k', n) \parallel \{p\}$.*

Proof: Case a): the labelling of p is a bijection onto $I \cup O$.

We construct a net $N(p)$ as follows: $T_{N(p)} = I \cup O$. For all $e, e' \in E$ with $e < e'$ we add an empty place s with ${}^\bullet s = \{l(e)\}$ and $s^\bullet = \{l(e')\}$; for every $e \in E$ which is minimal with respect to $<$, we add a place s with one token and ${}^\bullet s = \emptyset, s^\bullet = \{l(e)\}$. All capacities are infinite. Obviously, $N(p)$ is 1-bounded; $PWA(N(p))$, and thus by Theorem 7.1.16 $Sem(N(p))$, is the set of prefixes of p.

Now let us consider the net $N(p) \parallel (I, O, k, n)$ and let s be the additional place. By construction, every firing sequence is a linearization of a prefix of p; thus, if we have a reachable marking M reached via $w \in (I \cup O)^*$ and a transition t with enough tokens in its preset under M, then either $t \in O$ and trivially $M(s) + W(t, s) \leq k$ or $t \in I$ and $\#(wt, I) - \#(wt, O) \leq k - n$ (since $p \in PW(O, I, \omega, k - n)$ and $wt \in L(O, I, \omega, k - n)$), thus $M(s) + W(t, s) = n + \#(wt, I) - \#(wt, O) \leq k$. Hence s is contact-free and $N(p) \parallel (I, O, k', n)$ is a capacity-modification of $N(p) \parallel (I, O, k, n)$. Now the result follows from compositonality.

Case b): the labelling of p is arbitrary.

We can extend or reduce I to I' and O to O' such that $|I'| = |\{e \in E \mid l(e) \in I\}|$, $|O'| = |\{e \in E \mid l(e) \in O\}|$, $l(E) \cap I \subseteq I'$ and $l(E) \cap O \subseteq O'$. Let $p' = (E_p, <_p, l')$ be labelled over $I' \cup O'$ such that l' is bijective and $l'(e) \in I' \Leftrightarrow l_p(e) \in I$ for all $e \in E$.

Obviously we have $p' \in PW(O', I', \omega, k - n)$; thus, we can conclude from case a) that $Sem(I', O', k, n) \parallel \{p'\} = Sem(I', O', k', n) \parallel \{p'\}$. If we have an element r of $q \parallel p$ with $q \in Sem(I, O, k, n)$, then we can change the labelling of q and r according to p' with results q' and r' and have $r' \in q' \parallel p'$. From uniformity it follows that $r' \in Sem(I', O', k, n) \parallel \{p'\}$. Hence, there exists $s' \in Sem(I', O', k', n)$ with $r' \in s' \parallel p'$. Changing back the labelling according to p, we obtain by uniformity some $s \in Sem(I, O, k', n)$ with $r \in s \parallel p$, hence $r \in Sem(I, O, k', n) \parallel \{p\}$. The other inclusion is symmetric.   □

It can also be shown that the necessary condition of Proposition 7.2.15 is sufficient. But we aim at another characterization which is easier to handle. Let us call a semantics *well-behaved* if it is consistent, prefix-closed, compositional, uniform, capacity-oriented and complementation-invariant. With the above result, we are able to show that a well-behaved semantics must be flowseq-closed – we have already seen that it does not have to be seq-closed on atomic nets. For this result we need a number of lemmas; in these lemmas, *Sem* is supposed to be well-behaved.

Lemma 7.2.16 *For all atomic nets* (I, O, k, n) *with* $k < \omega$, $m \geq n$ *and* $k' \geq m - n + k$ *we have* $Sem(I, O, k, n) = Sem(I, O, k, n) \parallel Sem(I, O, k', m)$.

Proof:

$$
\begin{aligned}
Sem(I, O, k, n) &= Sem(I, O, k, n) \parallel Sem(I, O, k, n) &\text{(7.2.4)}\\
&= Sem(I, O, k, n) \parallel Sem(O, I, k, k - n) &\text{(7.2.4)}\\
&= Sem(I, O, k, n) \parallel Sem(O, I, k + m - n, k - n) &\\
&\qquad\qquad\qquad\qquad\qquad\text{(capacity-modifaction)}\\
&= Sem(I, O, k, n) \parallel Sem(I, O, k + m - n, m) &\text{(7.2.4)}\\
&= Sem(I, O, k, n) \parallel Sem(I, O, k', m) &\\
&\qquad\qquad\qquad\qquad\qquad\text{(capacity-modification)}
\end{aligned}
$$

$\square$

Lemma 7.2.17 *For all* $q \in Sem(I, O, \omega, n)$ *there exists* $k \in \mathbb{N}$ *such that for all finite* $k' \geq k$ *we have* $q \in Sem(I, O, k', n)$.

Proof: There exists $k \in \mathbb{N}$ such that $q \in PW(I, O, k, n)$, hence for all finite $k' \geq k$ we have $q \in PW(I, O, k', n)$ and $q \in PW(O, I, \omega, k' - n)$. By Proposition 7.2.15, this implies $Sem(I, O, k', n) \parallel \{q\} = Sem(I, O, \omega, n) \parallel \{q\}$. Since q is contained in the right-hand side, we can find $p \in Sem(I, O, k', n)$ with $p \preceq q$. Since by Lemma 7.2.16 $Sem(I, O, k', n) = Sem(I, O, k', n) \parallel Sem(I, O, \omega, n)$, we conclude that $q \in p \parallel q$ is in $Sem(I, O, k', n)$. $\square$

Lemma 7.2.18 *If for some partial word* q *there exists a* $k \in \mathbb{N}$ *such that for all finite* $k' \geq k$ *we have* $q \in Sem(I, O, k', n)$, *then* $q \in Sem(I, O, \omega, n)$.

Proof: Let $q = (E, <, l)$. For sufficiently large k' we have $(E, \emptyset, l) \in PW(O, I, \omega, k' - n)$. By Proposition 7.2.15 we have $Sem(I, O, k', n) \parallel \{(E, \emptyset, l)\} = Sem(I, O, \omega, n) \parallel \{(E, \emptyset, l)\}$. Since q is contained in the left-hand side, we conclude that $q \in Sem(I, O, \omega, n)$. $\square$

Lemma 7.2.19 *Let* $(\{i\}, \{o\}, k, n)$, $k < \omega$, *be an atomic net, and let* $p = (E, <, l)$ *be a partial order labelled over* $\{i, o\}$ *such that* $\#(p, i) \leq k - n$, $\#(p, o) \leq n$ *and* $<= \{(e, e')\}$ *with* $\{l(e), l(e')\} = \{i, o\}$. *Then* $p \in Sem(\{i\}, \{o\}, k, n)$. *(Figure 7.5 shows the two forms* p *may have.)*

Proof: First, let us assume we are given some partial order $p = (E, <, l)$ labelled over $\{i, o\}$ such that $\#(p, i) = k - n + 1$, $\#(p, o) = n$, $<= \{(e, e')\}$, $l(e) = o$ and $l(e') = i$. We have $p \in PW(\{o\}, \{i\}, \omega, k - n)$, hence by Proposition 7.2.15

$$(*) \qquad Sem(\{i\}, \{o\}, k, n) \parallel \{p\} = Sem(\{i\}, \{o\}, k + 1, n) \parallel \{p\}.$$

Let $p' = (E, \emptyset, l)$. By consistency $p' \in Sem(\{i\}, \{o\}, k + 1, n)$, hence $p \in p' \parallel p$ is in the right-hand side of $(*)$. Thus, p must be a sequentialization of some element of $Sem(\{i\}, \{o\}, k, n)$ by Lemma 7.1.8 i), and since $p' \notin Sem(\{i\}, \{o\}, k, n)$, we conclude that $p \in Sem(\{i\}, \{o\}, k, n)$. By prefix-closure, this shows the lemma for the case $l(e) = i$ and $l(e') = o$. In the case $l(e) = o$ and $l(e') = i$, we have $p \in Sem(\{o\}, \{i\}, k, k - n)$ by the first case, and thus are done by Theorem 7.2.4. □

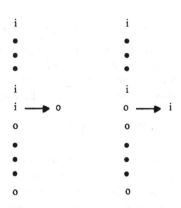

Figure 7.5

Theorem 7.2.20 *Let Sem be a well-behaved semantics. Then Sem is flowseq-closed, i.e. for all atomic nets (I, O, k, n) we have $FlowseqSem(I, O, k, n) = Sem(I, O, k, n)$.*

Proof: If $I = \emptyset$ or $O = \emptyset$, then $FlowseqSem(I, O, k, n) = Sem(I, O, k, n)$ by definition. Hence assume $I \neq \emptyset \neq O$.

Case a): k is finite.

Let $q = (E, <, l) \in Sem(I, O, k, n)$ and q' be a flow-sequentialization of q. First, let us assume that $l(E) \subseteq \{i, o\}$ for some $i \in I, o \in O$. Choose k', m according to Lemma 7.2.16 such that $\#(q, i) \leq k' - m$ and $\#(q, o) \leq m$. Let $p_1, p_2 \in Sem(\{i\}, \{o\}, k', m)$ be the two partial words according to Lemma 7.2.19 with $\#(p_1, i) = \#(p_2, i) = \#(q, i)$ and $\#(p_1, o) = \#(p_2, o) = \#(q, o)$. Then, q' can be obtained from q by repeatedly $\parallel$-composing q with $id^*_{I \cup O}(p_1)$ or $id^*_{I \cup O}(p_2)$, and thus $q \in Sem(I, O, k, n)$ by Lemma 7.2.16. (Observe that $Sem(\{i\}, \{o\}, k', m)$ is a subset of $Sem(I, O, k', m)$ up to the alphabet by uniformity.)

In the general case, we change in q and q' every label from I to some fixed $i \in I$ and every label from O to some fixed $o \in O$ and obtain q_{io} and q'_{io}. By uniformity

$q_{io} \in Sem(I, O, k, n); q'_{io}$ is a flow-sequentialization of q_{io} and thus in $Sem(I, O, k, n)$ by the special case. Again by uniformity we conclude that $q' \in Sem(I, O, k, n)$.

Case b): $k = \omega$.

Let $q \in Sem(I, O, \omega, n)$ and q' be a flow-sequentialization of q. Choose k according to Lemma 7.2.17. Then, by case a), we have $q' \in Sem(I, O, k', n)$ for all finite $k' \geq k$. Hence $q' \in Sem(I, O, \omega, n)$ by Lemma 7.2.18.   □

Now we can give a characterization of capacity-orientation:

Theorem 7.2.21 *Let Sem be a semantics which is uniform, consistent, prefix-closed, compositional and complementation-invariant. Then Sem is capacity-oriented if and only if Sem is flowseq-closed and for all atomic nets (I, O, k, n) and $k' \geq k$ we have $Sem(I, O, k, n) = Sem(I, O, k', n) \cap PW(I, O, k, n)$.*

Proof: '$\Rightarrow$' By Theorem 7.2.20, *Sem* must be flowseq-closed. We only show one inclusion of the above equality, the proof for the other is analogous. Let $q \in Sem(I, O, k, n)$ and let p be the flow-reduction of q. Since $q \in PW(I, O, k, n) = PW(O, I, k, k - n)$, we can apply Proposition 7.2.15 for p and conclude that there exists $q' \in Sem(I, O, k', n)$ sucht that $q \in q' \parallel p$. By Proposition 7.2.8, q is a flow-sequentialization of q', hence $q \in Sem(I, O, k', n)$.

'$\Leftarrow$' follows immediately from Theorem 7.2.9.   □

This theorem allows to check a compositional semantics for well-behavedness by considering atomic nets only. By Theorem 7.1.10, consistency and prefix-closure carry over from atomic nets to nets in general. Uniformity is based on atomic nets by definition. Theorem 7.2.4 gives a criterion for complementation-invariance which refers to atomic nets. Finally, if all these criteria are satisfied, capacity-orientation can be checked by applying Theorem 7.2.21.

For example, let us consider the compositional semantics Sem_4 which is defined on atomic nets (I, O, k, n) as the seq-closure of $SL(I, O, k, n)$. In [HRT89], the question is raised whether Sem_4 is well-behaved. Of course, uniformity is not considered in [HRT89], but Sem_4 is easily seen to be uniform. It also follows directly from the definition that Sem_4 is consistent, prefix-closed and compositional. Furthermore we have $SL(I, O, k, n) = SL(O, I, k, k - n)$, hence $Sem_4(I, O, k, n) = Sem_4(O, I, k, k - n)$. From the seq-closure we get $Sem_4(I, O, k, n) = Sem_4(I, O, k, n) \parallel Sem_4(I, O, k, n)$ and this implies complementation-invariance by Theorem 7.2.4. Figure 7.6 shows the atomic net $(\{a, b\}, \{c\}, 3, 1)$ and some p in $Sem_4(\{a, b\}, \{c\}, 3, 1) \cap PW(\{a, b\}, \{c\}, 2, 1)$ which is not in $Sem_4(\{a, b\}, \{c\}, 2, 1)$. Hence Sem_4 is not capacity-oriented by Theorem 7.2.21.

We have already seen that *Flow* is a well-behaved semantics. With Theorem 7.2.21, we can now prove that *Flow* is the unique minimal well-behaved semantics, i.e. that *Flow* is contained in any well-behaved semantics.

Theorem 7.2.22 *Let Sem be a well-behaved semantics and N an arbitrary net. Then $Flow(N) \subseteq Sem(N)$.*

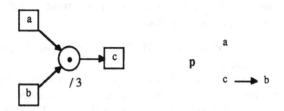

Figure 7.6

Proof: By compositionality, it is enough to consider the case that N is an atomic net (I, O, k, n). Let $p = (E, <, l) \in Flow(I, O, k, n) \subseteq PW(I, O, k, n)$. For some large $m \geq k - n$ we have $(E, \emptyset, l) \in PW(O, I, \omega, m)$, thus $(E, \emptyset, l) \in Sem(O, I, \omega, m)$ by consistency. From the flowseq-closure of Sem we conclude $p \in Sem(O, I, \omega, m)$.

Since $p \in PW(I, O, k, n)$, we have for any prefix p' of p that $\#(p', O) - \#(p', I) \leq n$, which implies $p \in PW(O, I, m+n, m)$. Hence $p \in Sem(O, I, m+n, m)$ by Theorem 7.2.21. Now Theorem 7.2.4 gives $p \in Sem(I, O, m + n, n)$, and Theorem 7.2.21 finally implies $p \in Sem(I, O, k, n)$ since $p \in PW(I, O, k, n)$. $\square$

One may expect from a semantics Sem that the dependences in some $p \in Sem(I, O, k, n)$ should model causality, and that causality can only exist between a transition of I and a transition of O, since the latter may need a token from the other or provide 'open space' for it. If we adopt this view, then Theorem 7.2.22 shows that $Flow$ is the only well-behaved semantics that could be called causal.

It remains an open problem to find out whether there exists a well-behaved semantics different from $Flow$ and PWA.

7.3 S-modification

In this section, we will take a critical view at compositionality by considering another net modification which one might expect to preserve the behaviour of a net. With complementation-modification, we have already met a net modification which allows under certain circumstances to add or remove a place. This is also possible with S-modification, which was defined in [Ber87].

Definition 7.3.1 Let N be a net. $s \in S$ is called *structurally redundant* if

i) $K(s) = \omega,$

ii) $\exists R \subseteq S - \{s\} : (M_N(s) \geq \sum_{r \in R} M_N(r) \wedge \ \forall t \in T :$
     $(W(s,t) \leq \sum_{r \in R} W(r,t) \ \wedge \ W(t,s) - W(s,t) \geq \sum_{r \in R} W(t,r) - W(r,t)))$

iii) $^\bullet s \subseteq {}^\bullet(S - \{s\}) \cup (S - \{s\})^\bullet$

$N' = (S', T_N, W', K', M_{N'})$ is called *S-modification* of N if $s \in S_N$ is structurally redundant, $S' = S_N - \{s\}$, $W' = W_N|_{S' \times T_N \cup T_N \times S'}$, $K' = K_N|_{S'}$ and $M_{N'} = M_N|_{S'}$.

An admissible semantics is called *S-modification invariant* if each net N and every S-modification of N have the same semantics.

This definition is transferred from [Ber87]. There, it is shown for nets without capacities that L is S-modification invariant.

The idea behind Definition 7.3.1 is the following. If a transition is enabled under some marking and we delete some place, then the transition is still enabled in the modified net. In this sense, a place may only restrict the behaviour af a net. Definition 7.3.1 gives static conditions on s which ensure that s does in fact not restrict the behaviour.

A place possibly restricts the firing of the transitions in its postset by the number of tokens it provides, and the firing of the transitions in its preset by the capacity. Condition i) ensures that the latter is not the case for s. According to ii), s initially carries at least as many tokens as the places in R together, and this does not change according to the last part of ii). Hence $W(s,t) \leq \sum_{r \in R} W(r,t)$ says: if there are too few tokens on s to fire t, then there are also too few in some place of R.

In particular, this implies that s cannot be the single place in the preset of some transition with empty postset. Symmetrically, s cannot be the single place in the postset of some transition with empty preset by iii). Therefore, N' is a well-defined net in the sense of this chapter, since it does not have an isolated transition.

Thus, S-modification-invariance seems to be a desirable property of a semantics. For example, S-modification allows to delete a place s if some other place has the same pre- and postset and the same initial marking. (Especially, for atomic nets N and an S-modification invariant semantics Sem, we have $Sem(N) = Sem(N \parallel N)$.)

Unfortunately, S-modification-invariance and compositionality exclude each other.

Theorem 7.3.2 *Let Sem be a compositional semantics. Then Sem is not S-modification-invariant.*

Proof: Let us assume that Sem is S-modification invariant. Consider the nets N, N' and N'' depicted in Figure 7.7.

By admissibility, we have that one of the partial words shown in Figure 7.8 is in $Sem(N'')$.

From compositionality, we easily get $p_1 \in Sem(N)$ or $p_2 \in Sem(N)$ if one of $p_1, \ldots, p_4$ is in $Sem(N'')$. On the other hand, for the S-modification N' of N we have $p_1 \notin Sem(N')$ and $p_2 \notin Sem(N')$, since by Theorem 7.1.16 $Sem(N') = PWA(N')$. This is a contradiction.

If only p_5 is in $Sem(N'')$, consider the nets $\bar{N}$ and $\bar{N}'$ that are obtained by exchanging the transitions a and b in N, N' resp. By compositionality we have $p_2 \in Sem(\bar{N})$. But p_2 is not in $Sem(\bar{N}')$. $\square$

This result depends on the operator $\parallel$. It raises the question whether compositionality with respect to $\parallel$ really is desirable. We will close this section by showing that a semantics can be well-behaved and S-modification invariant, if we work with a different operator introduced in [Gra81], which we have already used in Theorem 3.1.7. This operator is designed to be used with partial words in general; one could also adapt this operator to the use of interval semiwords.

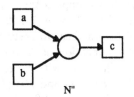

N"

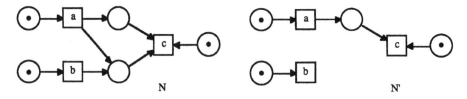

N   N'

Figure 7.7

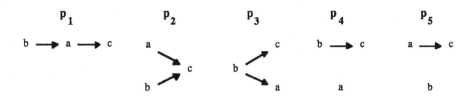

Figure 7.8

Definition 7.3.3 Let p_i be a partial order labelled over Z_i, $i = 1, 2$. Then for a partial order p labelled over $Z_1 \cup Z_2$ we have $p \in p_1 \parallel \preceq p_2$ if $p_i \preceq id^*_{Z_i}(p)$ for $i = 1, 2$. The definition of $\parallel \preceq$ is extended to sets of labelled partial orders as usual.

Let us remark that $p \in p_1 \parallel \preceq p_2$ if and only if $p \succeq p'$ for some $p' \in p_1 \parallel p_2$. Now it is obvious from Proposition 7.1.9 and Lemma 7.1.14 that PW is compositional with respect to $\parallel \preceq$. For compositionality with respect to $\parallel \preceq$, one can easily show an analogous result to Theorem 7.2.4 and conclude that PW is complementation-invariant. We have seen in Proposition 7.2.6 that PW is capacity-oriented; uniformity, consistency and prefix-closure are clear from the definition anyway. Now we conclude this chapter by showing that PW is the only natural choice if we are interested in compositionality w.r.t. $\parallel \preceq$.

Theorem 7.3.4 *The semantics PW is the only semantics which is consistent, compositional with respect to $\parallel \preceq$, uniform, complementation-invariant and capacity-oriented. It is also prefix-closed and S-modification invariant.*

Proof: If a semantics Sem is compositional with respect to $\parallel \preceq$ and complementation-invariant, we have analogously to Theorem 7.2.4 that for all atomic nets $Sem(I, O, k, n) =$

$Sem(I,O,k,n) \parallel \preceq Sem(I,O,k,n)$, which implies the seq-closure of *Sem*. Now it is not too hard to repeat the proofs that led to Theorem 7.2.22, and one gets the uniqueness of *PW*. It remains to show that *PW* has the last property mentioned in the theorem.

Let N be a net and N' an S-modification of N w.r.t. a place $s \in S$. Let $\bar{N}$ and $\bar{N}'$ be the complementations w.r.t. all places with finite capacity followed by capacity-modification to ω. Then $PW(\bar{N}) = PW(N)$ and $PW(\bar{N}') = PW(N')$. The place s is structurally redundant in $\bar{N}$, and $\bar{N}'$ is an S-modification of $\bar{N}$ w.r.t. s. Since $\bar{N}$ and $\bar{N}'$ have infinite capacities, it follows that $L(\bar{N}) = L(\bar{N}')$. Hence $PW(\bar{N}) = PW(\bar{N}')$ and $PW(N) = PW(N')$. □

Concluding Remarks

In this book, a considerable number of behaviour descriptions have been presented that are suitable to support modular construction. We have demonstrated their adequateness further by showing that they are fully abstract or characterize an external equivalence. The choice between these descriptions depends on the construction method one wants to use and on the requirements for correctness. As construction methods we have studied parallel composition with synchronous and with asynchronous communication and several forms of refinement. A very fruitful idea of correctness has been to simply require deadlock- or deadlock/divergence-freeness; for the case that the branching behaviour has to be considered in detail, we have also required that some form of bisimulation be used.

Many of the presented behaviour descriptions can be grouped under the heading 'failure semantics'. The two basic forms – $\mathcal{F}$- and $\mathcal{FD}$-semantics – can be justified by the requirement to build deadlock-free or deadlock/divergence-free nets by parallel composition with synchronous communication, as we have shown in Chapter 3. Thus, the exchange of failure-equivalent building blocks in a parallel composition is required to preserve deadlock-freeness, but in fact it preserves behaviour in a much stronger sense, i.e. it preserves failure semantics. In Chapter 4, we have looked at $\mathcal{F}$-deterministic nets; here, the behaviour preservation is even stronger. Furthermore, $\mathcal{F}$-deterministic nets have provided us with a general framework for the behaviour-preserving refinement of places.

The basic forms of failure semantics can easily be adapted to various related situations, for example to additional fairness considerations or to the important case that only safe nets should be constructed. A more involved further development has been described in Chapter 5. If we want to use action refinement as an additional construction method, we have to combine the failure idea with some form of partial order semantics; we have obtained full abstractness for a partial order semantics based on interval orders.

These results establish quite a finished picture, although research on action refinement is still very active. For the dual methods, the approach presented in Chapter 4 is rather a beginning. We have studied parallel composition with asynchronous communication and have obtained some promising results – but only for the restricted case that one component is a P-daughter-net. We have defined I, O-nets as a suitable form of deterministic P-daughter-nets; they give strong results on behaviour preservation and provide a general framework for the behaviour-preserving refinement of transitions. It would be interesting to generalize these results and to combine them with the results for the synchronous case. It is also challenging to develop a method for the equivalence-preserving refinement of places and a suitable semantics. Furthermore, we have studied parallel composition with synchronous communication for nets with capacities in Chapter 7; so far, we have only considered unlabelled nets in these studies; in the future, these results should be generalized to labelled nets and combined with the other results of this book.

Although this book devotes much effort to showing the necessity of certain distinctions, there might be good reasons to use a semantics that is more discriminating than necessary. What is necessary depends on the purpose we have in mind in the first place; if we are confident that two systems or building blocks are equivalent in a stronger sense than necessary, then proving this stronger equivalence might be helpful for other purposes later on. Also testing a stronger equivalence might be more efficient as shown for failure equivalence and bisimulation in [KS90].

Still, we should consider whether we have good reasons for the distinctions we make. Especially, our results in Chapters 5 and 6 show that in order to deal with action refinement we need interval orders or the ST-idea (which are closely related), and both have the flavour of the temporal rather than that of causality. Although other arguments have been put forward in favour of causality-based partial order semantics, see e.g. [Bes88b,Rei88], it is still open and in my view an important and challenging question whether causality based semantics can be proven to be necessary for practical reasons.

In practical applications, very often some forms of high-level net are used. As far as these are abbreviations for place/transition-nets, our results should carry over easily. A more difficult, but important task will be to translate our results to the symbolic level of high-level nets. Finally, it is not expected that a specification describes, for example, a required failure semantics for a parallel system; instead, it would be more practical to specify the desired behaviour by means of temporal logic formulae. For such an approach, the behaviour descriptions given in this book are just a first step. Hopefully, they will be an essential tool for the development of a modular design method where nets are specified by temporal logic.

Bibliography

[Abr87]   S. Abramsky: Observation equivalence as a testing equivalence. Theor. Comp. Sci. 53:225–241, 1987

[Ace89]   L. Aceto: On relating concurrency and nondeterminism. Technical Report 6/89, Dept. Comp. Sci. Univ. of Sussex, Brighton, 1989

[Ace90]   L. Aceto: Full abstractions for series-parallel pomsets. Technical Report 1/90, Dept. Comp. Sci. Univ. of Sussex, Brighton, 1990

[AH89]   L. Aceto, M. Hennessy: Towards action-refinement in process algebras. In: Proc. 4th LICS, 138–145. IEEE Computer Society Press, 1989. A full version has appeared as: Computer Science Report 3/88, Dept. Comp. Sci. Univ. of Sussex, Brighton, 1988

[AH91]   L. Aceto, M. Hennessy: Adding action refinement to a finite process algebra. In: J. Leach Albert, B. Monien, M. Rodríguez Artalejo(eds.): ICALP 91, Lect. Notes Comp. Sci. 510, 506–519. Springer, 1991

[And83]   C. André: The behaviour of a Petri net on a subset of transitions. R.A.I.R.O. 17:5–21, 1983

[AR88]   I.J. Aalbersberg, G. Rozenberg: Theory of traces. Theoret. Comput. Sci. 60:1–82, 1988

[Bau88]   B. Baumgarten: On internal and external characterizations of PT-net building block behaviour. In: G. Rozenberg(ed.): Advances in Petri Nets 1988, Lect. Notes Comp. Sci. 340, 44–61. Springer, 1988

[Bau90]   B. Baumgarten: Petri-Netze. Grundlagen und Anwendungen. BI-Wissenschaftsverlag, 1990

[BC87]   G. Boudol, I. Castellani: On the semantics of concurrency: Partial orders and transition systems. In: H. Ehrig et al.(eds.): TAPSOFT 87, Vol. I, Lect. Notes Comp. Sci. 249, 123–137. Springer, 1987

[BD87]   E. Best, R. Devillers: Sequential and concurrent behaviour in Petri net theory. Theoret. Comput. Sci. 55:87–136, 1987

[BDC92]   L. Bernardinello, F. De Cindio: A survey of basic net models and modular
            net classes. In: G. Rozenberg(ed.): Advances in Petri Nets, Lect. Notes
            Comp. Sci., Springer, 1992. Submitted

[BDKP91]   E. Best, R. Devillers, A. Kiehn, L. Pomello: Concurrent bisimulations in
            Petri nets. Acta Informatica 28:231–264, 1991

[Ber87]   G. Berthelot: Transformations and decompositions of nets. In: W. Brauer
            et al.(eds.): Petri Nets: Central Models and Their Properties, Lect. Notes
            Comp. Sci. 254, 359–376. Springer, 1987

[Bes88a]   E. Best: Kausale Semantik nichtsequentieller Programme. Habilitation-
            sschrift, Rhein. Friedr.-Wilh.-Univ. Bonn, 1988. Also available as GMD-
            Bericht Nr. 174, Oldenbourg Verlag, 1989

[Bes88b]   E. Best: Weighted basic Petri nets. In: F.H. Vogt(ed.): Concurrency 88,
            Lect. Notes Comp. Sci. 335, 257–276. Springer, 1988

[BF88]   E. Best, C. Fernandéz: Nonsequential Processes. A Petri Net View.
            EATCS Monographs on Theor. Comput. Sci. 13. Springer, 1988

[BHR84]   S.D. Brookes, C.A.R. Hoare, A.W. Roscoe: A theory of communicating
            sequential processes. J. ACM 31:560–599, 1984

[BK84]   J.A. Bergstra, J.W. Klop: Process algebra for synchronous communica-
            tion. Information and Control 60:109–137, 1984

[BKO87]   J.A. Bergstra, J.W. Klop, E.R. Olderog: Failures without chaos: A new
            process semantics for fair abstraction. In: M. Wirsing(ed.): Formal De-
            scription of Programming Concepts III, 77–103. North-Holland, 1987

[Bol90]   B. Bollobàs: Graph Theory. An Introductory Course. Graduate Texts in
            Mathematics 63. Springer, 1990

[BR84]   S.D. Brookes, A.W. Roscoe: An improved failures model for communicat-
            ing processes. In: S.D. Brookes, A.W. Roscoe, G. Winskel(eds.): Seminar
            on Concurrency, Lect. Notes Comp. Sci. 197, 281–305. Springer, 1984

[BR88]   A. Bourguet-Rouger: External behaviour equivalence between two Petri
            nets. In: F.H. Vogt(ed.): Concurrency 88, Lect. Notes Comp. Sci. 335,
            237–256. Springer, 1988

[Bra84]   W. Brauer: How to play the token game. Petri Net Newsletter 16:3–13,
            1984

[Bro83]   S.D. Brookes: A model for communicating sequential processes. Technical
            Report CMU-CS-83-149, Carnegie-Mellon-Univ., Pittsburgh, 1983

[Car86]   H. Carstensen: Fairneß bei nebenläufigen Systemen. PhD thesis, Univ.
            Hamburg, 1986

[CDMP87]   L. Castellano, G. De Michelis, L. Pomello: Concurrency vs. interleaving: An instructive example. Bull. EATCS 31:12–15, 1987

[Che91]   G. Chehaibar: Replacement of open interface subnets and stable state transformation equivalence. In: Proc. 12th Int. Conf. Applications and Theory of Petri Nets, Gjern, 390–409, 1991

[Dar82]   P. Darondeau: An enlarged definition and complete axiomatisation of observational congruence of finite processes. In: M. Dezani-Ciancaglini, U. Montanari(eds.): International Symposium on Programming, Lect. Notes Comp. Sci. 137, 47–62. Springer, 1982

[DCDMPS88] F. De Cindio, G. De Michelis, L. Pomello, C. Simone: A state transformation equivalence for concurrent systems: Exhibited functionality-equivalence. In: F.H. Vogt(ed.): Concurrency 88, Lect. Notes Comp. Sci. 335, 222–236. Springer, 1988

[DCDMS87]   F. De Cindio, G. De Michelis, C. Simone: GAMERU: A language for the analysis and design of human communication pragmatics within organizational systems. In: G. Rozenberg(ed.)· Advances in Petri Nets 1987, Lect. Notes Comp. Sci. 266, 21–44. Springer, 1987

[DD89]   P. Darondeau, P. Degano: Causal trees. In: G. Ausiello et al.(eds.): ICALP 89, Lect. Notes Comp. Sci. 372, 234–248. Springer, 1989

[DDNM88]   P. Degano, R. De Nicola, U. Montanari: A distributed operational semantics for CCS based on condition/event systems. Acta Informatica 26:59–91, 1988

[DDNM89]   P. Degano, R. De Nicola, U. Montanari: Partial orderings descriptions and observations of nondeterministic concurrent processes. In: J.W. de Bakker et al.(eds.): Proc. REX School / Workshop Linear Time, Branching Time and Partial Order in Logic and Models of Concurrency. Noordwijkerhout, 1988, Lect. Notes Comp. Sci. 354, 438–466. Springer, 1989

[Dev90a]   R. Devillers: The semantics of capacities in P/T-nets. In: G. Rozenberg(ed.): Advances in Petri Nets 1989, Lect. Notes Comp. Sci. 424, 128–150. Springer, 1990

[Dev90b]   R. Devillers: Maximality preserving bisimulation. Technical Report LIT-214, Univ. Bruxelles, 1990. To appear in Theoret. Comput. Sci.

[Dev91]   R. Devillers: Maximality preservation and the ST-idea for action refinement. Technical Report LIT-242, Univ. Bruxelles, 1991

[Die90]   V. Diekert: Combinatorics on Traces. Lect. Notes Comp. Sci. 454. Springer, 1990

[Dij71]   E.W. Dijkstra: Hierarchical ordering of sequential processes. Acta Informatica 1:115–138, 1971

[DM87]   P. Degano, U. Montanari: Concurrent histories: A basis for observing distributed systems. J. Comp. Sys. Sci. 34:422–461, 1987

[DNH84]   R. De Nicola, M.C.B. Hennessy: Testing equivalence for processes. Theoret. Comput. Sci. 34:83–133, 1984

[DV89]   V. Diekert, W. Vogler: On the synchronization of traces. Math. Systems Theory 22:161–175, 1989

[Eng85]   J. Engelfriet: Determinacy $\rightarrow$ (observation equivalence = trace equivalence). Theoret. Comput. Sci. 36:21–25, 1985

[Eng90]   U. Engberg: Partial Orders and Fully Abstract Models for Concurrency. PhD thesis, Computer Science Dept. Aarhus Univ., 1990. Technical Report DAIMI PB-307

[Fis70]   P.C. Fishburn: Intransitive indifference with unequal indifference intervals. J. Math. Psych. 7:144–149, 1970

[Gen87]   H. J. Genrich: Predicate/transition nets. In: W. Brauer et al.(eds.): Petri Nets: Central Models and Their Properties, Lect. Notes Comp. Sci. 254, 207–247. Springer, 1987

[GG89a]   R.J. v. Glabbeek, U. Goltz: Partial order semantics for refinement of actions - neither necessary nor always sufficient, but appropriate when used with care. EATCS Bulletin 38:154–163, 1989

[GG89b]   R.J. v. Glabbeek, U. Goltz: Equivalence notions for concurrent systems and refinement of actions. In: A. Kreczmar, G. Mirkowska(eds.): MFCS 89, Lect. Notes Comp. Sci. 379, 237–248. Springer, 1989

[GG90]   R.J. v. Glabbeek, U. Goltz: Refinement of actions in causality based models. In: J.W. de Bakker, W.P. de Roever, G. Rozenberg(eds.): Proc. REX Workshop on Stepwise Refinement of Distrib. Systems 1989, Lect. Notes Comp. Sci. 430, 267–300. Springer, 1990

[GL91]   R. Gorrieri, C. Laneve: The limit of $split_n$-bisimulations for CCS agents. In: A. Tarlecki(ed.): MFCS 91, Lect. Notes Comp. Sci. 520, 170–180. Springer, 1991

[Gla90a]   R.J. v. Glabbeek: private communication, 1990

[Gla90b]   R.J. v. Glabbeek: Comparative Concurrency Semantics and Refinement of Actions. PhD thesis, Univ. Amsterdam, 1990

[Gla90c] R.J. v. Glabbeek: The refinement theorem for ST-bisimulation semantics. In: M. Broy, C.B. Jones(eds.): Proc. IFIP Working Conference on Programming Concepts and Methods, Sea of Galilee, Israel, 1990. To appear

[GMM88] R. Gorrieri, S. Marchetti, U. Montanari: A2CCS: A simple extension of CCS for handling atomic actions. In: M. Dauchet, M. Nivat(eds.): CAAP '88, Lect. Notes Comp. Sci. 299, 258–270. Springer, 1988

[Gol87] U. Goltz: Über die Darstellung von CCS-Programmen durch Petrinetze. Dissertation, RWTH Aachen, 1987. Also available as GMD-Bericht Nr. 172, Oldenbourg Verlag 1988

[Gol88a] R. Gold: Verklemmungsfreiheit bei modularer Konstruktion fairer Petrinetze. Diplomarbeit, Techn. Univ. München, 1988

[Gol88b] U. Goltz: On representing CCS programs by finite Petri nets. In: M.P. Chytil et al.(eds.): Proc. MFCS 1988, Lect. Notes Comp. Sci. 324, 339–350. Springer, 1988

[Gra81] J. Grabowski: On partial languages. Fundamenta Informaticae IV.2:428–498, 1981

[GV87] R.J. v. Glabbeek, F. Vaandrager: Petri net models for algebraic theories of concurrency. In: J.W. de Bakker et al.(eds.): PARLE Vol. II, Lect. Notes Comp. Sci. 259, 224–242. Springer, 1987

[GV90] R. Gold, W. Vogler: Quality criteria for partial order semantics of place/transition nets. In: B. Rovan(ed.): MFCS 90, Lect. Notes Comp. Sci. 452, 306–312. Springer, 1990

[GV91] R.J. v. Glabbeek, F. Vaandrager: The difference in splitting into n and n+1. 1991. Unpublished

[GW89a] R.J. v. Glabbeek, W.P. Weijland: Refinement in branching time semantics. In: Proc. AMAST Conf., 197–201, 1989

[GW89b] R.J. v. Glabbeek, W.P. Weijland: Branching time and abstraction in bisimulation semantics. In: G.X. Ritter(ed.): Information Processing 89, 613–618. North-Holland, 1989

[Hac76a] M. H. T. Hack: The equality problem for vector addition systems is undecidable. Theoret. Comput. Sci. 2:77–95, 1976

[Hac76b] M. H. T. Hack: Petri net languages. Technical Report TR-159, MIT, Boston, 1976

[Hoa85] C.A.R. Hoare: Communicating Sequential Processes. Prentice-Hall, 1985

[HRT89]   J. Hirshfeld, A. Rabinovich, B.A. Trakhtenbrot: Discerning causality in
           interleaving behaviour. In: A.R. Meyer, M.A. Taitslin(eds.): Logic at
           Botik '89, Lect. Notes Comp. Sci. 363, 146–162. Springer, 1989

[Jen87]   K. Jensen: Coloured Petri nets. In: W. Brauer et al.(eds.): Petri Nets:
           Central Models and Their Properties, Lect. Notes Comp. Sci. 254, 248–299.
           Springer, 1987

[JK90]   R. Janicki, M. Koutny: Observing concurrent histories. In: H.M.S.
           Zedan(ed.): Real-Time Systems, Theory and Applications, 133–142.
           North-Holland, 1990

[JM91]   L.A. Jategaonkar, A.R. Meyer: Testing equivalence for Petri nets with
           action refinement. Unpublished Manuscript, 1991

[Kie88]   A. Kiehn: On the interrelationship between synchronized and non-syn-
           chronized behaviour of Petri nets. J. Inf. Process. Cybern. EIK 24:3–18,
           1988

[Kos82]   S.R. Kosaraju: Decidability of reachability in vector addition systems.
           In: Proc.14th Annual ACM Symposium on Theory of Computing, San
           Francisco, 267–281, 1982

[KS90]   P.C. Kanellakis, S.A. Smolka: CCS expressions, finite state processes, and
           three problems of equivalence. Information and Computation 86:43–68,
           1990

[Lam86]   L. Lamport: On interprocess communication I. Distributed Comp. 1:77–
           85, 1986

[May81]   E.W. Mayr: An algorithm for the general Petri net reachability problem.
           In: Proc 13th Annual ACM Symposium on Theory of Computing, 238–
           246, 1981

[Maz84]   A. Mazurkiewicz: Traces, histories, graphs: Instances of a process monoid.
           In: M.P. Chytil et al.(eds.): Proceeding of the 11th Symposium on Math-
           ematical Foundations of Computer Science (MFCS), Lect. Notes Comp.
           Sci. 176, 115–133. Springer, 1984

[Maz87]   A. Mazurkiewicz: Trace theory. In: W. Brauer et al.(eds.): Petri Nets: Ap-
           plications and Relationships to other Models of Concurrency, Lect. Notes
           Comp. Sci. 255, 279–324. Springer, 1987

[Maz88]   A. Mazurkiewicz: Compositional semantics of pure place/transition sys-
           tems. Fundamenta Informaticae XI:331–356, 1988

[Mil77]   R. Milner: Fully abstract models of typed λ-calculi. Theor. Comput. Sci.
           4:1–22, 1977

[Mil80]   R. Milner: A Calculus for Communicating Systems. Lect. Notes Comp. Sci. 92. Springer, 1980

[Mil81]   R. Milner: A modal characterisation of observable machine-behaviour. In: G. Astesiano, C. Böhm(eds.): CAAP 81, Lect. Notes Comp. Sci. 112, 267–310. Springer, 1981

[Mil83]   R. Milner: Calculi for synchrony and asynchrony.   Theor. Comput. Sci. 25:267–310, 1983

[Mil89]   R. Milner: Communication and Concurrency. Prentice Hall, 1989

[Mül85]   K. Müller: Constructable Petri nets.   Elektr. Inf. Kybern. 21:171–199, 1985

[Mur90]   D.V.J.. Murphy: Time, Causality and Concurrency. PhD thesis, Univ. of Surrey, 1990

[NEL89]   M. Nielsen, U. Engberg, K. Larsen: Partial order semantics for concurrency. In: J.W. de Bakker et al.(eds.): Proc. REX School / Workshop Linear Time, Branching Time and Partial Order in Logic and Models of Concurrency. Noordwijkerhout, 1988, Lect. Notes Comp. Sci. 354, 523–548. Springer, 1989

[NPW81]   M. Nielsen, G.D. Plotkin, G. Winskel: Petri nets, event structures and domains I. Theor. Comput. Sci. 13:85–108, 1981

[OH86]   E.R. Olderog, C.A.R. Hoare: Specification-oriented semantics for communicating processes. Acta Informatica 23:9–66, 1986

[Old89]   E.R. Olderog: Nets, terms and formulas: Three views of concurrent processes and their relationship. Habilitationsschrift, Univ. Kiel, 1989

[Par81]   D. Park: Concurrency and automata on infinite sequences.   In: P. Deussen(ed.): Proc. 5th GI Conf. on Theoretical Comp. Sci., Lect. Notes Comp. Sci. 104, 167–183. Springer, 1981

[Pel87]   E. Pelz: Closure properties of deterministic Petri nets. In: F.J. Brandenburg(ed.): STACS 87, Lect. Notes Comp. Sci. 247, 371-382. Springer, 1987

[Pet81]   J.L. Peterson: Petri Net Theory. Prentice-Hall, 1981

[Plo81]   G. Plotkin: A structural approach to operational semantics. Technical Report DAIMI FN-19, Aarhus Univ., 1981

[Pom86]   L. Pomello: Some equivalence notions for concurrent systems - an overview. In: G. Rozenberg(ed.): Advances in Petri Nets 1985, Lect. Notes Comp. Sci. 222, 381–400. Springer, 1986

[Pom88]   L. Pomello: Osservatore, Reti di Petri, Processi. PhD thesis, Univ. of
            Milano and Torino, 1988

[Pra86]   V. Pratt: Modelling concurrency with partial orders. Int. J. Parallel Prog.
            15:33–71, 1986

[PS90]   L. Pomello, C. Simone: Preorders of concurrent systems. Internal report,
            DSI, Milano, 1990

[PS92]   L. Pomello, C. Simone: A survey of equivalence notions for net based
            systems. In: G. Rozenberg(ed.): Advances in Petri Nets, Lect. Notes
            Comp. Sci., 1992. Submitted

[PT87]   R. Paige, R. Tarjan: Three partition refinement algorithms.   SIAM J.
            Computing 16:973–989, 1987

[RB81]   W.C. Rounds, S.D. Brookes: Possible futures, acceptances, refusals and
            communicating processes. In: Proc. 22nd Ann. Symp. on Foundations of
            Comp. Sci., 140–149. IEEE, 1981

[Rei85]   W. Reisig: Petri Nets. EATCS Monographs on Theoretical Computer
            Science 4. Springer, 1985

[Rei88]   W. Reisig: Concurrency is more fundamental than interleaving. EATCS
            Bulletin 3:181–185, 1988

[Rei90]   W. Reisig: Petri nets and algebraic specification. Technical Report SFB-
            Bericht Nr. 342/1/90B, Inst. Informatik, Techn. Univ. München, 1990

[RT88]   A. Rabinovich, B.A. Trakhtenbrot: Behaviour structures and nets. Fun-
            damenta Informaticae 11:357–404, 1988

[SM83]   I. Suzuki, T. Murata: A method for stepwise refinement and abstraction
            of Petri nets. J. Comp. Sys. Sci. 27:51–76, 1983

[Sou91]   Y. Souissi: On liveness preservation by composition of nets via a set of
            places. In: G. Rozenberg(ed.): Advances in Petri Nets 1991, Lect. Notes
            Comp. Sci. 524, 277–295. Springer, 1991

[Sta81]   P.H. Starke: Processes in Petri nets. J. Inf. Process. Cybern. EIK 17:389–
            416, 1981

[Tau89]   D. Taubner: Finite Representations of CCS and TCSP Programs by
            Automata and Petri Nets. Lect. Notes Comp. Sci. 369. Springer, 1989

[TV89]   D. Taubner, W. Vogler: Step failures semantics and a complete proof
            system. Acta Informatica 27:125–156, 1989

[Val79]   R. Valette: Analysis of Petri nets by stepwise refinement. J. Comp. Sys.
            Sci. 18:35–46, 1979

[VJ85]   R. Valk, M. Jantzen: The residue of vector sets with application to decid-
           ability problems in Petri nets. Acta Informatica 21:643–674, 1985

[VN82]   G. Vidal-Naquet: Deterministic languages for Petri nets. In: C. Gi-
           rault, W. Reisig(eds.): Application and Theory of Petri Nets, Informatik-
           Fachber. 52, 198–202. Springer, 1982

[Vog87]   W. Vogler: Behaviour preserving refinements of Petri nets. In: G. Tin-
           hofer, G. Schmidt(eds.): Graph-Theoretic Concepts in Computer Science,
           Proc. WG 86, Bernried, Lect. Notes Comp. Sci. 246, 82–93. Springer, 1987

[Vog89]   W. Vogler: Failures semantics and deadlocking of modular Petri nets.
           Acta Informatica 26:333–348, 1989

[Vog90]   W. Vogler: Representation of a swapping class by one net. In: G. Rozen-
           berg(ed.): Advances in Petri Nets 1989, Lect. Notes Comp. Sci. 424, 467–
           486. Springer, 1990

[Vog91a]   W. Vogler: Executions: A new partial order semantics of Petri nets.
           Theoret. Comput. Sci. 91:205–238, 1991

[Vog91b]   W. Vogler: Failures semantics based on interval semiwords is a congruence
           for refinement. Distributed Computing 4:139–162, 1991

[Vog91c]   W. Vogler: Bisimulation and action refinement. In: C. Choffrut,
           M. Jantzen(eds.): STACS 91, Lect. Notes Comp. Sci. 480, 309–321.
           Springer, 1991

[Vog91d]   W. Vogler: Deciding history preserving bisimilarity. In: J. Leach Albert,
           B. Monien, M. Rodríguez Artalejo(eds.): ICALP 91, Lect. Notes Comp.
           Sci. 510, 495–505. Springer, 1991

[Vos87]   K. Voss: Interface as a basic concept for systems specification and verifi-
           cation. In: K. Voss(ed.): Concurrency and Nets, 585–604. Springer, 1987

Index

Lecture Notes in Computer Science

For information about Vols. 1–535
please contact your bookseller or Springer-Verlag